For over 35 years, biological scientists have come to rely on the research protocols and methodologies in the critically acclaimed *Methods in Molecular Biology* series. The series was the first to introduce the step-by-step protocols approach that has become the standard in all biomedical protocol publishing. Each protocol is provided in readily-reproducible step-by-step fashion, opening with an introductory overview, a list of the materials and reagents needed to complete the experiment, and followed by a detailed procedure that is supported with a helpful notes section offering tips and tricks of the trade as well as troubleshooting advice. These hallmark features were introduced by series editor Dr. John Walker and constitute the key ingredient in each and every volume of the *Methods in Molecular Biology* series. Tested and trusted, comprehensive and reliable, all protocols from the series are indexed in PubMed.

# Theranostics

## Methods and Protocols

Edited by

**Jyotsna Batra and Srilakshmi Srinivasan**

*School of Biomedical Sciences, Institute of Health and Biomedical Innovation (IHBI), Translational Research Institute, Queensland University of Technology (QUT), Brisbane, QLD, Australia*

Humana Press

*Editors*
Jyotsna Batra
School of Biomedical Sciences
Institute of Health and Biomedical Innovation (IHBI), Translational Research Institute
Queensland University of Technology (QUT)
Brisbane, QLD, Australia

Srilakshmi Srinivasan
School of Biomedical Sciences
Institute of Health and Biomedical Innovation (IHBI), Translational Research Institute
Queensland University of Technology (QUT)
Brisbane, QLD, Australia

ISSN 1064-3745 ISSN 1940-6029 (electronic)
Methods in Molecular Biology
ISBN 978-1-4939-9768-8 ISBN 978-1-4939-9769-5 (eBook)
https://doi.org/10.1007/978-1-4939-9769-5



This Humana imprint is published by the registered company Springer Science+Business Media, LLC, part of Springer Nature.
The registered company address is: 233 Spring Street, New York, NY 10013, U.S.A.

# Preface

The term "theranostics" implies to a combination of therapy and diagnostics and has been used by scientific community in a variety of contexts. Over the past decade, the field of theranostics has advanced considerably with respect to advances in biomarker identification, new molecular imaging probes, genetics and high-throughput techniques, imaging-guided molecular therapy, and new nanotheranostics. The field is appropriate to explain the advancement in science to establish accurate and personalized therapies for various diseases and to combine diagnostic and therapeutic applications into a single agent for the development of a promising paradigm involving diagnosis, drug delivery, and monitoring of disease response to treatment.

New clinicians, researchers, and students working in this multidisciplinary field often ask how to obtain well-optimized and well-detailed protocols to achieve their research outcome in this field. This book is designed specifically to meet that demand and is published in time to meet the needs of medical researchers in a comprehensive manner encompassing bioengineering, diagnostics, in vivo imaging and its use for image-guided therapy, and a variety of other miscellaneous subjects. To accomplish this daunting task, we had the good fortune to recruit nearly 52 leaders in the field worldwide contributing to the 18 chapters.

Given the multidisciplinary nature of the field, the book is broken into five different sections.

**Part I (*Bioengineering*) consists of three chapters (Chapters 1–3).** Many theranostic approaches has been limited due to the lack of representative laboratory models that mimic the biological processes occurring in vivo and the restrictive access and limited live imaging capabilities that in vivo models provide. Synthetic scaffold materials allow to engineer bone-like microenvironments derived from primary human cells, which can further be co-cultured with metastatic cell lines. Chapter 1 reviews the several easy-to-follow methods for the characterization of cells grown in hydrogels and also describes an enzymatic approach for an improved cell recovery for subsequent analysis. Chapter 2 reviews the application of melt electrowriting technology (MEW) to provide 3D microfiber scaffolds suitable for this purpose. Using primary human cells, MEW scaffolds support the reproducible formation of human bone-like 3D microenvironments. Chapter 3 describes imaging techniques to obtain multidimensional real-time data of cancer cells in the microenvironment context. This novel preclinical platform will contribute to the better understanding of the nature of the dynamic interactions between bone and cancer, as well as serve to measure the efficiency of anticancer drugs.

**Part II (Molecular Diagnostics) consists of Chapters 4–9.** The current diagnostic methods, such as biopsy sampling, tumor tissue staining, and imaging techniques, which are invasive, painful, time-consuming, and expensive require skilled personnel. Therefore, there is an unmet clinical need to develop noninvasive, easy-to-use tools to identify patients before clinical manifestation. Furthermore, the quality of the samples for analysis remains an undeniable challenge which was also discussed in these chapters. Chapter 4 describes exosome-based therapeutics that represent a most promising next-generation approach and are considered valid diagnostic biomarkers and potential therapeutic tools. Exosomes are widely disseminated, heterogenous entities, and the isolation of large and high-quality exosomes is daunting. In this chapter, we reveal the protocol and key insights into the isolation, purification, and characterization of exosomes using ultracentrifugation method. MicroRNAs (miRNAs) are small endogenous noncoding RNA molecules which are powerful regulators of the different cellular processes involved in the pathogenesis of various diseases. The use of miRNA-based therapies has been proposed with the onset of early-phase clinical trials to assess the therapeutic efficacy of miRNAs. The stratification of patients based on differential expression of miRNAs and the therapeutic targeting of such miRNAs would enable patient-specific tailored intervention. Chapter 5 describes experimental methodology for the detection of miRNAs in plasma samples by RT-qPCR. Chapter 6 describes a robust and cost-effective protocol to isolate and enrich miRNAs from saliva samples. Similarly, circulating tumor cells (CTCs) have shown promising potential as liquid biopsies that facilitate early detection, prognosis, therapeutic target selection, and monitoring treatment response. Chapter 7 describes a CTC isolation and analysis method by Tianyu Guo et al. for cancer detection and therapeutic response monitoring. Chapter 8 explains the isolation of CTCs using the spiral microfluidic technology for the efficient sorting of CTCs from patient blood samples for targeted therapy. Both these two chapters are based on CTC isolation methods using different technologies. While Chapter 7 is a marker-independent CTC enrichment method based on the cell size and deformability and the immunofluorescence staining method to detect CTCs, Chapter 8 details the spiral microfluidic technology which is also a marker-independent technique but utilizes hydrodynamic forces for size-based efficient sorting of CTCs from patient blood samples.

Pharmacogenetics involves the search for genetic variations that lead to interindividual differences in drug response. This genotyping approach have paved the way to new opportunities to deliver a better quality of care through more precise characterization of the individual's genetic makeup that will, in turn, contribute to the interindividual variations in drug response. Chapter 9 provides perspectives from the identification of such causative genetic variations—to illustrate how to analyze the raw data obtained by different sequencing techniques while indicating the potential challenges that may arise at each step. Chapter 10 explains the use of automatic Idylla™ system for the analysis of specific mutations in cancer patients.

**Part III (Molecular Imaging) consists of Chapters 11–13**. Numerous advances have been made in recent years in exogenous probes and nanoparticles that allow precise and specific imaging in situ as well as label-free clinical imaging approaches. Chapter 11 summarizes a simple method of generating fluorescent probes using bacterial artificial chromosomes employed in in situ genetic analysis of cells in response to treatment. Chapter 12 focuses on the label-free digital live cell imaging technique and has expanded from its use in the laboratory to the clinical setting, and currently, it is being developed for use in theranostics. Chapter 13 provides a concise review and process of utilizing porous silicon nanoparticles for an improved permeability into the target tissues and for precise deep tissue imaging and diagnostics.

**Part IV (Imaging-Guided Therapy) is composed of Chapters 14–18**. The chapters in this section reveal how advanced imaging techniques now make highly precise clinical intervention possible, further providing accurate and efficient delivery of drug to the target delivery by crossing the barrier and escape clearance through the reticuloendothelial system to ensure sustained presence at effective concentrations. Nanoparticle-based drug delivery systems are emerging as a promising drug delivery platform, due to their distinct advantages leading to various biological actions in the body. Chapter 14 reviews generalized methods for the use of aptamers in diagnostics using optical and electrochemical platforms of detection and in delivering drug to the cancer cells and details a method for the in vitro selection of DNA aptamers against a protein target by SELEX, a combinatorial single-stranded oligonucleotide synthesis technique that specifically bind to a target ligand. Chapter 15 highlights how aptamers can be utilized that has the potential to enhance the precision of molecular medicine and targeted therapeutics. Chapter 16 reviews a simple method of generating fluorescence in situ hybridization probes using bacterial artificial chromosomes that have unlimited possibilities for the analysis of any genomic regions. Chapter 17 explores the method for developing a library of nanoparticles for the efficient delivery of therapeutic agents across the blood-brain barrier. Chapter 18 introduces the biomimetic synthetic strategy and procedures for preparing genetically engineered nanovesicles for disease cell-specific targeting.

It is impossible to describe all the areas that encompass theranostics or will have an impact on the science and practice of personalized medicine. Our aim is to offer a reasonable solution towards disease diagnosis and therapy by giving the readers a sense of new methods and challenges associated with developing theranostics. It has been a great pleasure working with the authors of this book. Without their enthusiasm, encouragement, and professional delivery of their contributions in a timely manner, it would not have been possible to make this book a reality. We hope that this book may form a foundation for further advances in the field of theranostics.

***Brisbane, QLD, Australia*** ***Jyotsna Batra***
***Woolloongabba, QLD, Australia*** ***Srilakshmi Srinivasan***

# Contents

# Contributors

TARIQ ALTALHI • *Department of Chemistry, Faculty of Science, Taif University, Taif, Saudi Arabia*

SRI RAM ARUNACHALAM • *School of Chemistry and Molecular Biosciences, University of Queensland, St Lucia, QLD, Australia; The School of Biomedical Sciences, Institute of Health and Biomedical Innovation, Queensland University of Technology, Kelvin Grove, QLD, Australia*

SHAHNAWAZ AHMAD BABA • *Chemical Biology Lab, Department of Biotechnology, Indian Institute of Technology, Roorkee, Roorkee, India*

JYOTSNA BATRA • *School of Biomedical Sciences, Institute of Health and Biomedical Innovation (IHBI), Translational Research Institute, Queensland University of Technology (QUT), Brisbane, QLD, Australia*

NATHALIE BOCK • *Faculty of Health, School of Biomedical Sciences, Australian Prostate Cancer Research Centre (APCRC-Q), Institute of Health and Biomedical Innovation (IHBI), Queensland University of Technology (QUT), Brisbane, QLD, Australia; Translational Research Institute (TRI), Queensland University of Technology (QUT), Woolloongabba, QLD, Australia; Centre in Regenerative Medicine, IHBI, Queensland University of Technology (QUT), Kelvin Grove, QLD, Australia*

JULIANA B. CANDIDO • *Centre for Cancer and Inflammation, Barts Cancer Institute, Queen Mary University of London, London, UK*

HU CHEN • *State Key Laboratory of Molecular Vaccinology and Molecular Diagnostics & Center for Molecular Imaging and Translational Medicine, School of Public Health, Xiamen University, Xiamen, China*

ROGER CHUNG • *Department of Biomedical Sciences, Faculty of Medicine & Health Sciences, Macquarie University, Sydney, NSW, Australia*

ROSS CRAWFORD • *Faculty of Science and Engineering, Institute of Health and Biomedical Innovation, Queensland University of Technology, Brisbane, QLD, Australia*

ALEXANDRE SANTOS CRISTINO • *UQ Diamantina Institute, The University of Queensland, Woolloongabba, QLD, Australia; Griffith Institute for Drug Discovery, Griffith University, Brisbane, QLD, Australia*

LAURA V. CROFT • *Faculty of Health, School of Biomedical Sciences, Institute of Health and Biomedical Innovation, Cancer and Ageing Research Program, Translational Research Institute, Queensland University of Technology, Woolloongabba, QLD, Australia*

LIBING FU • *Department of Biomedical Sciences, Faculty of Medicine & Health Sciences, Macquarie University, Sydney, NSW, Australia*

TIANYU GUO • *Centre for Molecular Oncology, Barts Cancer Institute, Queen Mary University of London, London, UK*

SIMON HEEKE • *Institute of Research on Cancer and Ageing of Nice (IRCAN), Inserm U1081, CNRS UMR7284, Université Côte d'Azur, Nice, France; Laboratory of Clinical and Experimental Pathology, CHU de Nice, University Hospital Federation OncoAge, Nice, France*

MICHELLE MEI CHIH HILL • *QIMR Berghofer Medical Research Institute, Brisbane, QLD, Australia; UQ Diamantina Institute, The University of Queensland, Woolloongabba, QLD, Australia*

PAUL HOFMAN • *Institute of Research on Cancer and Ageing of Nice (IRCAN), Inserm U1081, CNRS UMR7284, Université Côte d'Azur, Nice, France; Laboratory of Clinical and Experimental Pathology, CHU de Nice, University Hospital Federation OncoAge, Nice, France; Hospital-Integrated Biobank (BB-0033-00025), Université Côte d'Azur, CHU de Nice, Nice, France*

ARUTHA KULASINGHE • *The School of Biomedical Sciences, Institute of Health and Biomedical Innovation, Queensland University of Technology (QUT), Brisbane, QLD, Australia; Translational Research Institute, Brisbane, QLD, Australia*

TUSHAR KUMERIA • *School of Pharmacy, Pharmacy Australia Center of Excellence, The University of Queensland, Woolloongabba, QLD, Australia; Mater Research Institute, Translational Research Institute, The University of Queensland, Woolloongabba, QLD, Australia*

GANG LIU • *State Key Laboratory of Molecular Vaccinology and Molecular Diagnostics & Center for Molecular Imaging and Translational Medicine, School of Public Health, Xiamen University, Xiamen, China; State Key Laboratory of Cellular Stress Biology, Innovation Center for Cell Biology, School of Life Sciences, Xiamen University, Xiamen, China; The MOE Key Laboratory of Spectrochemical Analysis & Instrumentation, College of Chemistry and Chemical Engineering, Xiamen University, Xiamen, China*

JINGYI LIU • *State Key Laboratory of Molecular Vaccinology and Molecular Diagnostics & Center for Molecular Imaging and Translational Medicine, School of Public Health, Xiamen University, Xiamen, China*

DANIELA LOESSNER • *Centre for Cancer and Inflammation, Barts Cancer Institute, Queen Mary University of London, London, UK; Queensland University of Technology, Brisbane, QLD, Australia*

YONG-JIE LU • *Centre for Molecular Oncology, Barts Cancer Institute, Queen Mary University of London, London, UK*

XINZHAN MAO • *Department of Orthopedic Surgery, The Second Xiangya Hospital, Central South University, Changsha, Hunan, China*

XUEYING MAO • *Centre for Molecular Oncology, Barts Cancer Institute, Queen Mary University of London, London, UK*

FARHANA MATIN • *School of Biomedical Sciences, Institute of Health and Biomedical Innovation (IHBI), Translational Research Institute, Queensland University of Technology (QUT), Brisbane, QLD, Australia*

JAIMIE A. MULDERS • *TrendBio Pty Ltd., Alphington, VIC, Australia*

RUCHI MUTREJA • *Chemical Biology Lab, Department of Biotechnology, Indian Institute of Technology, Roorkee, Roorkee, India*

NAVEEN KUMAR NAVANI • *Chemical Biology Lab, Department of Biotechnology, Indian Institute of Technology, Roorkee, Roorkee, India*

KENNETH O'BYRNE • *Faculty of Health, School of Biomedical Sciences, Institute of Health and Biomedical Innovation, Cancer and Ageing Research Program, Translational Research Institute, Queensland University of Technology, Woolloongabba, QLD, Australia; Cancer Services, Princess Alexandra Hospital, Brisbane, QLD, Australia*

NUR OZTEN • *Institute of Life Sciences and Biotechnology, Bezmialem Vakif University, Istanbul, Turkey; Department of Pharmaceutical Toxicology, Faculty of Pharmacy, Bezmialem Vakif University, Istanbul, Turkey*

JONG Y. PARK • *Department of Cancer Epidemiology, Moffitt Cancer Center, Tampa, FL, USA*

ELEONORA PEERANI • *Centre for Cancer and Inflammation, Barts Cancer Institute, Queen Mary University of London, London, UK*

AMIRALI POPAT • *School of Pharmacy, Pharmacy Australia Center of Excellence, The University of Queensland, Woolloongabba, QLD, Australia; Mater Research Institute, Translational Research Institute, The University of Queensland, Woolloongabba, QLD, Australia*

INDIRA PRASADAM • *Faculty of Science and Engineering, Institute of Health and Biomedical Innovation, Queensland University of Technology, Brisbane, QLD, Australia*

CHAMINDIE PUNYADEERA • *The School of Biomedical Sciences, Institute of Health and Biomedical Innovation, Queensland University of Technology, Kelvin Grove, QLD, Australia; Translational Research Institute, Brisbane, QLD, Australia*

ZHI QU • *School of Pharmacy, Pharmacy Australia Center of Excellence, The University of Queensland, Woolloongabba, QLD, Australia; Mater Research Institute, Translational Research Institute, The University of Queensland, Woolloongabba, QLD, Australia*

DEREK J. RICHARD • *Faculty of Health, School of Biomedical Sciences, Institute of Health and Biomedical Innovation, Cancer and Ageing Research Program, Translational Research Institute, Queensland University of Technology, Woolloongabba, QLD, Australia*

HARLEY ROSE ROBINSON • *QIMR Berghofer Medical Research Institute, Brisbane, QLD, Australia; UQ Diamantina Institute, The University of Queensland, Woolloongabba, QLD, Australia*

JOAN RÖHL • *Faculty of Health, School of Biomedical Sciences, Australian Prostate Cancer Research Centre (APCRC-Q), Institute of Health and Biomedical Innovation (IHBI), Queensland University of Technology (QUT), Brisbane, QLD, Australia; Translational Research Institute (TRI), QUT, Woolloongabba, QLD, Australia*

ABEL SANTOS • *School of Chemical Engineering and Advanced Materials, The University of Adelaide, Adelaide, SA, Australia; Institute for Photonics and Advanced Sensing (IPAS), The University of Adelaide, Adelaide, SA, Australia; ARC Centre of Excellence for Nanoscale BioPhotonics (CNBP), The University of Adelaide, Adelaide, SA, Australia*

BINGYANG SHI • *Department of Biomedical Sciences, Faculty of Medicine & Health Sciences, Macquarie University, Sydney, NSW, Australia*

SALAH ALI A. SHOWIHEEN • *Faculty of Science and Engineering, Institute of Health and Biomedical Innovation, Queensland University of Technology, Brisbane, QLD, Australia*

ELZBIETA STANKIEWICZ • *Centre for Molecular Oncology, Barts Cancer Institute, Queen Mary University of London, London, UK*

KIMBERLY D. STEWART • *Department of Chemistry and Physiology and Functional Genomics, Center for Research at the Bio/Nano Interface, Shands Cancer Center, UF Genetics Institute, McKnight Brain Institute, University of Florida, Gainesville, FL, USA*

ANTONIA RUJIA SUN • *Faculty of Science and Engineering , Institute of Health and Biomedical Innovation, Queensland University of Technology, Brisbane, QLD, Australia*

WEIHONG TAN • *Department of Chemistry and Physiology and Functional Genomics, Center for Research at the Bio/Nano Interface, Shands Cancer Center, UF Genetics Institute, McKnight Brain Institute, University of Florida, Gainesville, FL, USA; Molecular Science and Biomedicine Laboratory, State Key Laboratory of Chemo/Bio-Sensing and Chemometrics, College of Chemistry and Chemical Engineering, College of Life Sciences, Aptamer Engineering Center of Hunan Province, Hunan University, Changsha, People's Republic of China*

KAI DUN TANG • *The School of Biomedical Sciences, Institute of Health and Biomedical Innovation, Queensland University of Technology, Kelvin Grove, QLD, Australia*

MAJID EBRAHIMI WARKIANI • *The School of Biomedical Engineering, University of Technology Sydney, Sydney, NSW, Australia*

XIAOXIN WU • *Department of Orthopedic Surgery, The Second Xiangya Hospital, Central South University, Changsha, Hunan, China; Faculty of Science and Engineering, Institute of Health and Biomedical Innovation, Queensland University of Technology, Brisbane, QLD, Australia*

YIN XIAO • *Faculty of Science and Engineering , Institute of Health and Biomedical Innovation, Queensland University of Technology, Brisbane, QLD, Australia*

EMRAH YUCESAN • *Institute of Life Sciences and Biotechnology, Bezmialem Vakif University, Istanbul, Turkey*

PENGFEI ZHANG • *State Key Laboratory of Molecular Vaccinology and Molecular Diagnostics & Center for Molecular Imaging and Translational Medicine, School of Public Health, Xiamen University, Xiamen, China*

# Part I

## Bioengineering

# Chapter 1

# Cell Recovery of Hydrogel-Encapsulated Cells for Molecular Analysis


**Eleonora Peerani, Juliana B. Candido, and Daniela Loessner**



## Abstract

Tissue engineering technologies have produced controllable and reproducible three-dimensional (3D) models that mimic the architecture and complexity of native tissues. In particular cell biology-based research is driven by the development of cell culture platforms and techniques that allow the analysis of cells cultured in 3D. Here we summarize several easy-to-follow methods for the characterization of cells that have been encapsulated and grown in hydrogels to measure their cell viability, metabolic activity, and mechanical properties of cell-containing hydrogels. We also describe an enzymatic approach for the digestion of cell-containing hydrogels and cell recovery thereby maintaining high cell viability for subsequent analysis.



**Key words** 3D cell culture, Biomaterials, Hydrogel, Spheroids, Stiffness, Degradation


## 1 Introduction

In the last decade we have witnessed a shift from two-dimensional (2D) cell cultures toward the use of 3D cell culture approaches. Routinely used 2D culture practices include the growth of cell monolayers on tissue culture plastic, a simplistic system that does not mimic the dimensionality of the extracellular microenvironment. However, the increasing need for biologically representative model systems that mirror the native extracellular matrix (ECM) and cellular processes in 3D has led to the development of an array of tissue-engineered microenvironments [1].

Hydrogel-based 3D cell culture models mimic the biochemical and mechanical properties of the ECM found in tissues and organs. In cancer research, naturally occurring materials are often implemented because they are biocompatible and retain cell binding and protease cleavage sites. Examples include collagen [2, 3], fibrin [4], gelatin [5], and Matrigel, which is the most widely used protein mixture of murine ECM proteins and growth factors [6]. Materials are not restricted to proteins derivatives; alginate [7] and


Jyotsna Batra and Srilakshmi Srinivasan (eds.), *Theranostics: Methods and Protocols*, Methods in Molecular Biology, vol. 2054, https://doi.org/10.1007/978-1-4939-9769-5_1,

hyaluronic acid hydrogels [8] are commonly used examples of naturally occurring polymers and glycosaminoglycans respectively as well as composite approaches [9–11]. However, these materials are limited in their range of mechanical properties, batch-to-batch consistency, and incorporation of biofunctional peptides or molecules [12].

To overcome these limitations, synthetic materials with modular and reproducible properties have been developed. Examples include polyethylene glycol [13, 14] and self-assembling peptides [15]. However, the synthetic origin of these materials can result in a lack of cell-instructive properties, thus, incorporation of cell degradation and integrin binding domains is often necessary to allow for bioactive characteristics [16]. The combination of the biocompatible features of naturally occurring materials with the modular properties of synthetic materials represents a desired strategy to use the advantages of both hydrogel approaches as 3D cell culture platforms [17].

Gelatin methacryloyl (GelMa) hydrogels consist of gelatin that has been functionalized with methacryloyl (methacrylamide and methacrylate) side groups resulting in a photo-crosslinkable polymerization reaction (Fig. 1). Gelatin is a naturally occurring material mainly composed of denatured collagen type-I and therefore

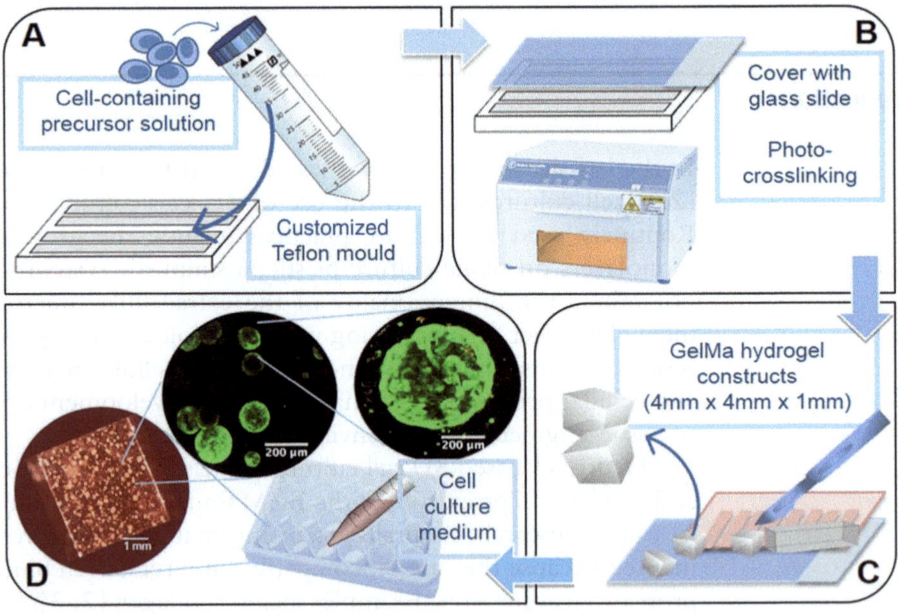


**Fig. 1** Schematic of GelMa hydrogel preparation. (**a**, **b**) The cell-containing precursor solution is casted into a customized Teflon mould, which is covered with a glass slide, and polymerized via photo-crosslinking. (**c**, **d**) Then, hydrogels are cut to obtain identical-sized constructs, transferred into a multiwell plate with cell culture medium, in which they are free-floating to allow cell proliferation and formation of multicellular spheroids over time (live cells stained with green-fluorescent calcein-AM; dead cells stained with red-fluorescent ethidium homodimer-1)

presents with low antigenicity and batch-to-batch variation, while retaining bioactive sites, such as matrix metalloproteinases degradation and arginine-glycine-aspartic acid (RGD) integrin binding domains [5]. This is crucial in a 3D setting, where cells exhibit altered cell surface protein, protease, and integrin expression levels compared to 2D monolayers [16]. Key cell functions, such as migration, proliferation, and matrix degradation, are regulated through cell–cell and cell–matrix interactions [17]. The ability of cells to communicate with each other and their surrounding matrix in a biologically representative system is one reason in the shift toward the use of 3D cell cultures. GelMa hydrogels have generated significant interest as they combine both the inherent bioactivity of its ECM component, collagen type-I, with the biochemical, mechanical and crosslinking modularity of synthetic matrices [18].

In tumor biology, one of the key advantages of 3D platforms is their ability to mimic the mechanical properties of cancerous tissues. GelMa hydrogel stiffness is affected by a number of parameters, including polymer and photoinitiator concentration, UV intensity and exposure time, temperature, and degree of functionalization [19]. By modifying these different parameters the matrix stiffness is controlled without compromising cell viability [20, 21]. In addition, the flexible GelMa design of predefined hydrogel shapes and its transparency allow for molecular and microscopic analyses of encapsulated cells. Its biocompatible and thixotropic properties and ability to polymerize at room temperature make it an attractive biomaterial for 3D bioprinting [22].

However, routinely used molecular and microscopic analyses of multicellular 3D constructs often require optimization or modifications of established procedures used for 2D monolayers. It is crucial to recover cells from 3D matrices without affecting their viability and expression profile. One strategy is the on-demand degradation of synthetic cell-containing hydrogels, which have been modified with a microbial protease-sensitive peptide, through the addition of a microbial protease, with minimal impact on human cell-secreted signaling proteins [23]. Due to the increased popularity of 3D models, there is a need for the development of standardized and reproducible techniques for the recovery of cells cultured in 3D.

Here we describe several cost-effective and accessible methods for the characterization of cell viability and mechanical properties of cell-containing hydrogels followed by an enzymatic approach for cell recovery from these 3D matrices. The metabolic activity is measured using an Alamar Blue cell viability assay. The active component of the Alamar Blue reagent is resazurin that permeates living cells and is reduced to resorufin. Viable cells will constantly convert resazurin to resorufin, which is then used as a measure for cell survival and proliferation. Resazurin is a blue and nonfluorescent, while resofurin is a red and fluorescent compound, and its fluorescence intensity is detected by spectrophotometry. The non-cytotoxic Alamar Blue reagent diffuses cell-containing GelMa hydrogels and is suitable for noninvasive analysis of cells grown within or on top of hydrogels [17, 24]. We use a standard curve made of a serial dilution of cells and cell-free controls for calibration of fluorescence intensity values and calculations of cell numbers. Furthermore, the stiffness of GelMa hydrogels is examined using an unconfined compression testing method previously developed in our laboratory [25]. Determining the mechanical properties of 3D matrices is important for a range of applications that study changes in the microenvironmental biomechanics related to disease development and progression or for the development of bioengineered tissue analogs and 3D platforms.

Finally, we analyze the efficiency of five different enzymatic approaches for the degradation of cell-containing GelMa hydrogels performing flow cytometry analysis of the retrieved cells. This procedure is particularly important for recovering cells from their 3D matrices for subsequent analytical approaches, including flow cytometry that is used to determine cell viability, frequency, and the expression of cell-specific markers [10]. The ability to degrade hydrogels used for 3D cell culture and to extract encapsulated cells without compromising their viability or surface receptor expression is critical for downstream molecular analysis using the retrieved cells. All tested enzymatic approaches resulted in complete hydrogel degradation and can be used depending on the application and type of subsequent cellular and molecular analysis (*see* **Note 1**). As collagen is the main component of GelMa [26], the efficiency of different collagenases is assessed in this study. In addition, gelatin retains a number of protease-sensitive and cell binding sites implicated in a number of physiological functions, including cell proliferation, migration and differentiation, disease development and progression, and are therefore key components of the bioactive niche analog [5]. Enzymes, such as trypsin, are known to interfere with integrins and other cell surface molecules [27], hence this approach may alter the expression of cell surface markers or receptors of the retrieved cells (*see* **Note 1**). Here we use trypsin/EDTA, collagenase I and V, dispase II, and a combination of collagenase XI and dispase II (Table 1) and analyze their efficiency in terms of degradation time, viability, and frequency of retrieved cells.

## 2 Materials

All reagents and materials should be stored in a dry place at room temperature unless otherwise stated. For usage, prewarm reagents at 37 °C in a sterile water bath. Follow strictly all waste disposal regulations when discarding used reagents. Perform all GelMa and cell-related experimental work in a biological safety cabinet.

**Table 1**
**Enzymes tested and respective details**

| Approach | Name | Working concentration | Company; catalog number |
|---|---|---|---|
| 1 | Collagenase I from *Clostridium histolyticum* | 1 mg/mL | ThermoFisher; 17018029 |
| 2 | Trypsin/EDTA | 0.5%/0.2% | Sigma-Aldrich; 59418C |
| 3 | Collagenase XI from *Clostridium histolyticum* and Dispase II from *Bacillus polymyxa* | 0.125 mg/mL and 0.625 U/mL | Sigma-Aldrich; C9407; STEMCELL Technologies; 7913 |
| 4 | Collagenase V from *Clostridium histolyticum* | 1 mg/mL | Sigma-Aldrich; C9263 |
| 5 | Dispase II from *Bacillus polymyxa* | 2.5 U/mL | STEMCELL Technologies; 7913 |

### 2.1 Analysis of Cell Viability and Microscopy

1. Live/dead viability kit for mammalian cells (ThermoFisher). Store at −20 °C and protected from light.
2. PBS (ThermoFisher).
3. Sterile 24-well plate.
4. Sterilized metal spatula.
5. Aluminum foil.
6. Microscope glass slides.
7. Standard fluorescence or confocal laser scanning microscope.

### 2.2 Analysis of Metabolic Activity

1. Alamar Blue (ThermoFisher). Store at 4 °C and protected from light.
2. High glucose Dulbecco's Modified Eagle Medium (DMEM; ThermoFisher) supplemented with 10% FBS (HyClone) and 1% penicillin–streptomycin (ThermoFisher). Store at 4 °C (*see* **Note 2**).
3. Phenol red-free DMEM (ThermoFisher). Store at 4 °C (*see* **Note 2**).
4. PBS (as above).
5. Sterile 48-well and 24-well plates.
6. Sterilized metal spatula.
7. Black 96-well plate with clear bottom (ThermoFisher).
8. Fluorescence microplate reader.

### 2.3 Analysis of Mechanical Properties

1. PBS (as above).
2. Sterile 24-well plate.
3. Sterilized metal spatula.
4. Instron 3342 machine with a 10 N load cell (adaptor type 00-0, rigid) or a similar mechanical tester.

### 2.4 Hydrogel Digestion

For reproducibility of experiments, enzyme working solutions are made fresh on the day of analysis as the activity of some enzymes decreases with storage and over time, while others are sensitive to repeated thawing and freezing cycles (Table 1; *see* **Note 3**).

1. High glucose DMEM (as above). Store at 4 °C (*see* **Note 2**).
2. High glucose DMEM, serum-free. Store at 4 °C (*see* **Note 2**).
3. Collagenase I. Store at 4 °C. Store aliquots of the stock solution at −20 °C.
4. Trypsin/EDTA, 10×. Store at −20 °C.
5. Collagenase XI. Store at −20 °C. Store aliquots of the stock solution at −20 °C.
6. Hank's Balanced Salt solution (HBSS; Sigma-Aldrich).
7. Dispase II in HBSS. Store at −20 °C.
8. Collagenase V. Store at 4 °C. Store aliquots of the stock solution at −20 °C.
9. PBS (as above; *see* **Note 4**).
10. Sterile 24-well plate.
11. Sterile 15 mL tubes.
12. Water bath.
13. Shaker with adjustable temperature (*see* **Note 5**).
14. Laboratory benchtop centrifuge.

### 2.5 Flow Cytometry

1. Flow cytometry buffer containing 1% FCS and 2 mM EDTA in PBS. Store at 4 °C (*see* **Note 6**).
2. DAPI (ThermoFisher). Store at room temperature and protected from light. Store aliquots of the stock solution at −20 °C.
3. Live/dead viability kit for mammalian cells (ThermoFisher). Store at −20 °C and protected from light.
4. PBS (as above).
5. 100% Ethanol (ThermoFisher).
6. V-bottom polypropylene plates (Sigma-Aldrich).
7. Flow cytometry tubes.
8. Aluminum foil.
9. Laboratory microplate centrifuge.
10. Standard flow cytometer.

## 3 Methods

All procedures are carried out at room temperature unless otherwise specified. When handling chemicals and biological material ensure to wear appropriate personal protective equipment and work in a biological safety cabinet. GelMa hydrogels are prepared as previously reported [18]. Ovcar-4 (kindly provided by Dr. Michelle Lockley, Barts Cancer Institute) or MIACaPa-2 (ATCC; CRL-1420) cells ($0.35 \times 10^6$ cells/mL hydrogel) are encapsulated in 5% (wt/vol) GelMa (50 mg/mL). Briefly, the precursor solution is crosslinked for 10 min and cut into identical-sized constructs (Fig. 1). Each cell-containing hydrogel is placed and maintained free-floating in an individual well of a 24-well plate in 1.5 mL, or 48-well plate in 0.5 mL, in high glucose DMEM culture medium supplemented with 10% FBS and 1% penicillin–streptomycin, replaced twice per week, for up to 28 days.

### 3.1 Analysis of Cell Viability and Microscopy

We recommend assessing cell viability one day post cell encapsulation and any other desired time point over the duration of the experiment. Additionally, brightfield microscopy may be performed to image cell morphology and formation of multicellular spheroids or cellular networks depending on the cell type. Cell-containing GelMa hydrogels were assessed at days 1, 4, 7, and 14 post cell encapsulation (Fig. 2).

1. Prepare a 0.5 μL/mL of calcein-AM and 2 μL/mL of ethidium homodimer-1 in PBS working solution and keep protected from light.
2. Use a sterilized metal spatula and transfer each hydrogel sample into an individual well of a 24-well plate.
3. Wash samples with 1 mL PBS for 15 min at 37 °C.
4. Remove the PBS and add 0.5 mL of the dye-containing PBS solution into each well.
5. Incubate for 20 min at 37 °C protected from light using aluminum foil.
6. Transfer sample on a microscope glass slide and image with a standard fluorescence or confocal laser scanning microscope using a 10/20× air objective.

### 3.2 Analysis of Metabolic Activity

The Alamar Blue reagent is light-sensitive, and we therefore recommend minimizing light exposure over the duration of the experiment. Cell-containing GelMa hydrogels were assessed at days 1, 14, and 28 post cell encapsulation (Fig. 3a). The Alamar Blue reagent is noncytotoxic, therefore, the same hydrogel samples may be analyzed for each time point.

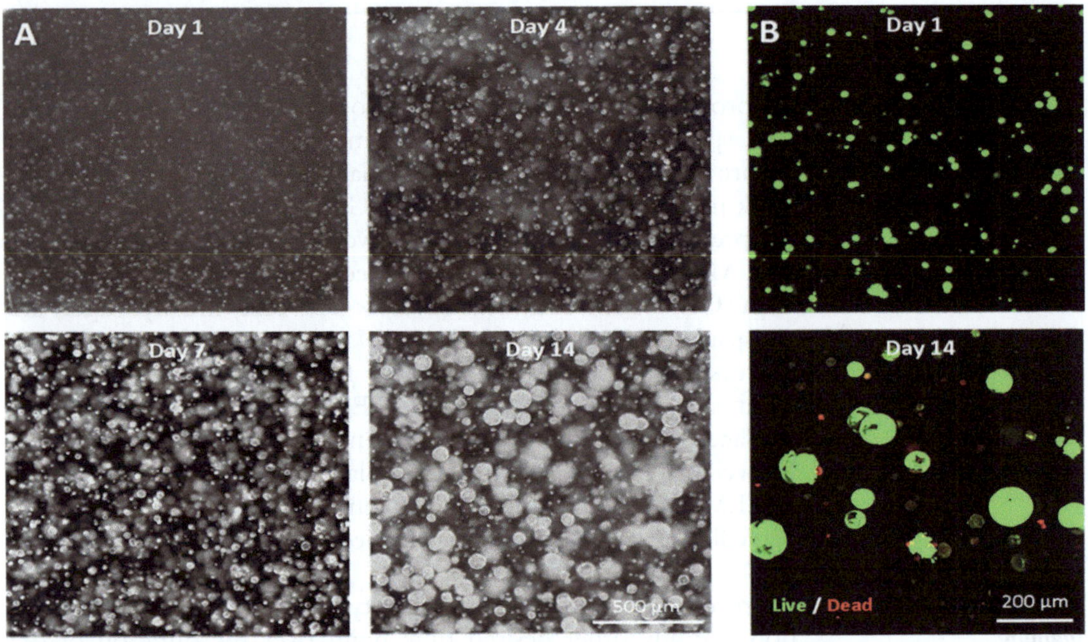


**Fig. 2** Cell growth and morphology in GelMa hydrogels. (**a**) Brightfield images indicate that GelMa hydrogels support cancer cell growth post cell encapsulation and formation of multicellular spheroids over 14 days. (**b**) Confocal micrographs indicate a high cell viability on days 1 (top) and 14 (bottom); live cells stained with green-fluorescent calcein-AM; dead cells stained with red-fluorescent ethidium homodimer-1

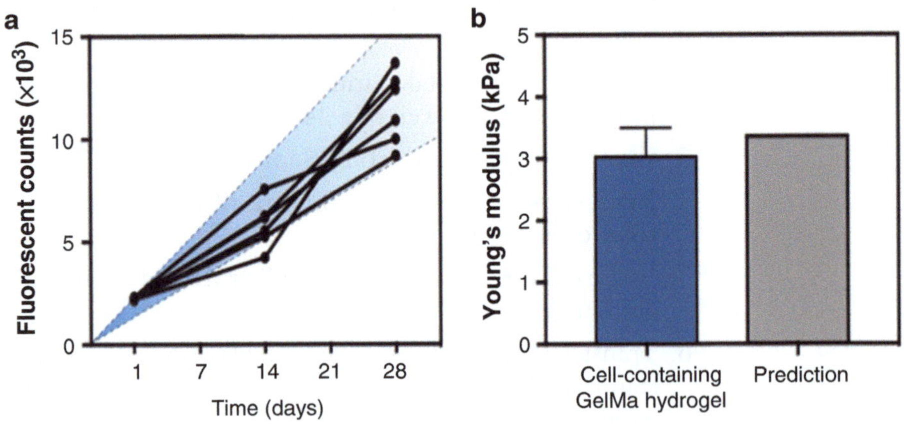


**Fig. 3** Metabolic activity and mechanical properties of cells grown encapsulated within GelMa hydrogels. (**a**) An increase in fluorescence counts, indicative of increased metabolic activity and cell proliferation, was observed over 28 days ($n = 6$). Data were corrected to fluorescence counts from culture medium without cells and normalized to day 1. (**b**) Mechanical properties of cell-containing GelMa hydrogels at day 14 of 3D culture are in good agreement with the empirical prediction using a revised version of the equation previously reported [19]

1. Plate a 1:2 serial dilution of cells into duplicate wells of a 48-well plate by consecutively mixing 500 μL of cell solution with 500 μL of culture medium (*see* **Note 7**). Place the 48-well plate in an incubator overnight at 37 °C for cells to attach.
2. Prepare the Alamar Blue solution by adding 10% Alamar Blue reagent in phenol red-free culture medium as per manufacturer's instruction. A minimum of 400 μL per well is recommended. We previously established a 4% Alamar Blue working solution for cells grown encapsulated within GelMa hydrogels [18].
3. Use a sterilized metal spatula and transfer each hydrogel sample into an individual well of the 48-well plate that contains the serial cell dilution.
4. Remove culture medium from the wells that contain the serial cell dilutions for the standard curve.
5. Add 400 μL of the 10% Alamar Blue working solution to all cell-containing hydrogel and cell-seeded wells. A cell-free hydrogel and culture medium without cells serve as negative controls.
6. Incubate the 48-well plate protected from light for 6 h at 37 °C.
7. Then, transfer 100 μL of the Alamar Blue reaction from each well into a black 96-well plate with clear bottom. Measurements are therefore performed in triplicates.
8. Read fluorescence signals using a fluorescence microplate reader with an excitation wavelength of 544 nm and an emission at 590 nm.
9. Use a sterilized metal spatula and transfer the tested hydrogels into an individual well of a 24-well plate filled with 1 mL PBS.
10. Incubate for 20–30 min or until the hydrogels appear decolored.
11. Then, place cell-containing hydrogels into a new 24-well plate filled with 1.5 mL culture medium and maintain in culture for additional time points and subsequent analysis (*see* **Note 8**).

### 3.3 Analysis of Mechanical Properties

The testing procedure may vary depending on the machine used. Cell-containing GelMa hydrogels were assessed on day 14 post cell encapsulation (Fig. 3b).

1. Use a sterilized metal spatula and transfer hydrogel samples into a 24-well plate filled with 1 mL PBS.
2. Mount the 10 N load cell onto the Instron 3342 machine and calibrate.
3. Place the sample in the center of the test pad and gently lower the load cell until it touches the surface of the hydrogel. This

should be the point where the force starts significantly deviating from zero. During this process, keep the sample hydrated but not submerged in PBS.

4. Record the anvil length of the individual sample (depending on the software, you may then be required to reset the extension to zero before proceeding to compression).
5. Reset the load cell to zero.
6. Begin sample compression (*see* **Note 9**).
7. Upon test completion, elevate the load cell and remove the tested sample. This should not be kept for further analysis as the hydrogel architecture is likely compromised.
8. To determine the Young's modulus, use natural logarithms of the stress and strain values to obtain a stress-strain curve for each hydrogel, following (*see* **Note 10**):

$$E = \text{stress}/\text{strain} = (\text{force}/\text{area})/(\text{extension}/\text{length})$$

The Young's modulus of each hydrogel is calculated as the slope of the best-fitted trendline. In Fig. 3, empirical predictions are calculated for 5% (wt/vol) GelMa with 10 min UV exposure time following this equation:

$$E = E_{\max}\left(\frac{C}{100}\right)^n\left(1 - e^{-(t-A)/\tau}\right)$$

where the compressive modulus ($E$, kPa) of GelMa hydrogels is calculated based on the GelMa concentration ($C$, % wt/vol) and UV exposure time ($t$, min), with $E_{\max}$ corresponding to 7995, $A$ equals 4.2, $\tau$ equals 11.5.

### 3.4 Hydrogel Digestion

Each digestion protocol was tested on a minimum of five cell-containing GelMa hydrogels, with three biological replicates on day 14 post cell encapsulation (Fig. 4).

1. Prepare the collagenase I working solution (Table 1) by weighing collagenase I powder in serum-free high glucose DMEM (*see* **Notes 2**, **3** and **11**).
2. Prepare the trypsin/EDTA working solution (Table 1) by thawing the stock solution at 37 °C using a water bath.
3. Prepare the collagenase XI and dispase II working solution (Table 1) by adding 25 mg of collagenase XI and 25 mL of dispase II to 200 mL serum-free high glucose DMEM (*see* **Notes 2** and **3**).
4. Prepare the collagenase V working solution (Table 1) by diluting collagen V in HBSS.

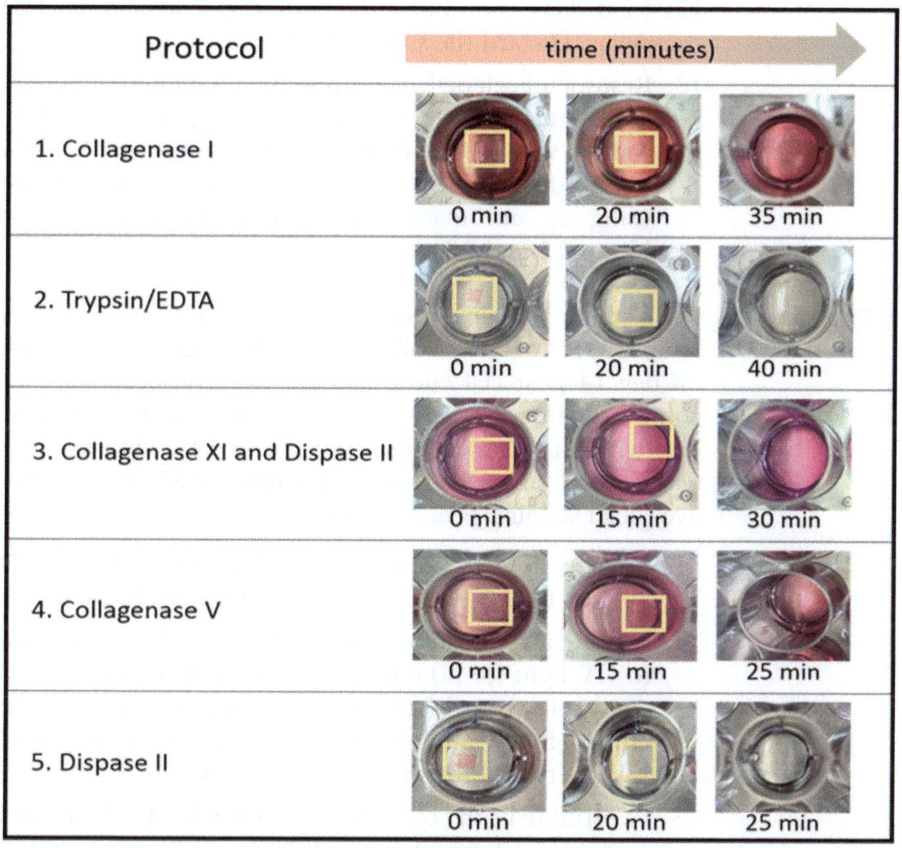


**Fig. 4** GelMa hydrogel digestion and relative time of digestion using five different enzymatic approaches. Hydrogel digestion was completed in 25–40 min. Collagenases I and V powders confer a brown coloration to the digestion solution, which is also caused by the presence of phenol red in the culture medium

5. Prepare the dispase II working solution (Table 1) in PBS (*see* **Note 3**).
6. Place each hydrogel sample in an individual well of a new 24-well plate and wash with 1 mL PBS for 20–30 min.
7. Replace the PBS with 1 mL of each digestion working solution to each sample.
8. Incubate using a shaker at 37 °C and 100 rpm until the hydrogel is completely dissolved; usually after 25–40 min (Fig. 4).
9. After 10 min, gently pipette the digest up and down to allow mechanical dissolution.
10. Gently pipette the solution up and down and transfer the content of each well to a 15 mL tube.
11. Add 2 mL FBS-containing high glucose DMEM to stop the reaction (*see* **Note 2**).

12. Centrifuge the solution for 5 min at 250 × *g* to obtain a cell pellet. Discard the supernatant.
13. Resuspend cell pellets in FBS-containing high glucose DMEM (*see* **Note 2**). Depending on the application, cells may be washed with PBS and centrifuged again to ensure the removal of the digestion solution. Cell counting may be performed using an automated cell counter or hemocytometer (*see* **Notes 12** and **13**).

### 3.5 Flow Cytometry

Perform the staining procedure on ice and lower the centrifuge temperature to 4 °C. Following the hydrogel digestion procedure on day 14 post cell encapsulation, the live/dead staining, DAPI counterstaining and flow cytometry were performed on the same day, without fixation (Fig. 5; *see* **Note 14**). Different conditions were tested on cells recovered from GelMa hydrogels for the live/dead staining procedure (Table 2).

1. Prepare flow cytometry buffer by mixing 1.25 g BSA and 200 μL EDTA (0.5 M) in 50 mL PBS (*see* **Note 6**).
2. As UV-treated control, expose 200 μL cell suspension in a tube to UV light for 30 min. For an ethanol-treated control, resuspend cells from Methods 3.4.12 in 1 mL of prechilled 100% ethanol for 1 h and then mix with 1 mL cell suspension without UV or ethanol treatment (50:50 ratio).
3. Centrifuge both controls from step 3.5.2 for 5 min at 250 × *g* at 4 °C and resuspend cell pellets including samples from Subheading 3.4, **step 12** in 200 μL culture medium and transfer them into V-bottom polypropylene plates. For a medium control, resuspend a cell pellet in 200 μL flow cytometry buffer (Table 2); here nine wells for nine conditions tested (Table 2).
4. Centrifuge the plate for 5 min at 250 × *g* at 4 °C.
5. Turn the plate over to remove supernatant. Then, flick the plate and blot once against clean paper towel.
6. Wash with 200 μL flow cytometry buffer and centrifuge the plate for 5 min at 250 × *g* at 4 °C.
7. Thaw live/dead reagents and DAPI.
8. Prepare a working solution of calcein-AM (50 μM) in DMSO. An ethidium homodimer-1 stock solution (2 mM) in DMSO is provided with the kit. Add 2 μL of calcein-AM solution and 4 μL of the ethidium homodimer-1 stock solution to 1 mL of PBS to create a master mix.
9. Resuspend cell pellets in 100 μL master mix, or single green (calcein-AM) or red (ethidium homodimer-1) dye only (Table 2), and incubate for at least 15–20 min protected from light using aluminum foil.

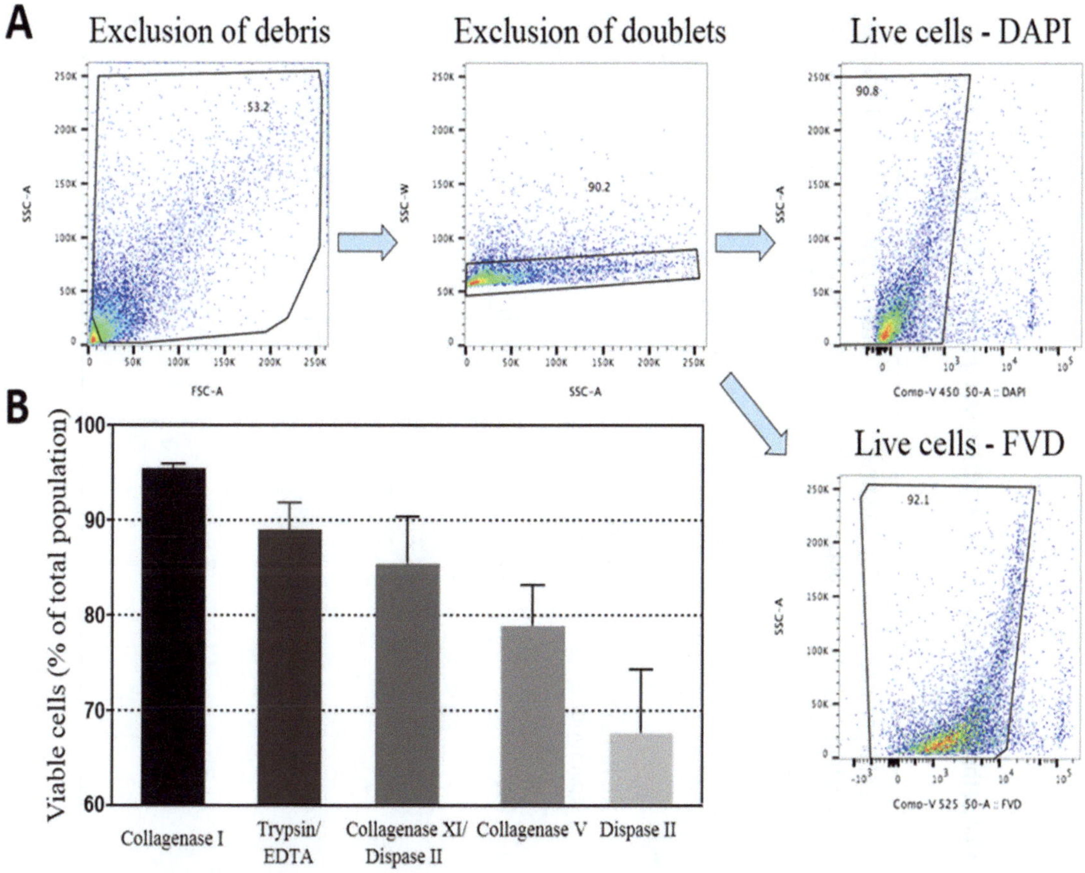


**Fig. 5** Flow cytometry and detection of cell viability. (**a**) Representative flow cytometry gating strategy for exclusion of cell debris and doublets and detection of viable cells using DAPI (450/50, *X*-axis) or alternatively fixable viability dye (FVD; 525/50, *X*-axis). (**b**) High cell viability across the five different enzymatic approaches tested, with collagenase I: 95.5 $\pm$ 0.5%, trypsin/EDTA: 89.0 $\pm$ 2.6%, collagenase XI and dispase II: 85.5 $\pm$ 4.4%, collagenase V: 78.9 $\pm$ 3.7% and dispase II: 67.7 $\pm$ 6.0% cell viability (mean $\pm$ SEM of five biological replicates with five gels each). Collagenase I and trypsin/EDTA-containing digestion procedures resulted in minor cell loss, while dispase II-containing procedures resulted in increased cell loss

10. Transfer the content of each well into an individual flow cytometry tube.
11. Prepare a working solution of DAPI (1 μg/mL) in PBS and add 100 μL to each sample and incubate for at least 15–20 min protected from light using aluminum foil.
12. Perform flow cytometry and ensure to include a laser that covers the detection wavelengths for calcein-AM (530/30 nm), ethidium homodimer-1 (610/20 nm) and DAPI (450/50 nm). Signals from cell debris were eliminated during data acquisition by gating (Fig. 5). Use single color stained cells for compensation. The population separates into two groups, with live cells stained in green and dead cells stained in red and blue.

**Table 2**
**Test conditions for flow cytometry and percentages of calcein-AM (CaAM)/ethidium homodimer-1 (EtH1)/DAPI staining of GelMa-retrieved cells using collagenase I digestion**

| Sample | Description | Staining | CaAM-positive population (%) | EtH- positive population (%) | DAPI-negative population (%) | Live/dead detection |
|---|---|---|---|---|---|---|
| 1 | Manufacturer's instruction; without DAPI | 0.1 μM CaAM, 8 μM EtH1 | 84.5 | 7.0 | 0.3 | Live |
| 2 | Manufacturer's instruction; with DAPI | 0.1 μM CaAM, 8 μM EtH1, 1 μg/mL DAPI | 90.5 | 7.3 | 91.2 | Live |
| 3 | Double concentration | 0.2 μM CaAM, 16 μM EtH1, 1 μg/mL DAPI | No distinct separation | No distinct separation | 91.7 | Live |
| 4 | Green dye only | 0.2 μM CaAM | 89.9 | – | 91.0 | Live |
| 5 | Red dye only | 16 μM EtH1 | – | 18.5 | 0.5 | Live |
| 6 | Manufacturer's instruction; UV-treated | 0.1 μM CaAM, 8 μM EtH1, 1 μg/mL DAPI | 3.4 | 81.7 | 5.8 | Dead |
| 7 | Manufacturer's instruction; Ethanol-treated: live cells (50:50) | 0.1 μM CaAM, 8 μM EtH1, 1 μg/mL DAPI | 49.6 | 58.0 | 47.6 | 50:50 |
| 8 | Manufacturer's instruction; phenol red-free DMEM | 0.1 μM CaAM, 8 μM EtH1, 1 μg/mL DAPI | 81.3 | 7.8 | 93.5 | Live |
| 9 | Unstained control | Unstained control | – | – | – | – |

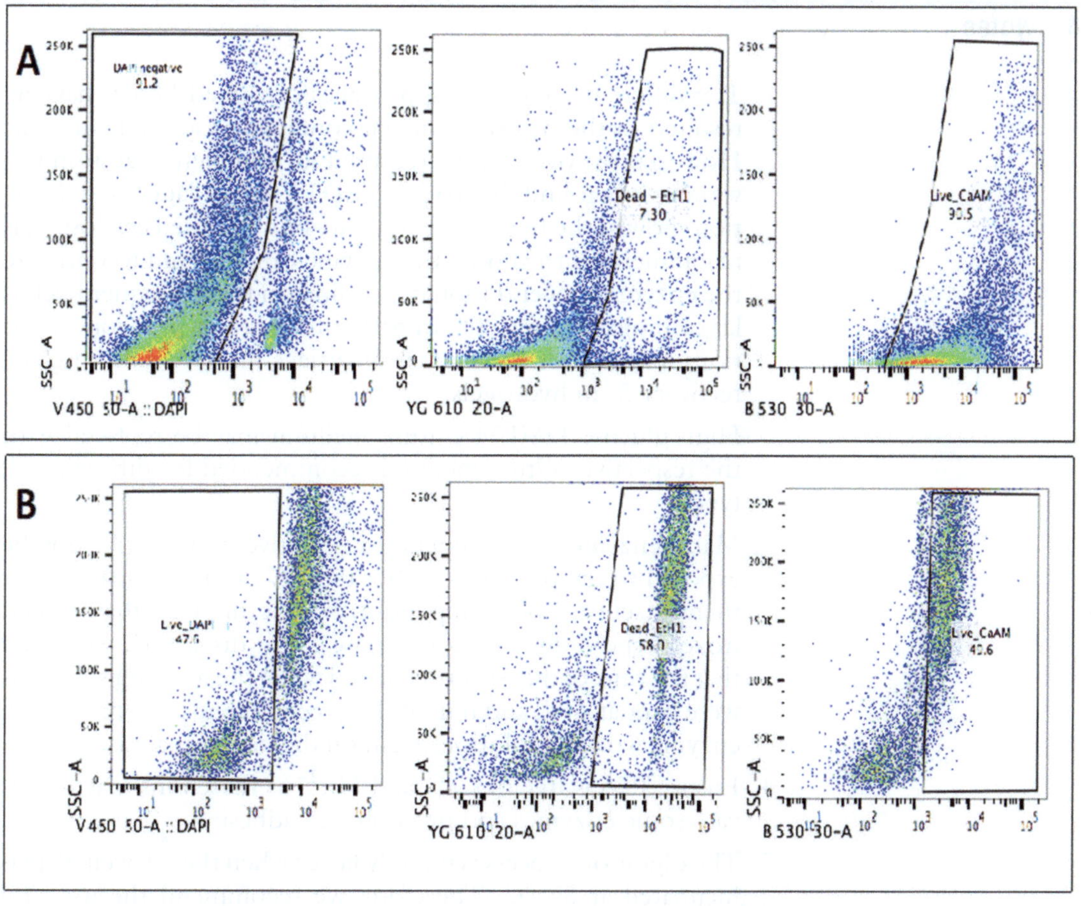


**Fig. 6** Calcein-AM (CaAM)/ethidium homodimer-1 (EtH1)/DAPI staining of GelMa-retrieved cells. Representative plots are shown from live/dead staining detection of samples 2 (**a**; Table 2; DAPI 450/50, EtH1 610/20, CaAM 530/30, *X*-axis) and 7 (**b**; Table 2; DAPI 450/50, EtH1 610/20, CaAM 530/30, *X*-axis)

Following the manufacturer's instruction (samples 1 and 2; Fig. 6) resulted in a clear separation of two cell populations, whereas double-concentrated calcein-AM and ethidium homodimer-1 (sample 3) showed no clear cell separation. A distinct separation of cells was observed with ethidium homodimer-1 and DAPI counterstaining, while calcein-AM occasionally yielded in a heterogeneous population with no clear separation (sample 3). The phenol red-free condition, whereby hydrogels were cultured and digested with phenol red-free DMEM, had no impact on the staining efficacy (sample 8). Our results indicate that including DAPI counterstaining to the live/dead protocol yielded the clearest cell separation of both live and dead cell populations (sample 3). An increased dye concentration is recommended when either the green or red dye only is used (samples 4 and 5).

## 4 Notes

1. The different enzymatic approaches described herein have all resulted in the digestion of cell-containing GelMa hydrogels. However, the overall time required for complete digestion was variable. We found that trypsin/EDTA and collagenase I digestion showed the highest cell viability but increased cell loss. It is recommended to choose the enzyme carefully and to optimize the enzyme concentration depending on the subsequent cellular or molecular analysis. In particular, high trypsin concentrations may impact the analysis of surface markers following cell recovery from hydrogels.
2. High glucose DMEM culture medium may be replaced with the respective culture medium recommended for different cell types.
3. Many enzymes are temperature-sensitive and should not be exposed to temperatures higher than 37 °C as this may render them inactive. For example dispase II cannot be thawed and frozen repeatedly. For collagenase I and dispase II, we noted that the time of digestion decreased when the working solutions were kept at 4 °C and reused. It is recommended to prepare all enzyme working solutions fresh on the day of analysis.
4. Ensure PBS is free of $Ca^{2+}$ and $Mg^{2+}$ as these ions may inactivate some enzymes and promote cell adhesion.
5. The digestion process is notably faster when the 24-well plate is incubated at 37 °C. Therefore, we recommend the use of a shaker with an incubator chamber with adjustable temperature.
6. The flow cytometry buffer should be made fresh every 2–4 weeks. Note that higher BSA concentrations may cause auto-fluorescence and need to be optimized accordingly. Alternatively, 2% FBS can be used instead of 2.5% BSA.
7. The serial dilution of cells needs to cover the range of cell numbers that are expected in the 3D culture over time. In our experiments, we have approximately $14.5 \times 10^3$ cells per hydrogel construct post cell encapsulation and therefore set up a serial dilution ranging from $0–2.0 \times 10^5$ cells.
8. The Alamar Blue reagent is nontoxic and allows noninvasive and continuous measurements of the same sample [28]. Alternatively, samples can be stored at −80 °C in 2 mL tubes for quantification of the DNA content using CyQuant or PicoGreen assays [18, 20].
9. The Bluehill 3 software was used for data collection with the stop threshold set at 50% of the total sample extension. Note that this in an arbitrary setting and may be changed according to the sample thickness and load cell compression speed.

10. To predict the Young's modulus of each hydrogel sample, the anvil length (height) and surface area of the sample need to be determined with a digital caliper. The anvil length of each individual sample should be recorded prior to compression testing.
11. Collagenase I used herein is a highly static powder and difficult to weigh accurately. We recommend weighing an excess amount and consequently adjusting to a final 1 mg/mL concentration in serum-free high glucose DMEM.
12. Caution when using an automated cell counter. Unless it is suitable to detect small cell numbers, counting may not be accurate to quantify cell retrieval from a hydrogel sample on day 1 of 3D cell culture. In our experiments, we have approximately $14.5 \times 10^3$ cells per hydrogel construct on day 1 of 3D cell culture, which was too small to be accurately detected using an automated cell counter.
13. Samples containing high cell numbers and debris as a result from the hydrogel digestion may increase the time for flow cytometry analysis or even block the flow cytometer. It is recommended to pass the retrieved cells through a nylon mesh or filter (70 μm) prior to proceeding with the live/dead staining.
14. The live/dead staining and DAPI counterstaining were carried out immediately following cell recovery from GelMa hydrogels to ensure a high cell viability. Thus, it is recommended to perform the hydrogel digestion, staining and flow cytometry analysis on the same day. When using primary cells, staining and flow cytometry analysis should be performed within a couple of hours to prevent clumping or cell death.

## Acknowledgements

We thank Dr. Michelle Lockley for providing Ovcar-4 cells, Dr. Laura Lecker for her assistance and guidance with the GelMa synthesis, and acknowledge the technical assistance of the microscopy and flow cytometry facilities.

### References

1. Loessner D, Holzapfel BM, Clements JA (2014) Engineered microenvironments provide new insights into ovarian and prostate cancer progression and drug responses. Adv Drug Deliv Rev 79-80:193–213. https://doi.org/10.1016/j.addr.2014.06.001
2. Vennin C, Chin VT, Warren SC, Lucas MC, Herrmann D, Magenau A et al (2017) Transient tissue priming via ROCK inhibition uncouples pancreatic cancer progression, sensitivity to chemotherapy, and metastasis. Sci Transl Med 9(384). https://doi.org/10.1126/scitranslmed.aai8504
3. De Jaeghere E, De Vlieghere E, Van Hoorick J, Van Vlierberghe S, Wagemans G, Pieters L et al (2018) Heterocellular 3D scaffolds as

biomimetic to recapitulate the tumor microenvironment of peritoneal metastases in vitro and in vivo. Biomaterials 158:95–105. https://doi.org/10.1016/j.biomaterials.2017.12.017

4. Liu J, Tan Y, Zhang H, Zhang Y, Xu P, Chen J et al (2012) Soft fibrin gels promote selection and growth of tumorigenic cells. Nat Mater 11 (8):734–741. https://doi.org/10.1038/nmat3361
5. Klotz BJ, Gawlitta D, Rosenberg A, Malda J, Melchels FPW (2016) Gelatin-methacryloyl hydrogels: towards biofabrication-based tissue repair. Trends Biotechnol 34(5):394–407. https://doi.org/10.1016/j.tibtech.2016.01.002
6. Hughes CS, Postovit LM, Lajoie GA (2010) Matrigel: a complex protein mixture required for optimal growth of cell culture. Proteomics 10(9):1886–1890. https://doi.org/10.1002/pmic.200900758
7. Tabriz AG, Hermida MA, Leslie NR, Shu W (2015) Three-dimensional bioprinting of complex cell laden alginate hydrogel structures. Biofabrication 7(4):045012. https://doi.org/10.1088/1758-5090/7/4/045012
8. Highley CB, Prestwich GD, Burdick JA (2016) Recent advances in hyaluronic acid hydrogels for biomedical applications. Curr Opin Biotechnol 40:35–40. https://doi.org/10.1016/j.copbio.2016.02.008
9. Cavo M, Caria M, Pulsoni I, Beltrame F, Fato M, Scaglione S (2018) A new cell-laden 3D Alginate-Matrigel hydrogel resembles human breast cancer cell malignant morphology, spread and invasion capability observed "in vivo". Sci Rep 8(1):5333. https://doi.org/10.1038/s41598-018-23250-4
10. Rebelo SP, Pinto C, Martins TR, Harrer N, Estrada MF, Loza-Alvarez P et al (2018) 3D-3-culture: A tool to unveil macrophage plasticity in the tumour microenvironment. Biomaterials 163:185–197. https://doi.org/10.1016/j.biomaterials.2018.02.030
11. Tang Y, Huang B, Dong Y, Wang W, Zheng X, Zhou W et al (2017) Three-dimensional prostate tumor model based on a hyaluronic acid-alginate hydrogel for evaluation of anti-cancer drug efficacy. J Biomater Sci Polym Ed 28 (14):1603–1616. https://doi.org/10.1080/09205063.2017.1338502
12. Caliari SR, Burdick JA (2016) A practical guide to hydrogels for cell culture. Nat Methods 13 (5):405–414
13. Studle C, Vallmajo-Martin Q, Haumer A, Guerrero J, Centola M, Mehrkens A et al (2018) Spatially confined induction of endochondral ossification by functionalized hydrogels for ectopic engineering of osteochondral tissues. Biomaterials 171:219–229. https://doi.org/10.1016/j.biomaterials.2018.04.025
14. Yue X, Nguyen TD, Zellmer V, Zhang S, Zorlutuna P (2018) Stromal cell-laden 3D hydrogel microwell arrays as tumor microenvironment model for studying stiffness dependent stromal cell-cancer interactions. Biomaterials 170:37–48. https://doi.org/10.1016/j.biomaterials.2018.04.001
15. Hedegaard CL, Collin EC, Redondo-Gómez C, Nguyen LTH, Ng KW, Castrejón-Pita AA et al (2018) Hydrodynamically guided hierarchical self-assembly of peptide–protein bioinks. Adv Funct Mater 28(16). https://doi.org/10.1002/adfm.201703716
16. Loessner D, Stok KS, Lutolf MP, Hutmacher DW, Clements JA, Rizzi SC (2010) Bioengineered 3D platform to explore cell-ECM interactions and drug resistance of epithelial ovarian cancer cells. Biomaterials 31(32):8494–8506. https://doi.org/10.1016/j.biomaterials.2010.07.064
17. Kaemmerer E, Melchels FP, Holzapfel BM, Meckel T, Hutmacher DW, Loessner D (2014) Gelatine methacrylamide-based hydrogels: an alternative three-dimensional cancer cell culture system. Acta Biomater 10 (6):2551–2562. https://doi.org/10.1016/j.actbio.2014.02.035
18. Loessner D, Meinert C, Kaemmerer E, Martine LC, Yue K, Levett PA et al (2016) Functionalization, preparation and use of cell-laden gelatin methacryloyl-based hydrogels as modular tissue culture platforms. Nat Protoc 11 (4):727–746. https://doi.org/10.1038/nprot.2016.037
19. Schuurman W, Levett PA, Pot MW, van Weeren PR, Dhert WJ, Hutmacher DW et al (2013) Gelatin-methacrylamide hydrogels as potential biomaterials for fabrication of tissue-engineered cartilage constructs. Macromol Biosci 13(5):551–561. https://doi.org/10.1002/mabi.201200471
20. Bartnikowski M, Bartnikowski NJ, Woodruff MA, Schrobback K, Klein TJ (2015) Protective effects of reactive functional groups on chondrocytes in photocrosslinkable hydrogel systems. Acta Biomater 27:66–76. https://doi.org/10.1016/j.actbio.2015.08.038
21. Erkoc P, Seker F, Bagci-Onder T, Kizilel S (2018) Gelatin methacryloyl hydrogels in the absence of a crosslinker as 3D glioblastoma multiforme (GBM)-mimetic microenvironment. Macromol Biosci 18(3). https://doi.org/10.1002/mabi.201700369

22. McBeth C, Lauer J, Ottersbach M, Campbell J, Sharon A, Sauer-Budge AF (2017) 3D bioprinting of GelMA scaffolds triggers mineral deposition by primary human osteoblasts. Biofabrication 9(1):015009. https://doi.org/10.1088/1758-5090/aa53bd

23. Valdez J, Cook CD, Ahrens CC, Wang AJ, Brown A, Kumar M et al (2017) On-demand dissolution of modular, synthetic extracellular matrix reveals local epithelial-stromal communication networks. Biomaterials 130:90–103. https://doi.org/10.1016/j.biomaterials.2017.03.030

24. Bonnier F, Keating ME, Wrobel TP, Majzner K, Baranska M, Garcia-Munoz A et al (2015) Cell viability assessment using the Alamar blue assay: a comparison of 2D and 3D cell culture models. Toxicol In Vitro 29(1):124–131. https://doi.org/10.1016/j.tiv.2014.09.014

25. Delaine-Smith RM, Burney S, Balkwill FR, Knight MM (2016) Experimental validation of a flat punch indentation methodology calibrated against unconfined compression tests for determination of soft tissue biomechanics. J Mech Behav Biomed Mater 60:401–415. https://doi.org/10.1016/j.jmbbm.2016.02.019

26. Young S, Wong M, Tabata Y, Mikos AG (2005) Gelatin as a delivery vehicle for the controlled release of bioactive molecules. J Control Release 109(1–3):256–274. https://doi.org/10.1016/j.jconrel.2005.09.023

27. Brown MA, Wallace CS, Anamelechi CC, Clermont E, Reichert WM, Truskey GA (2007) The use of mild trypsinization conditions in the detachment of endothelial cells to promote subsequent endothelialization on synthetic surfaces. Biomaterials 28(27):3928–3935. https://doi.org/10.1016/j.biomaterials.2007.05.009

28. Quent VM, Loessner D, Friis T, Reichert JC, Hutmacher DW (2010) Discrepancies between metabolic activity and DNA content as tool to assess cell proliferation in cancer research. J Cell Mol Med 14(4):1003–1013. https://doi.org/10.1111/j.1582-4934.2010.01013.x

# Chapter 2

# Bioengineered Microtissue Models of the Human Bone Metastatic Microenvironment: A Novel In Vitro Theranostics Platform for Cancer Research

**Nathalie Bock**

## Abstract

One of the major limitations of studying cancer in distant sites is the lack of representative laboratory models that mimic the biological processes occurring in vivo. In this protocol, we demonstrate the application of melt electrowriting technology (MEW) to provide 3D microfiber scaffolds suitable for this purpose. Using primary human cells, MEW scaffolds support the reproducible formation of human bone-like 3D microenvironments. Co-culture with human cancer cells provides an in vitro bioengineered model of metastases in bone, suitable for investigating cell–cell and cell–matrix interactions between bone and cancer cells. By proposing variations to standard tissue histology, immunohistochemistry, immunofluorescence, and 3D imaging techniques, we show how to characterize cell morphology and protein expression in a reproducibly engineered bone metastatic microtissue.

**Key words** Bone, Metastasis, Prostate cancer, Cancer model, Osteoblasts, Scaffold, Tissue engineering, Melt electrowriting, Microtissue, Tumor microenvironment

## 1 Introduction

Hormonal-dependent cancers such as prostate cancer and breast cancer preferentially metastasize to bone. This condition is associated with poor quality of life, poor prognosis, and a major cause of mortality in these patients, due to very few treatment options available [1]. The mechanisms associated with bone metastasis and disease progression are still poorly understood, due to the complexity of recapitulating human in vivo cancer–bone interactions in laboratory models. With a growing consensus that the tumor microenvironment is a key modulator of response to therapy [2], three-dimensional (3D) in vitro models represent a promising platform to bridge the gap between simplistic two-dimensional (2D) models of cells grown in monolayers, and complex human cancer xenografts in animal hosts [3]. This approach is essential to understand biological processes in a more physiological context and

Jyotsna Batra and Srilakshmi Srinivasan (eds.), *Theranostics: Methods and Protocols*, Methods in Molecular Biology, vol. 2054, https://doi.org/10.1007/978-1-4939-9769-5_2, © Springer Science+Business Media, LLC, part of Springer Nature 2019

warrants the discovery and validation of novel contributing factors that facilitate cancer progression in the bone compartment. Finally, platforms that are simple and inexpensive will allow scalability for their translation into high-throughput settings for personalized therapies.

Recent 3D in vitro bone metastasis models include rotary culture systems [4], microfluidic devices [5] and bioreactors [6]. These devices however face reproducibility issues and do not provide 3D cell–matrix interactions [3]. Conversely, recent advances in tissue engineering technologies have enabled to engineer metastatic microenvironments able to recapitulate some of the features of native metastatic tissues [7–10]. Synthetic scaffold materials allow to engineer bone-like microenvironments derived from primary human bone cells, which can further be co-cultured with metastatic cell lines in a direct mode [11–13]. Direct contact models are more relevant; however, difficulties arise in data analysis, as often cell populations cannot be separated, preventing deep mechanistic analysis. Technical difficulties also arise with 3D analysis, due to the interference of synthetic scaffold materials, which challenges downstream analysis via traditional imaging technologies.

Here we demonstrate how additive manufacturing using melt electrowriting (MEW) technology provides a versatile synthetic material, which can be reproducibly cultured with primary human cells. Following osteogenic differentiation, these biomaterial/cellular constructs lead to patient-specific mineralized 3D microtissues that partly simulate the native bone microenvironment. Further co-culture with prostate cancer cells generate an appropriate in vitro 3D model of metastases found in the bone. Specifically, this model is suitable to study the final step of the metastatic cascade, when cancer cells have reached the bone. This model of established metastases can be analyzed efficiently with modified versions of standard histology, immunohistochemistry, immunofluorescence techniques, and the use of 3D imaging technologies (Fig. 1) [14–16]. Ultimately, our microtissue model and the standards defined in this protocol will be able to provide a suitable theranostics platform that will lead to faster discovery and implementation of novel therapies against bone metastasis [17, 18].

## 2 Materials

### 2.1 Scaffold Manufacturing

#### 2.1.1 Materials and Consumables

1. Medical-grade polycaprolactone (mPCL, PURASORB PC12, 2 mm pellets, molecular weight 95–140 kDa, Corbion Purac, The Netherlands).
2. Laboratory wipes.
3. Ziploc bags.

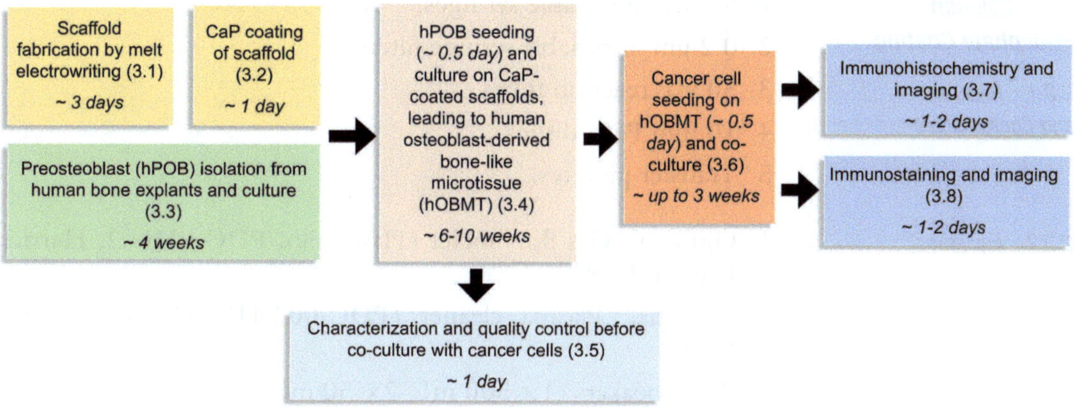


**Fig. 1** Work flow and expected durations for each step of the protocol

*2.1.2 Equipment*

1. Melt electrowriting apparatus, built in-house according to [19].
2. 3CC clear Luer-Lock plastic syringe.
3. 23 G tapered needle.
4. Laser cutter.
5. Laboratory bench oven.
6. Corel Draw software.
7. Manually written G-Code (incremental type) provided below. This code will generate the layer-by-layer manufacturing of a microfiber scaffold with overall size 90 × 70 mm with a spacing of 150 μm in between the fibers, comprising 80 layers in the *z* direction in a '0°–90°' pattern, yielding approximately 600 μm in the *z* direction.

   Please note that in order to save space, '*' in the code means a new line, i.e. pressing 'Enter'. Make sure to replace all '*' by pressing the 'Enter' key in the final code. Any comments found after the percentage sign (%) are notes for the user that do not impact the code and can be removed.

   G17 G21 G40 G49 G54 G80 G94 F450 *%G91 is relative coordinates!* *F1000 *G91 *G1 x10 *M98 p1 l20 *% trail print for jet stabilization* *G1 y5 *M98 p1236 l40 *% number of layers in overall scaffold construct--40X and 40Y* *G1 x-10 **M30 **o1236 *M98 p1237 l300 *% loop size/FD* *G1 x90 *G1 y-90 *M98 p1238 l300 *% loop size/FD* *M99 **o1237 *%x loops* *G1 x90 *G1 y0.15 *G1 x-90 *G1 y0.15 *M99 **o1238 *%y loops* *G1 x-0.15 *G1 y90 *G1 x-0.15 *G1 y-90 *M99 **o1 *% trail loop* x100 *y0.2 *x-100 *y0.2 *M99 *M2
8. Mach3 CNC control software (Artsoft, USA).

### 2.2 Calcium Phosphate Coating

#### 2.2.1 Materials and Consumables

1. 50 mL disposable syringes.
2. 0.2 μm disposable syringe filters.
3. 50 mL reaction tubes.
4. 10 mm Petri dishes.
5. Transfer pipettes.

#### 2.2.2 Equipment

1. Optional: Gas flow mixer (PlasmaFlo, PDC-FMG-2, Harrick Plasma, USA).
2. Optional: Plasma cleaner (PDC-002-HP, Harrick Plasma, USA).
3. Pyrex beakers (1× 500 mL, 2× 50 mL).
4. 500 mL measuring cylinder.
5. 500 mL Schott glass bottle.
6. Two medium magnetic stirring bars.
7. Magnetic stirrer.
8. Metal spatula and metal tweezers.
9. Calibrated pH meter.
10. Optional: pH strips.
11. Vacuum desiccator (vacuum power 0.01 bar).
12. Water bath set to 37 °C.

#### 2.2.3 Reagents

1. Sterile ultrapure water ($ddH_2O$): obtained by purifying deionized water, and attaining sensitivity of 18 MΩ-cm at 25 °C.
2. Ethanol: 70% v/v solution in water.
3. Hydrochloric acid: ≥32% solution in water.
4. 10× Simulated Body Fluid (SBF) solution: Rinse all glassware and magnetic stirring bars with $ddH_2O$ prior to start (critical). Prepare *Solution A* by dissolving, in the exact order, the chemicals listed in Table 1 in approximately 300 mL $ddH_2O$ in a 500 mL beaker, under magnetic stirring until dissolved. In a 50 mL beaker, prepare *Solution B* by dissolving 710.0 mg of

**Table 1**
**Reagent amounts and order of use for SBF 10×**

| Reagent | Amount (g) | Order |
|---|---|---|
| Sodium chloride (NaCl) | 29.215 | 1 |
| Potassium chloride (KCl) | 0.187 | 2 |
| Calcium chloride dihydrate ($CaCl_2 \cdot 2H_2O$) | 1.838 | 3 |
| Magnesium chloride hexahydrate ($MgCl_2 \cdot 6H_2O$) | 0.508 | 4 |

disodium hydrogen phosphate powder ($Na_2HPO_4$) in 15 mL dd$H_2O$ under magnetic stirring until dissolved (*see* **Note 1**). Place calibrated and clean pH meter probe in the beaker containing *Solution A* to monitor pH. Under stirring, carefully add *Solution B* to *Solution A* dropwise, ensuring that the pH remains between 1.5 and 3.5 (absolutely below 4) by HCl titration. When *Solution B* has been fully added, rinse the beaker with 5 mL dd$H_2O$. Add the 5 mL from last step to *Solution A* dropwise, while maintaining the pH at 3.7–3.9. Transfer the solution to a 500 mL measuring cylinder. Top up the volume to 500 mL with dd$H_2O$. Ensure that the pH is 4.0. This solution is referred to as SBF 10×. Transfer SBF 10× solution to a 500 mL glass bottle, store at 4 °C and use within 4 weeks (*see* **Note 2**). On the coating day, the use of SBF 10× will require the addition of sodium hydrogen carbonate ($NaHCO_3$) powder, as explained in Subheading 3.

5. 2 M NaOH: dissolve 4 g of sodium hydroxide (NaOH) in 50 mL dd$H_2O$ (needed only if the gas flow mixer and plasma cleaner are not available).
6. 0.5 M NaOH: dissolve 1 g of NaOH in 50 mL dd$H_2O$ (*see* **Note 3**).

### 2.3 In Vitro Cell Culture

#### 2.3.1 Materials and Consumables

1. Sterile reaction tubes (15, 50 mL).
2. Sterile Pasteur (glass) pipettes.
3. Sterile serological pipettes (5, 10, 25 mL).
4. Waste bottle connected to vacuum for liquid aspiration.
5. Biodine or bleach for aspirated liquid neutralization.
6. Sterile pipette tips (no filter needed, 1000, 200, 20 μL).
7. Cell counting slides for automated cell counter.

#### 2.3.2 Equipment

1. Tube racks for 15 and 50 mL tubes.
2. Pipettors (P1000, P200, P20).
3. Serological pipettors.
4. Sterile metal spatula and metal tweezers (autoclaved).
5. Biological safety cabinet class II (laminar flow).
6. Automated cell counter.
7. Swing-bucket centrifuge suitable for 15 mL reaction tubes.
8. Humidified cell culture incubator set to 37 °C and 95% air/5% $CO_2$.

#### 2.3.3 Reagents

1. Ethanol: 70% v/v solution in water.

### 2.4 Human Preosteoblast (hPOB) Isolation and Expansion

#### 2.4.1 Materials and Consumables

1. Human bone tissue obtained from donors undergoing knee or hip replacement surgery under informed consent and appropriate ethics documents, as per country regulations.
2. Tissue culture flask with reclosable lid 150 $cm^2$ (product 90552, TPP, Switzerland) or Tissue culture flasks 175 $cm^2$. *See* **Note 4**.
3. 1 mL cryovials.
4. Cryoboxes.

#### 2.4.2 Equipment

1. Bone curette.
2. Bone-cutting rongeurs.
3. Forceps.
4. Cell freezing container (such as Mr. Frosty™, Nalgene, USA), allowing 1 °C/min for rate of cooling once placed in −80 °C freezer.
5. −80 °C freezer.
6. Liquid nitrogen tank for long term cell storage.

#### 2.4.3 Reagents

1. Sterile phosphate buffer saline (PBS-P/S): 99% v/v divalent ion-free solution, 1% v/v penicillin–streptomycin (P/S).
2. Expansion medium (EM): alpha minimum essential medium (αMEM) with ribonucleosides, deoxyribonucleosides, sodium pyruvate, low glucose, phenol red, L-glutamine, 10% v/v fetal bovine serum (FBS), 1% v/v P/S.
3. Trypsin-ETDA (TE): 0.25% v/v Trypsin in PBS, 1 mM EDTA.
4. Freezing medium: 90% v/v FBS, 10% v/v dimethyl sulfoxide (DMSO).
5. Isopropanol for cell freezing container.
6. Liquid nitrogen.

### 2.5 hPOB Culture on CaP-mPCL Scaffolds

#### 2.5.1 Materials and Consumables

1. Low-attachment or cell-culture treated 12-well plates.
2. Autoclaved plastic reaction tubes (0.6 mL).
3. Sterile 50 mL disposable syringes.
4. Sterile 0.2 μm disposable syringe filters.
5. Specimen plastic containers (3 mL).

#### 2.5.2 Equipment

1. Optional: Rocking platform mixer. *See* **Note 5**.

#### 2.5.3 Reagents

1. Paraformaldehyde (PFA): 4% v/v solution in PBS.
2. 1 M β-glycerophosphate: Dissolve 21.6 g of β-glycerophosphate disodium salt hydrate powder suitable for cell culture, ≥99.0% titration, in 100 mL dd$H_2O$. Filter-

sterilize with disposable syringes and 0.2 μm syringe filters. Aliquot 17 × 5.65 mL in 15 mL labeled sterile reaction tubes and store at −20 °C.

3. Ascorbate-2-phosphate: 50 mg/mL solution in water. Dissolve 500 mg of L-Ascorbic acid 2-phosphate sesquimagnesium salt hydrate, ≥99.0%, in 10 mL dd$H_2O$. Filter-sterilize with disposable syringes and 0.2 μm syringe filters. Aliquot 17 × 565 μL in labeled sterile reaction tubes and store at −20 °C.
4. 0.1 mM dexamethasone: Dissolve 100 mg of dexamethasone-- cyclodextrin complex, water soluble, suitable for cell culture, in 2.55 mL dd$H_2O$ to prepare *Stock 1* (*S1*). Filter-sterilize with disposable syringes and 0.2 μm syringe filters. Aliquot in 25 × 100 μL sterile reaction tubes labeled *S1* and store at −20 °C. To prepare *Ready-to-use-stock* (0.1 mM), dilute 10 μL of *S1* in 10 mL dd$H_2O$ and aliquot in 17 × 565 μL in sterile reaction tubes, labeled *DEX* and store at −20 °C.
5. Osteogenic medium (OM): EM, 10 mM β-glycerophosphate, 0.17 mM ascorbate-2-phosphate, 100 nM dexamethasone. Thaw one aliquot of each osteogenic supplement and add the following amounts to a pre-made bottle of EM: 5.61 mL of 1 M β-glycerophosphate, 561 μL of ascorbate-2-phosphate 50 mg/mL solution, and 561 μL of 0.1 mM *Ready-to-use-stock* (*DEX*).

## 2.6 Biological Sample Preparation for Scanning Electron Microscopy (SEM)

### 2.6.1 Materials and Consumables

1. Specimen plastic containers (3 mL).
2. Small Petri dishes (5 mm).
3. Filter paper.

### 2.6.2 Equipment

1. Metal tweezers.
2. Fume hood laminar flow cabinet.
3. Waste Winchester bottles.
4. SEM 9 or 12 mm aluminum stubs.
5. Carbon conductive tape (double-sided).
6. Gold sputter coater.
7. Scanning electron microscope.

### 2.6.3 Reagents

1. Sterile PBS: 100% v/v divalent ion-free solution.
2. 0.1 M cacodylate buffer (pH 7.4).
3. Glutaraldehyde (Glut): 3% solution in 0.1 M cacodylate buffer (pH 7.4), 0.4 M sucrose, 1 M calcium chloride ($CaCl_2$).

4. Osmium tetroxide: 1% solution in 0.1 M cacodylate buffer.
5. Graded ethanol series: 50%, 70%, 90%, 100% v/v solutions.
6. Hexamethyldisilazane (HDMS): 100% solution.

### 2.7 Cancer Cell Co-culture with hOBMT

#### 2.7.1 Materials and Consumables

1. Metastatic human cancer cell lines. This protocol was tested and validated for standard unmodified human metastatic prostate cancer cell lines; LNCaP, C4-2B, PC3 (ATCC, USA), used in passages 15–40 (*see* **Note 6**).
2. Low-attachment or cell-culture treated 24-well plates (*see* **Note 7**).

#### 2.7.2 Equipment

1. Rocking platform mixer.
2. Optional: Inverted epifluorescence microscope, if using fluorescently labeled cells.

#### 2.7.3 Reagents

1. Sterile PBS: 100% v/v divalent ion-free solution.
2. PFA: 4% v/v solution in PBS.
3. Co-culture media (CCM): Roswell Park Memorial Institute 1640 medium, 10% v/v FBS, 1% v/v P/S.

### 2.8 Histology and Immunohistochemistry (IHC)

#### 2.8.1 Materials and Consumables

1. Disposable microtome blades.
2. Microscopy slides suitable for IHC. *See* **Note 8**.
3. Laboratory wipes.
4. Histology delineating pen.
5. Proteinase K ready-to-use solution.
6. Primary and secondary antibodies of interest suitable for IHC.
7. IgG controls.
8. DAKO liquid DAB + Substrate Chromogen System (Dako, Australia).
9. Mayer's hematoxylin.
10. Pertex® mounting media (HistoLab®, Sweden).
11. Transfer pipettes.
12. Tips (1000 μL, 200 L, 10 μL).
13. 1.6 mL plastic reaction tubes.
14. 8 mm plastic weighing trays.
15. Rectangular cover glasses (22 × 22 mm) for histology slide mounting.

#### 2.8.2 Equipment

1. Calibrated pH meter.
2. Histology cassettes.
3. Automated tissue processor.

4. Paraffin embedding machine.
5. Histology metal molds.
6. Rotary microtome.
7. Histology brushes.
8. Water bath set at 40 °C.
9. Bench oven such as food dehydrator set to 52 °C.
10. Plastic slide racks suitable for oven.
11. IHC glass jars.
12. De-cloaking chamber.
13. Glass slide metal racks fitting glass jars.
14. Fume hood laminar flow cabinet.
15. Pipettes (P1000, P200, P20).
16. Histology humidity incubation box.
17. Slide scanner or light microscope.
18. Histology boxes for storage.

#### 2.8.3 Reagents

1. Tap water.
2. Graded ethanol series (100%, 90%, 70%, 50% v/v).
3. Hydrochloric acid: $\geq$32% solution in water.
4. PFA: 4% v/v solution in PBS.
5. Xylene. *See* **Note 9**.
6. Sterile PBS: 100% v/v divalent ion-free solution.
7. 50 mM Tris–HCl (pH 7.4).
8. Hydrogen peroxide ($H_2O_2$): 3% solution in water.
9. Tri-sodium citrate buffer: Add 2.94 g Tri-sodium citrate dihydrate in 1 L dd$H_2O$. Adjust the pH to 6 and add 500 μL of Tween 20.
10. Tris-EDTA buffer: 10 mM Tris EDTA, 0.05% v/v Tween 20.
11. Bovine Serum Albumin (BSA): 2% w/v in PBS.
12. Ammonium hydroxide: 0.1% solution in water.

### 2.9 Immunofluorescence (IF)

#### 2.9.1 Materials and Consumables

1. Disposable 0.2 μm syringe filters.
2. Disposable 20 mL syringes.
3. Optional: 5 mm biopsy punches.
4. Cell-culture treated 48-well plates.
5. Pipette tips (1000, 200, 20 μL).
6. Pipettors (P1000, P200, P20).
7. 4′,6-Diamidino-2-Phenylindole (DAPI).

8. Fluorescently labeled phalloidin (such as CytoPainter from Abcam).
9. Primary antibodies of interest suitable for IF.
10. Fluorescently labeled secondary antibodies (such as Alexa-Fluor® from Abcam).
11. Paraffin film.
12. Aluminum foil.
13. Rectangular cover glasses (22 × 22 mm).

#### 2.9.2 Equipment

1. Reusable Polytetrafluoroethylene (PFTE) board (10 × 10 cm).
2. Sterilized metal tweezers.
3. Rocking platform mixer.
4. Inverted epifluorescence microscope (IX73, Olympus, Australia).
5. Confocal laser scanning microscope (CLSM), or spinning disc confocal microscope (SDCM).

#### 2.9.3 Reagents

1. Sterile PBS: 100% v/v divalent ion-free solution.
2. PFA: 4% v/v solution in PBS.
3. Triton X-100: 0.2% v/v solution in PBS.
4. BSA: 0.5% w/v solution in PBS. Prepare fresh and 0.2 μm filter-sterilized (store at 4 °C and use within 1 week).

## 3 Methods

### 3.1 Scaffold Manufacturing Using Melt Electrowriting

The following procedures apply for the house-built apparatus described in [19]. *See* **Note 10**.

1. Fill 3CC syringe with PCL pellets at 75% of the syringe volume. Insert the plunger and set vertically with the syringe head facing upwards, in an oven set to 60 °C for 8 h. This will allow the PCL pellets to melt and air bubbles to accumulate in the top area of the syringe.
2. Using the syringe plunger, slowly push the molten PCL to evacuate entrapped air bubbles and to pack the material inside the syringe. The syringe should be 50% filled with molten PCL.
3. Secure a 23 G needle tip to the syringe.
4. Turn on the MEW apparatus, set the syringe and needle block heaters to 74 and 85 °C respectively, to warm up the syringe and the needle area evenly (*see* **Note 11**). Wait for 2 h for homogeneous material distribution.
5. Set the extrusion pressure regulator to 2.2 bar (*see* **Note 11**).

6. Set the working distance between the needle tip and the aluminum collector (previously cleaned using 70% v/v ethanol) to 9 mm.
7. Load G Code in Mach 3CNC software.
8. Clean the syringe needle tip with 70% v/v ethanol-soaked laboratory wipes.
9. Close the MEW door and gradually increase the voltage to 10.1 kV (*see* **Note 11**). Wait for 15 min until the jet stabilizes.
10. Once jet is stable, start printing using Mach 3 CNC program (printing lasts for about 2 days and 16 h).
11. At the end of the printing, spray scaffold with 70% v/v ethanol. Using ethanol-sterilized tweezers, transfer the printed scaffold to a Ziploc bag.
12. Using Corel Draw, draw a 8 × 6 cm square that contains 48 squares (1 × 1 cm).
13. Turn on the laser cutter.
14. Place printed scaffold on the laser cutter tray.
15. Load Corel Draw program. Position the laser cutting path to match the position of printed scaffold on laser cutter tray (leave some margins on all sides).
16. In the settings, set the laser settings as 1.5% power, 20 mm/s for speed, PSI: 800 and 1 mm *z* axis height.
17. Start the cut. This will lead to 48 scaffolds with 150 μm fiber spacing and ~12 μm fiber diameter, which is ideal for preosteoblast cell seeding, differentiation and mineralization (*see* **Note 12**).
18. If not proceeding with CaP coating directly, store scaffolds in Petri dishes in a desiccator under vacuum and use up to 6 months after printing.

### 3.2 Calcium Phosphate Coating

In order to confer osteo-inductive properties to the MEW mPCL scaffolds used for differentiation of human preosteoblasts, scaffolds are coated with calcium phosphate (CaP) prior to use for cell culture [20, 21]. Treatment involves three key stages; surface activation, biomimetic mineralization, and post-treatment.

#### 3.2.1 Surface Activation

PCL is a hydrophobic polymer, which requires to be treated in order to improve hydrophilicity prior to CaP coating. Plasma treatment is ideal for this procedure (Option 1). If plasma treatment is not available, NaOH treatment can be used (Option 2).

##### Option 1. Plasma Treatment

1. Wipe clean the plasma chamber with ethanol 70% v/v.
2. Place the scaffolds in the chamber, ensuring sufficient space between the scaffolds (>5 mm).

3. Switch on the vacuum pump while holding the door shut for a few seconds for proper chamber seal.
4. Monitor pressure. When close to 200 Mtorr, open the oxygen, argon, and nitrogen gas supply taps (the valve to the chamber is still closed at this stage).
5. Slowly adjust the flowmeters to 10 mm on the scale reading, which corresponds to 7.3 mL/min argon, 6.3 mL/min nitrogen, 7.9 mL/min oxygen.
6. Once the chamber pressure drops below 200 Mtorr, open the three-way valve to chamber, to flood the treatment chamber with the appropriate gas mixture.
7. Wait till the vacuum pressure goes below 200 Mtorr again, then turn the RF power level to medium (MED) to ignite plasma.
8. Process the samples for desired duration depending on the scaffold thickness. For the scaffolds being generated here (1 × 1 cm) with a thickness between 500 μm and 1 mm, a minimum duration of 7 min is recommended. Ideally use 10 min.
9. At the end of the process, turn RF power off.
10. Close all gas supply taps and the three-way valve.
11. Turn off vacuum pump.
12. Very slowly open the three-way valve to flood chamber with ambient air.
13. Using tweezers, flip the scaffolds upside down to expose the nonactivated side upwards.
14. Repeat **steps 2–12**. Make sure to use the same coating duration as for the previous side.
15. Transfer the scaffolds to Petri dishes and proceed to the biomimetic mineralization.

Option 2. NaOH Treatment

1. Prepare a 2 M NaOH solution in a 50 mL reaction tube and prewarm it at 37 °C.
2. Place the scaffolds in 10 mm Petri dishes using tweezers, ensuring scaffolds are not overlapping.
3. Add 70% v/v ethanol to the dishes (just enough to cover the scaffolds), and place the dishes, without lids, under vacuum in a desiccator, to ensure homogenous scaffold prewetting. After 15 min, slowly release vacuum.
4. Discard ethanol and rinse twice with $ddH_2O$, inverting each scaffold two times to ensure even rinsing.
5. Immerse prewet scaffolds in prewarmed 2 M NaOH in a reaction tube (do not fully close tube), and perform a 5 min vacuum treatment in the desiccator. This step will ensure solution infiltration in the scaffolds.

6. Transfer the reaction tube into the water bath set to 37 °C and incubate for 45 min, while agitating manually from time to time.
7. Discard the NaOH solution and fill the tube with dd$H_2O$. Place the tube under gentle stirring for 5 min. Discard water and repeat this procedure at least five times or more, until the pH is approximately 7 using a pH meter or pH strips.
8. Transfer the scaffolds to Petri dishes and proceed to biomimetic mineralization.

*3.2.2 Biomimetic Mineralization*

1. Check that the SBF 10× pH is 4. Adjust the pH with HCl drops if necessary.
2. Rinse a 50 mL beaker and magnetic stirrer with dd$H_2O$.
3. Add 50 mL of SBF 10×, pH meter probe and magnetic stirrer to the clean beaker. Prepare *Solution C* by gradually adding small amounts of sodium $NaHCO_3$ powder, until the pH reaches 5.7. Perform slower $NaHCO_3$ additions to approach pH 6 (*see* **Note 13**), allowing for pH to adjust while the powder is dissolving.
4. In a new tube 50 mL reaction tube (do not use glass container to avoid potential etching and incorporation of undesirable species on the material's surface), filter *Solution C* using a disposable 50 mL syringe and 0.2 μm filter (*see* **Note 14**). Immerse the scaffolds in the filtered solution (do not fully close the tube), and perform a 5 min vacuum treatment in desiccator to ensure solution infiltration in scaffolds.
5. Before reaction tube transfer, check for entrapped air bubbles on the surface of the scaffolds, which will negatively affect coating homogeneity. Using a transfer pipette, displace any bubbles from the scaffolds' surface.
6. Transfer the Flacon tube with the scaffolds to a water bath set to 37 °C for 30 min (*see* **Note 15**).
7. Repeat **steps 4–6**, excluding vacuum treatment, in order to increase coating thickness and homogeneity (*see* **Note 16**) and rinse once with dd$H_2O$.
8. Prepare 0.5 M NaOH in reaction tube and prewarm at 37 °C.
9. Discard the water and fill the tube with the scaffolds with prewarmed 0.5 M NaOH in order to remove di-calcium phosphate dihydrate, a highly soluble CaP phase, detrimental for stable osteogenesis to occur. Perform a 5 min vacuum treatment in desiccator (do not fully close tube), to ensure solution infiltration within the scaffolds.
10. Transfer the reaction tube to the water bath set to 37 °C for 30 min.

#### 3.2.3 Post-treatment

1. Discard the NaOH solution and transfer the scaffolds to 10 mm Petri dishes.
2. Add $ddH_2O$ until the scaffolds are covered, and gently swirl the dishes for 5 s.
3. Using a transfer pipette, gently flush $ddH_2O$ over the scaffolds, minimizing direct contact between the transfer pipette and the coated scaffold surface.
4. Discard the rinsing solution and repeat five times. Using pH meter of pH strips, check that the last rinsing solution is pH 7 before final discarding.
5. Transfer the scaffolds in Petri dishes and in desiccator under vacuum overnight to fully dry the scaffolds.
6. Optional, but recommended: Image CaP-coated mPCL scaffolds for quality control using standard SEM (Subheading 3.5) and ensure coating is homogeneous in the depth of the scaffold (Fig. 2a) and has a cauliflower-appearance (Fig. 2b, c).
7. Store CaP-coated mPCL scaffolds in desiccator under vacuum until further use for cell culture, ideally the next day and no less than after 2 weeks (*see* **Note 17**).

### 3.3 Human Preosteoblast (hPOB) Cell Isolation and Expansion

Human preosteoblasts can be isolated from human bone fragments collected from trabecular bone from either hip or knee surgery. This procedure has to be performed under donor informed consent and appropriate institutional ethics approval. All steps must be performed in a biological safety cabinet class II (laminar flow), under aseptic conditions. Spray with 70% v/v ethanol all tools and equipment required during isolation.

1. According to the type of joint collected (hip or knee), collect healthy trabecular bone from the tibial plateau, acetabular ground, or femoral head (Fig. 3a).
2. Place the fresh human bone in 10 mm Petri dishes (*see* **Note 18**).
3. Use the bone cutters to hold the bone and the bone curette to chop the bone into small fragments, 2–5 mm in size (Fig. 3b).
4. Transfer the bone explants in a 50 mL reaction tube and record the mass.
5. Add 30 mL PBS-P/S to the tube, close and shake vigorously for a few seconds to wash out debris, loose connective tissues, and fat. Repeat this step 5–10 times until the PBS becomes clear (Fig. 3c).
6. Transfer 2–5 g of bone explants into cell culture-treated flask with reclosable lid (Fig. 3d), using a sterile spatula. Add expansion medium (17 mL). The bone explants should be covered by the medium, but not floating.

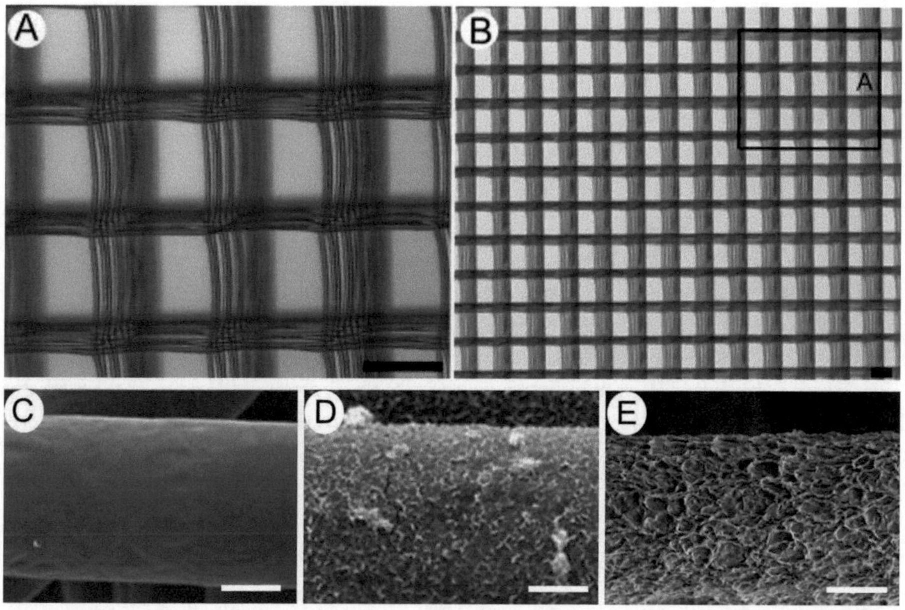


**Fig. 2** (**a**, **b**) Stereomicroscope images of melt electrowritten (MEW) scaffold [22]. (**c**, **d**) SEM images of MEW scaffold fiber; (**c**) before chemical treatment [19], (**d**) after NaOH treatment followed by CaP coating [19], (**e**) after plasma treatment followed by CaP coating. Scale bar are 100 μm (**a**, **b**) and 5 μm (**c**–**e**). Copyright Permission Sources: (**a**, **b**) [22], Copyright 2016 Elsevier; (**c**, **d**) [19], Copyright Springer Nature (2017); (**e**) Unpublished figure provided by Asha Mathew from QUT

7. Culture the explants in a humidified cell culture incubator at 37 °C and 95% air, 5% $CO_2$. Do not move the flasks for the first 2 days of isolation. Change media every 3–4 days. Prior to each media change until passage 1, use prewarmed PBS-P/S to wash the bone explants/cells first. Do no aspirate the bone explants. Typical cell outgrowth takes place after 5–7 days, displaying an elongated morphology (Fig. 3e).
8. When cells reach 70% confluence (typically 14 days), gently move bone explants to one corner of the flask, by tilting the flask. Aspirate the media using the opposite corner. Remove the reclosable lid of the TPP flask and using sterile tweezers, quickly transfer the bone explants to a Petri dish. Close the Petri lid and wash the cells with 15 mL prewarmed PBS-P/S *twice* (critical step) prior to adding 4 mL of 0.25% w/v Trypsin-EDTA. Place the flask back in the incubator for 5 min. Monitor cell detachment with a bench microscope. Once the cells are fully detached, use 8 mL of EM to inactivate Trypsin-EDTA (*see* **Note 19**).
9. Transfer the cell suspension in a 15 mL reaction tube and centrifuge at 400 × *g* for 5 min.

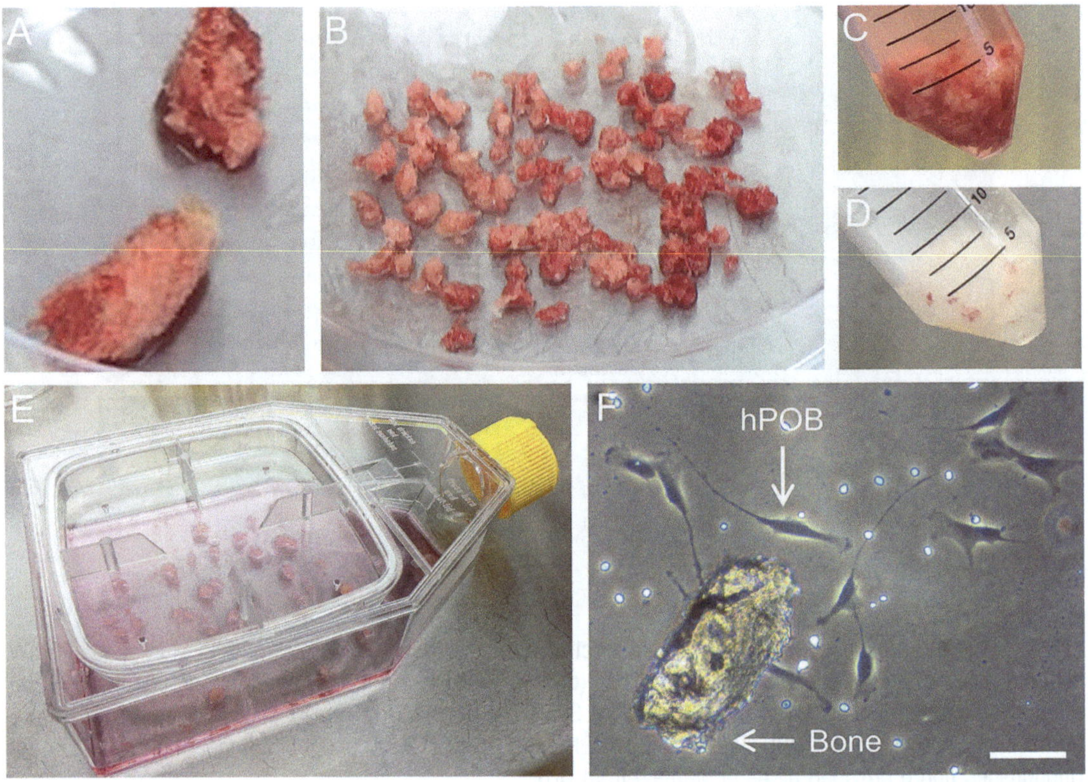


**Fig. 3** (**a**) Human bone tissue from the knee joint. (**b**) Morcellized bone after manual fragmentation. (**c**, **d**) Bone fragments before (**c**) and after (**d**) ten washes in PBS. (**d**) Bone fragments in culture in 175 $cm^2$ flask with reclosable lid. (**e**) Microscopy image of a bone fragment and human preosteoblasts (hPOB) outgrown from the bone fragment after 7 days of culture. Scale bar is 50 μm

10. Prepare the cell freezing container box (ensure it is at room temperature (RT)), label one cryovial and thaw the freezing media from −20 °C to RT.
11. After centrifugation, aspirate the media and resuspend the pellet in 2 mL of media.
12. Count the cells using the cell counter. At passage 1 and 70% confluence, the average yield of isolated cells is approximately $2.5 \times 10^5$ cells/g of bone tissue. When using 4 g of bone tissue, an average of one million cells is expected at passage 1.
13. Collect $5 \times 10^5$ cells (minimum) into a new 15 mL reaction tube and centrifuge at 400 × *g* for 5 min.
14. Aspirate the media and resuspend the pellet with 1 mL freezing media. Rapidly add the suspension to the cryovial, place the cryovial into the freezing box and immediately place the freezing box in a −80 °C freezer for at least 24 h, before long term storage in liquid nitrogen (<−190 °C).

15. Divide the other $5 \times 10^5$ cells into two new 175 $cm^2$ cell culture-treated flasks (approximately 1400 cells/$cm^2$) in expansion medium (25 mL each). Perform media change every 3–4 days. Culture until 80–90% confluence before harvesting with Trypsin-EDTA as mentioned previously or for experimental use (*see* **Note 20**).
16. Once cells are either passaged or frozen, place the bone explants back in the original TPP flask using reclosable lid and sterile tweezers, and add 17 mL of expansion medium. Isolation can be repeated up to four times using the same bone explants if bone explants are temporarily removed from the flask during the trypsinization step (*see* **Note 4**).
17. Place all leftover human tissue in a waste bottle containing biodine prior to autoclave and discarding. Autoclave all consumables used during isolation prior to discarding.

### 3.4 Seeding and Culture of hPOB on CaP-mPCL Scaffolds

The following steps must be performed in a biological safety cabinet class II (laminar flow), under aseptic conditions. Spray all tools and equipment with 70% v/v ethanol.

1. Prior to scaffold seeding, isolate primary human preosteoblasts from the bone tissue obtained from a donor according to Subheading 3.3 and expand using EM or thaw frozen stocks (*see* **Note 21**). Only use cells under passage 5 for scaffold seeding.
2. Place the CaP-mCPL scaffolds in 10 mm Petri dishes and fill with 70% v/v ethanol. After 20 min, remove the ethanol and let the scaffolds air-dry inside the laminar flow cabinet overnight.
3. Prior to seeding, expose the scaffolds to UV light from the laminar flow cabinet for 20 min on each side of the scaffolds and transfer the scaffolds to low-attachment 12-well plates (*see* **Note 22**), placing one scaffold per well.
4. Add PBS in between wells to maintain humidity.
5. After trypsinization, prepare a cell solution of $10 \times 10^6$ cells/mL primary hPOB in EM.
6. Seed the scaffolds by pipetting five drops (10 μL/drop) (Fig. 4a) of cell suspension onto each scaffold, leading to $5 \times 10^5$ cells per scaffold (*see* **Note 23**). Incubate for 4 h in a humidified incubator at 37 °C and 95% air, 5% $CO_2$ to allow cell attachment to scaffold fibers.
7. After 2 h, add 50 μL of EM per construct, to ensure the cell constructs remain wet. This amount will be just enough to cover the entire scaffold (*see* **Note 24**). Place the plate back into the incubator.
8. After 4 h, aspirate PBS in between wells. Gently add 0.5 mL of EM to hPOB/scaffold constructs. Change media every

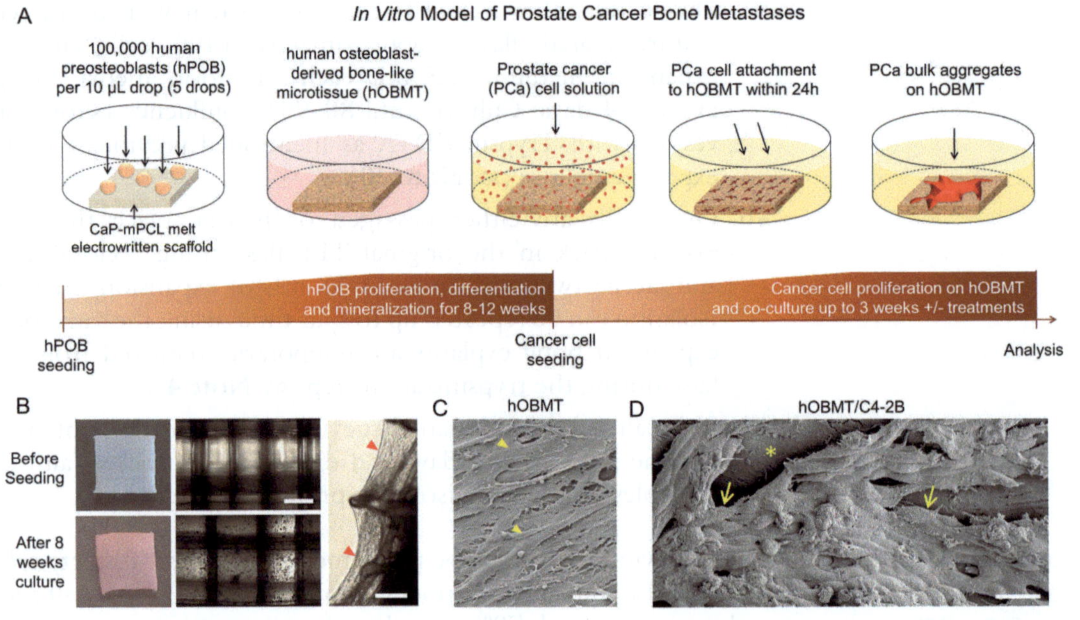


**Fig. 4** (**a**) Schematic of the processing steps leading to a 3D in vitro model of prostate cancer bone metastases. (**b**) Photos (left) and microscopy images (right) of the CaP-mPCL melt electrowritten scaffold before hPOB seeding (top) and after seeding +8 weeks of culture (bottom). Far right light microscope image shows hPOB confluence in scaffold after 2 weeks post-seeding, as evidenced by bridging hPOB monolayers around the scaffold edges (head arrows). Scale bar is 100 μm. (**c**) SEM image of hOBMT showing osteoblastic monolayers (head arrows) recovering scaffold, extracellular matrix, and mineral deposition. Scale bar is 10 μm. (**d**) SEM image of the hOBMT/C4-2B constructs after 21 days showing hOBMT (stars) and C4-2B areas (arrows). Scale bar is 25 μm

3–4 days (1 mL/well) and culture until confluence, as evidenced by light microscopy assessment (Fig. 4b) (approximately 2 weeks, depending of the initial amount seeded and the proliferative capacity of the cells).

9. When confluent, switch to OM (2 mL/well) and culture for a further 8–12 weeks depending on the degree of mineralization required and the mineralization capacity of the cells (*see* **Note 25**). The 3D live mineralized cell/scaffold construct obtained here is hereafter referred to as human osteoblast-derived bone-like microtissue (hOBMT). Cell morphology and mineralization can be assessed using SEM (Fig. 4c), explained in Subheading 3.5 and IF (Fig. 6c), explained in Subheading 3.7.

### 3.5 Scanning Electron Microscopy (SEM)

Scanning electron microscopy (SEM) is a useful tool to analyze the topography and the composition of a material's surface. It can be used here to perform quality control of the surface of mineralized hOBMT to ensure ECM and mineral deposition (*see* Fig. 4c), and to assess the interactions of cancer cells with hOBMT in the co-cultures (*see* Fig. 4d). SEM can also be used to assess the quality of the CaP coating prior to cell seeding (Fig. 2). Materials generally need a thin layer of a conductive coating (gold or carbon), prior to imaging. For biological samples, such as the microtissue constructs proposed here, the samples also need to be dehydrated prior to conductive coating and imaging. During these procedures, it is critical to not leave the specimens exposed to air. All steps are performed at RT.

1. At desired time points, in the biological safety hood cabinet, aspirate the media from the wells containing the hOBMT.
2. Wash the hOBMT with 3 mL prewarmed PBS, twice.
3. Add 1 mL of 3% Glut to specimen containers.
4. Aspirate the last PBS rinsing solution (one sample at a time) and immediately transfer the hOBMT to the specimen containers, with Glut. Incubate for at least 1 h. *See* **Note 26**.
5. Take the samples into the fume hood laminar flow cabinet.
6. Discard the Glut solution to an appropriate waste residue bottle.
7. Proceed to buffer rinse by adding 0.1 M cacodylate buffer until samples are fully covered. After 10 min, discard the solution in an appropriate waste residue bottle and repeat this step twice.
8. Proceed to osmium tetroxide post-fixation, by adding 1 mL of 1% osmium tetroxide in cacodylate buffer and incubate for 1 h.
9. Discard the solution in an appropriate waste residue bottle and proceed to rinsing. Add 1 mL of $ddH_2O$ for 10 min. Aspirate water and repeat this step once.
10. Proceed to sample dehydration using graded ethanol series (1 mL of each): ethanol 50% v/v 10 min, twice, ethanol 70% v/v 10 min, twice, ethanol 90% v/v twice 10 min, ethanol 100% v/v, 15 min, twice.
11. Add 1 mL of HDMS for 30 min.
12. Discard HDMS in an appropriate waste residue bottle and repeat **step 11**.
13. Prepare Petri dishes with filter paper.
14. Discard the last HDMS solution in an appropriate waste residue bottle.
15. Transfer the hOBMT to Petri dishes and air dry in fume hood for 1 h.
16. Remove the plastic protection from one side of the double-sided conductive tape, and stick onto the SEM stubs.
17. Slowly remove the plastic protection of the other side of the double-sided conductive tape and gently mount the hOBMT

onto the SEM stubs using tweezers. Ensure the surface of interest is uppermost.

18. Using the gold sputter coater, coat to a thickness of 10 nm.
19. Load stubs in SEM and image using a high voltage of 10 kV and an appropriate working distance (*see* **Note 27**). *See* Fig. 4c, d for expected images.

### 3.6 hOBMT Co-culture with Cancer Cells

The following steps must be performed in a biological safety cabinet class II (laminar flow), under aseptic conditions. Spray all tools and equipment with 70% v/v ethanol.

1. Culture the authenticated and mycoplasma-tested prostate cancer cells, below passage 30 (LNCaP, C4-2B, PC3), for at least 2 weeks prior to co-culture with the hOBMT.
2. On co-culture day, once the hOBMT are sufficiently mineralized (8–12 weeks culture in OM), select the control hOBMT samples that will not be co-cultured.
3. Wash the control hOBMT twice in PBS and fix in PFA 4% for 40 min or Glut 3% for 1 h. For PFA samples, rinse twice with PBS, add fresh PBS and store in 3 mL specimen containers at 4 °C until further characterization. These samples serve as a reference control and can be used for micro-computed tomography to assess mineralization, and for histology to assess ECM markers. The Glut samples can be used for SEM imaging according to Subheading 3.5.
4. Aspirate the media from the hOBMT, which will be used for co-culture, and wash with prewarmed co-culture media (CCM) three times, for 10 min each rinse (place back in the incubator during washes).
5. Prepare the cancer cell solutions in CCM; $1 \times 10^5$ cell/mL for metastatic cell lines such as C4-2B and PC3 and $2 \times 10^5$ cell/mL for nonmetastatic cell lines such as LNCaP (*see* **Note 28**).
6. Place one individual hOBMT in the well of a 24-well plate.
7. Dispense 500 μL of the desired cancer cell solution onto each hOBMT.
8. Place the well plate on the rocking platform mixer inside the incubator, on gentle rocking for at least 12 h (be consistent once you select one speed and attachment duration).
9. After 12 h, remove the plate from the incubator and aspirate CCM. Wash twice with 1 mL fresh prewarmed CCM. Do not invert the construct to ensure that the cancer cell side is uppermost.
10. Transfer the hOBMT/PCa co-culture microtissues in fresh 12-well plates making sure that seeded surface is uppermost.

11. Continue co-culture in CCM for up to 3 weeks or less, changing media every 2–3 days with 2.5 mL/well. Add any treatments that may be required to assess effects on the cancer cells in bone-like microenvironments, depending on the research questions.
12. If using fluorescently labeled cancer cells, successful and homogeneous attachment can be assessed with a standard inverted epifluorescence microscope, by temporarily inverting co-culture in culture wells and imaging (make sure to place the co-cultures back to normal after imaging). Regular epifluorescence imaging can help to provide information on cancer cell condition and morphologies on hOBMT over time (Fig. 6b).

### 3.7 Histology and Immunohistochemistry Staining of Microtissues

Tissue histology and immunohistochemistry (IHC) are used as standard techniques to assess the morphology and structure of cells and tissues using microscopy. By embedding tissues in a paraffin wax, thin sections of tissue can be collected and stained, while preserving the macrostructure of the tissue. This may however result difficult to achieve with bone tissues and alternative embedding materials, such as PMMA, may need to be used. One advantage of the bioengineered microtissues developed here is that they can be embedded successfully with traditional paraffin wax. Also, they do not need to undergo EDTA decalcification, a process which is usually required for in vivo bone tissues (ranging from 3 weeks to 6 months).

After histology, the use of IHC allows to stain specific cellular components such as proteins that are present intracellularly, as well as secreted in the matrix. By revealing antigens at the surface of proteins, primary antibodies can selectively bind to the antigens. Using secondary antibodies and color development further allows to visualize and localize the areas of cells and tissues containing the proteins of interest. These processes, used for clinical tissues, can be performed similarly on the bioengineered microtissues, with some modifications listed below. Quantitative data about protein concentration can be obtained with standard imaging software, informing on the effects of co-culture with various cancer cells, and/or impact of treatments, on protein expression in a human bone metastatic context.

All steps are performed at RT, except otherwise stated.

1. At desired time points, aspirate the media from the wells containing the hOBMT/cancer cell co-cultures (hereafter referred to as 'microtissues').
2. Wash with 3 mL PBS twice.
3. Fix the microtissues using 1 mL PFA 4% for 40 min.
4. Rinse the microtissues with 3 mL PBS twice.

**Table 2**
**Tissue processing steps with automated tissue processor**

| Reagent | Temperature | Time | Vacuum | Agitation |
|---|---|---|---|---|
| 70% v/v ethanol | Ambient | 15 min | Off | Yes |
| 70% v/v ethanol | Ambient | 15 min | Off | Yes |
| 70% v/v ethanol | Ambient | 30 min | On | Yes |
| 90% v/v ethanol | Ambient | 30 min | On | Yes |
| 90% v/v ethanol | Ambient | 30 min | On | Yes |
| 100% v/v ethanol | Ambient | 30 min | On | Yes |
| 100% v/v ethanol | Ambient | 45 min | On | Yes |
| 100% v/v ethanol | Ambient | 60 min | On | Yes |
| Xylene | Ambient | 45 min | On | Yes |
| Xylene | Ambient | 45 min | On | Yes |
| Xylene | Ambient | 60 min | On | Yes |
| Paraffin wax | 60 °C | 45 min | On | Yes |
| Paraffin wax | 60 °C | 60 min | On | Yes |
| Paraffin wax | 60 °C | 60 min | On | Yes |

5. Using tweezers, place the microtissues in histology cassettes and dehydrate the samples using an automated histology tissue processor with an overnight run that follows the steps listed in Table 2.
6. Remove the cassettes from the tissue processor and place the microtissues in histology molds. Embed in melted paraffin and let it harden.
7. Collect the solidified paraffin blocks and proceed to histology sectioning using a rotatory microtome and fresh metal blades, cutting 5 μm sections. Transfer the sections into a warm water bath set at 40 °C. Collect the sections soaking in the bath using microscopy slides (>3/slide).
8. Add the slides to a rack suitable for the oven. Insert the rack in the oven set to 52 °C for 4 h (*see* **Note 29**).
9. Prepare glass jars with the fresh solutions listed in Table 3. Place the dried slides in the glass slide rack. Deparaffinize and rehydrate sections by immersing the rack into the jars containing the various solutions, for the exact times listed in Table 3.
10. Depending on the primary antibody tested, there are three possible retrieval means:

**Table 3**
**Deparaffinization and rehydration of histology sections**

| Reagent | Time |
|---|---|
| Xylene I | 5 min |
| Xylene II | 5 min |
| 100% ethanol | 3 min |
| 100% ethanol | 3 min |
| 90% ethanol | 1 min |
| 70% ethanol | 1 min |
| 50% ethanol | 1 min |
| 50 mM Tris–HCl (pH 7.4) wash buffer | 1 min |

**Table 4**
**Bone markers validated for immunohistochemistry for bioengineered microtissues**

| Antibody | Product | Dilution | Incubation | Antigen Retrieval |
|---|---|---|---|---|
| Collagen I | Abcam ab23446 | 1:100 | Overnight 4 °C | Proteinase K, 15 min, RT |
| Osteocalcin | Abcam ab13418 | 1:300 | Overnight 4 °C | Proteinase K, 15 min, RT |
| Osteopontin | Abcam ab8448 | 1:500 | 1 h at RT | Tri-sodium citrate buffer pH 6 95 °C for 5 min |
| Fibronectin | Abcam ab299 | 1:1000 | 1 h at RT | Tri-sodium citrate buffer pH 6 95 °C for 5 min |

(a) Heat-mediated antigen retrieval with tri-sodium citrate buffer.

(b) Heat-mediated antigen retrieval with Tris-EDTA buffer.

(c) Proteolytic antigen retrieval.

Some antibodies suppliers mention the best method to be used for each antibody. If not, try all three methods in order to find out the best retrieval method for your specific antibody. *See* Table 4 for validated human bone related markers for the bioengineered microtissues.

For heat-mediated retrieval, places the slides in a chamber filled with tri-sodium citrate buffer or Tris-EDTA buffer. Incubate for 5 min at 95 °C in a decloaking chamber. For proteolytic retrieval, use a ready-to-use Proteinase K solution. Delineate the sections on the slide using histology delineating pen. Add 2–3 drops of Proteinase K on each section for 5–15 min.

11. Wash two times with Tris–HCl buffer for 2 min each wash using the glass jars.
12. Remove the slides from rack. Using laboratory wipes, absorb any residual liquid on the slides (do not touch the sections). Delineate the sections treated by heat with the histology delineating pen.
13. Using transfer pipettes, cover the slides with 3% $H_2O_2$ and incubate for 5 min, to block endogenous peroxidase activity.
14. Wash three times with Tris–HCl buffer for 2 min each wash.
15. Using transfer pipettes, add PBS-BSA 2% onto the sections for 1 h.
16. Prepare the primary antibody dilutions with PBS-BSA 2% in 1.6 mL reaction tubes.
17. Incubate the sections with the primary antibody solutions, with 'no primary' controls (use 2% BSA/PBS) and with IgG controls for 1 h RT or overnight at 4 °C (check the suppliers recommendation), in the histology incubating chamber.
18. Gently tap the slides onto the tissue paper to remove the antibody solutions and dry any excess liquid with laboratory wipes (do not directly touch the sections) (*see* **Note 30**).
19. Using the metal rack and the glass jars, wash three times in Tris–HCl buffer for 2 min each wash (three different jars). Agitate well between washes. Gently tap slides onto the tissue paper to remove excess buffer after the final wash.
20. Using transfer pipettes, incubate the sections with the secondary antibody solution for 30 min at RT.
21. Using the metal rack and glass jars, wash three times in Tris–HCl buffer for 2 min each wash (three different jars). Agitate well between washes. Tap slides onto tissue paper to remove excess buffer after the final wash.
22. Prepare the DAB solution by mixing 20 μL of Chromogen in 1 mL substrate.
23. Prepare a jar filled with tap water and a jar filled with Tris–HCl buffer.
24. Using transfer pipettes, fully cover the sections with a few drops of DAB solution for color development. As soon as color changes, place slide in the jar containing tap water. This step is critical and needs to be done one antibody (with their controls) at a time. In some cases, color development takes place within 10 s. If nothing is visible, do not leave more than 10 min. It is critical to use the same development time for all sections containing the tested antibody.

**Table 5**
**Dehydration of histology sections**

| Reagent | Time |
|---|---|
| 50% v/v ethanol | 1 min |
| 70% v/v ethanol | 1 min |
| 90% v/v ethanol | 1 min |
| 100% v/v ethanol | 3 min |
| 100% v/v ethanol | 3 min |
| Xylene I | 5 min |
| Xylene II | 5 min |

25. Wash once by immersing slides in a jar containing Tris–HCl buffer for 2 min.
26. Using transfer pipettes, fully cover the sections with a few drops of Mayer's hematoxylin for 2 min for counterstaining.
27. Fill a glass jar with 0.1% ammonium hydroxide.
28. Tap the slide onto the tissue paper to remove Mayer's hematoxylin and immerse the slide in the jar ammonium hydroxide for blueing, for 3–5 s, until blue color appears (do not overblue the sections).
29. Check under the microscope that the staining is visible. If necessary, repeat the hematoxylin and the blueing steps.
30. Prepare glass jars with the solutions listed in Table 5. Place the slides in a glass slide metal rack. Dehydrate the sections by immersing the rack into the jars containing the solutions for the exact time listed in Table 5. Make sure to work in the fume hood laminar flow cabinet for the last two steps in Xylene (*see* **Note 9**).
31. Using a transfer pipette, dispense the Pertex® mounting media on the center of the slide. Gently press a rectangular cover glass until the Pertex fully covers sections, avoiding bubbles.
32. Check that no air bubbles are entrapped in the mounted slide using a light microscope (*see* **Note 31**).
33. Let mounting agent harden overnight.
34. Using a slide scanner or a light microscope, view the slides and take images at high and low magnification, identifying the location of the proteins of interest (in cancer cells, in hOBMT, or in both) (Fig. 5a, b). *See* **Note 32**.

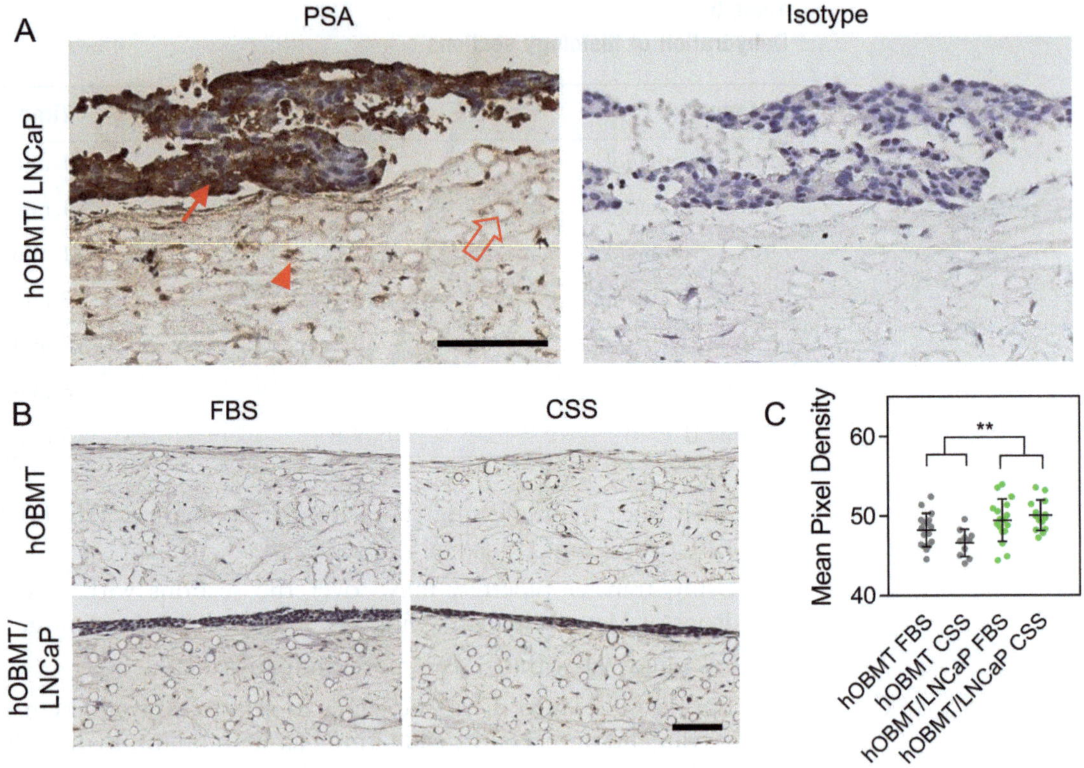


**Fig. 5** (**a**) IHC images of hOBMT/LNCaP after 10 days of co-culture in CCM, stained for PSA and isotype control (brown), with counterstained nuclei (purple). Solid arrow and arrow head show the staining in the LNCaP and in the hOBM, respectively. The open arrow shows a scaffold fiber. (**b**) IHC images of the hOBMT and hOBMT/LNCaP constructs, stained for Osterix, a transcription factor for osteoblast differentiation, in normal culture (using 10% FBS) conditions or in androgen-deprived conditions (using 10% charcoal stripped serum (CSS) instead of FBS. (**c**) Quantification of the mean pixel densities (non zero values) in hOBMT areas only (hOBMT alone and hOBMT in co-culture with LNCaP), showing increased staining in the hOBMT areas of hOBMT/LNCaP (scatter plot with means $\pm$ SD displayed). $^{**}P < 0.01$

### *3.8 Immunofluorescence Staining of Microtissues*

Similarly to histology and IHC, the use of immunofluorescence (IF) allows to visualize intracellular and extracellular proteins using fluorescence microscopy. IF provides a suitable complementary technique to IHC for the analysis of the bioengineered microtissues. IF involves a less labor-intensive processing of tissues, and the ability to maintain the tissue intact for staining (no sectioning required). There is hence decreased risk to damage the bioengineered tissues, as can be the case during histology and IHC processing. Furthermore, the use of confocal laser scanning microscopy allows to provide higher-resolution images with a level of tissue details that is not always possible with IHC. Visualization of nuclei, actin cytoskeleton, extracellular and cytoplasmic proteins can be achieved using dyes, specific primary antibodies, and fluorescently conjugated secondary antibodies. For the in vitro artificial metastatic tissue model proposed here, the use of IF/CLSM is especially suitable for assessing the morphology of cancer cells found at the surface of the hOBMT. However, while some degree of 3D reconstruction is possible with images taken with CLSM, the penetration of dyes and antibodies in the 3D structure is limited and staining may not be possible for the center of the microtissues. In this instance, histology and IHC will be more suited for analysis (*see* previous section). However, IF on full 3D microtissues and the use of CLSM will provide detailed 3D images of the superficial parts of the co-culture microtissues (~50 μm), which is ideal for detailing the nature of interactions between cancer cells and hOBM.

All steps are performed at RT, except otherwise stated.

1. At desired time points, aspirate the media from wells containing the microtissues.
2. Wash with 3 mL of PBS twice.
3. Fix the microtissues using 1 mL of PFA 4% for 40 min.
4. Rinse the microtissues with 3 mL of PBS twice.
5. Using the tweezers, place one microtissue onto the PTFE board. Using a biopsy punch, punch 5 mm tissue samples or continue with un-punched microtissues (*see* **Note 33**).
6. Transfer the punched microtissue samples to a 48-well plate.
7. Add 500 μL of 0.2% Triton X-100/PBS for exactly 10 min for cell permeabilization. Place the plate on a rocking platform (gentle rock).
8. Wash the samples three times with 1 mL of PBS (5 min/wash) on the rocking platform.
9. Using the tweezers, transfer the samples to new wells and add 500 μL of 0.5% w/v BSA/PBS. Incubate for 15–20 min on the rocking platform.
10. Prepare the primary antibody solutions diluted in 0.5% BSA/PBS (*see* **Note 34**). For 5 mm biopsy tissue punches, use 80 μL/tissue sample (*see* **Note 35**).
11. If staining for a specific protein, proceed to the next step. If staining for nuclei and/or F-actin only, proceed to **step 14** without the use of secondary antibody solutions.
12. Add the primary antibodies solutions and the controls with 0.5% BSA/PBS (no primary) to the tissue samples and incubate for 1 h on shaker (gentle shaking). After 30 min, use tips (new tip for each well) to gently flip the tissue sample on the other side to ensure maximum surface exposure for antibody binding.

13. Aspirate the antibody solutions (use fresh tips for each well) and wash the samples three times with 1 mL of PBS (5 min/wash) on the rocking platform (critical step).
14. Prepare the labeled secondary antibodies solutions in 0.5% BSA/PBS (*see* **Note 34** for concentrations), containing DAPI (5 μg/mL) and labeled-phalloidin (0.8 U/mL) (*see* **Note 36**), making sure that the labeled phalloidin and labeled secondary antibodies emit at different wavelengths allowing standard imaging using green and red/far red channels available on most fluorescence microscopes (*see* **Note 37**).
15. Aspirate the last PBS wash and incubate samples with 100 μL of secondary antibody solutions for 45 min on a rocking platform. Flip the samples on the other side at half time, using clean tweezers, to maximize exposure to antibody.
16. Aspirate the antibody solutions (use fresh tips for each well) and wash the samples three times with 1 mL PBS (10 min/wash) on the rocking platform (critical step).
17. Using the tweezers, transfer the stained samples to a new 48-well plate (*see* **Note 38**). Add 500 μL fresh PBS. Proceed to imaging or store at 4 °C until imaging, making sure to seal the plate with paraffin film and cover with aluminum foil (*see* **Note 39**).
18. On imaging day (epifluorescence or confocal), set a rectangular cover glass on the microscope stage. Using sterile tweezers, transfer the stained scaffold to cover glass. Add a few drops of PBS to keep moist or use a second cover slip to sandwich the scaffold and keep it moist. Make sure not to entrap air bubbles.
19. Assess whether the correct side with cancer cells is being imaged.
20. Proceed to imaging. Use epifluorescence to assess the initial quality of the staining (Fig. 6b). For confocal imaging, determine the *z*-stack depth by setting limits that encompass all the cancer cells present at the surface of the hOBMT (ranging from 50 to 100 μm depth). Use the maximum intensity projection to visualize the overall signal for intracellular information or use snapshots of the 3D reconstructed image for global information on cancer cell arrangement and morphology on hOBMT (Fig. 6d, e).

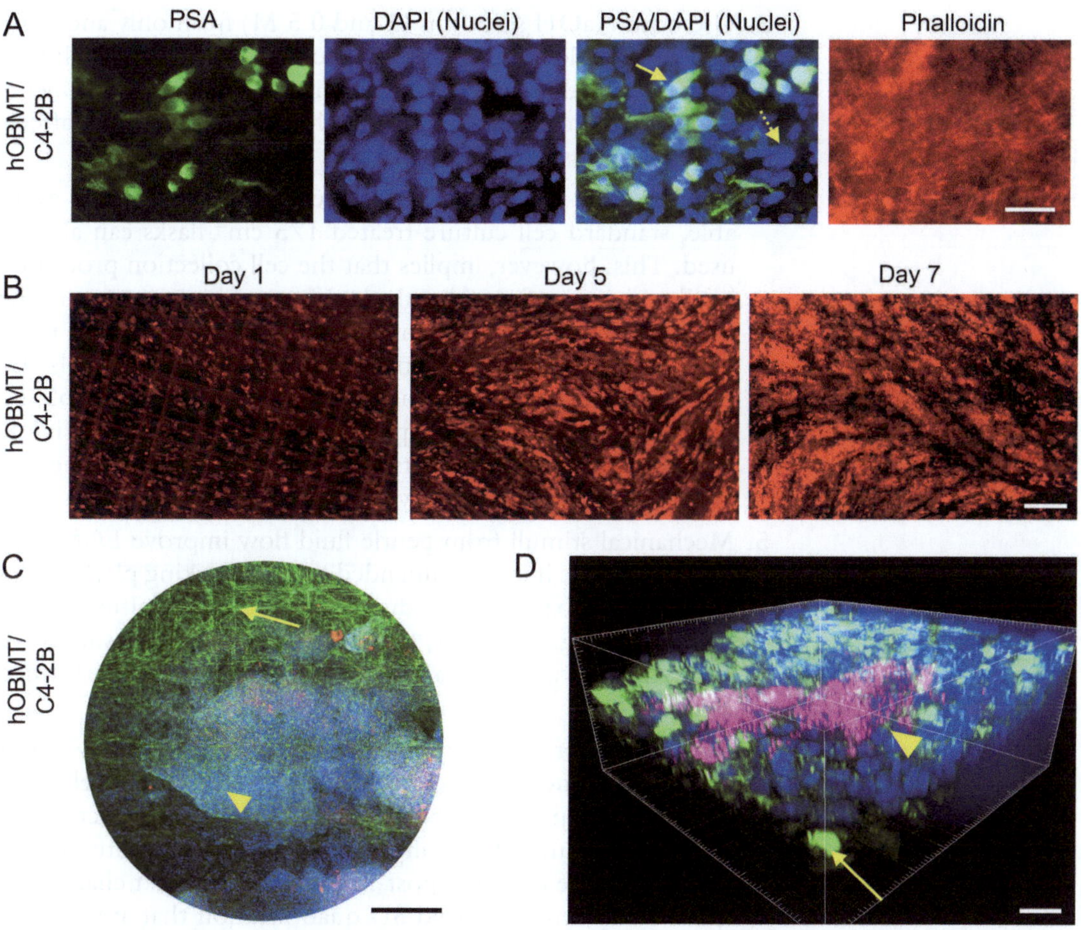


**Fig. 6** Various fluorescence images of the hOBMT in co-culture with unlabeled and mKO2 (red)-labeled cancer cells, referred to as 'm' as a prefix. (**a**) Epifluorescence of the hOBMT/C4-2B constructs after 3 h of co-culture, stained for PSA. Scale bar is 50 μm. The full arrow refers to a cancer cell and the dashed arrow refers to the nucleus of pHOB in the hOBM. (**b**) Epifluorescence of the hOBMT/mC4-2B constructs over time (red channel) showing single cancer cells growing in aggregates following hOBMT ECM. Scale bar is 200 μm. (**c**, **d**) Spinning disk confocal (SDC) maximal projections (**c**) and 3D reconstruction (**d**) of the hOBMT/mC4-2B constructs after 3 weeks of co-culture (blue: nuclei, green: F-actin, red: cancer cells). Scale bar is 500 μm (**c**) and 100 μm (**d**)

## 4 Notes

1. Efficient coating is reliant on exact matching of the weights listed in Table 1 and other weights listed in the reagents section, with a 1 mg precision. Weighing trays need to be rinsed with $ddH_2O$ so that the full amounts of chemicals are added to the solutions.
2. The pH of the SBF 10× solution may drop over time, but successful coatings can still be obtained with a starting pH of approximately 3.6–3.8. However, coating reproducibility will be ensured by using always the same pH (± 0.1).

3. Use both NaOH solutions (2 and 0.5 M) fresh only and store only in propylene reaction tubes, to avoid etching, which may happen with glass containers. This may contribute to the incorporation of undesirable species on the material's surface, affecting the success of coating.
4. If cell culture-treated flasks with reclosable lids are not available, standard cell culture-treated 175 $cm^2$ flasks can also be used. This, however, implies that the cell collection procedure by Trypsin is performed with the bone explants present in the flask. As a result, the trypsin step can only be performed twice using the same bone explants, as trypsin affects the cells still embedded in the bone tissue. When using the reclosable lid flasks, which allow bone explants removal prior to trypsinization, the same bone explants can be used up to four times for collection, maximizing the yield of isolated cells.
5. Mechanical stimuli from gentle fluid flow improve ECM production, hence it is recommended to use a rocking platform set inside an incubator for the duration of hOBMT culture, set on gentle rocking. However, this is not critical to the success of scaffold mineralization and static culture can be performed as well.
6. While it is not necessary for the analysis of hOBMT/cancer cells microtissues by histology, the ability to transfect cells with a fluorescent probe enables routine fluorescence microscopy during the course of co-culture. This strategy facilitates immunofluorescence imaging, post-fixation. *See* the next chapter for specific real-time and fixed 3D quantification that are possible with fluorescently labeled cancer cells on hOBMT.
7. The use of low-attachment plates is recommended to minimize cancer cell attachment to the bottom of wells, hence maximizing their attachment to hOBMT. However this is not critical and standard 24-well plates may be used as well.
8. The use of suitable coated glass slides is highly recommended for the histology of microtissues. Indeed, traditional microscopy glass slides have shown poor adherence of the histology sections of the microtissues, which partly detached during the subsequent deparaffinization, rehydration, and staining processes.
9. Xylene is a highly toxic organic solvent; ensure to work in the fume hood laminar flow cabinet.
10. If self-assembling the apparatus using the list of equipment specified in [19], it is recommended to seek electrician assessment to certify the high-voltage components and compliance with the safety regulations. The use of high voltage is associated with a risk of electrocution. Ensure to have magnetic interlocks on the enclosure box which enable instantaneous

voltage switch-off if the enclosure is opened during the course of electrowriting.

11. The electrowriting parameters may be subject to variations according to temperature and humidity. A 10% change from the values mentioned above (voltage, temperature, pressure) may be necessary to obtain the desired fiber diameter and spacing, to address changes in local conditions (room temperature and humidity).
12. Note that other dimensions, porosities, and fiber diameter may be obtained by varying the electrowriting parameters [22]. Note that porosities <100 μm may lead to less homogenous CaP coating in the depth of the scaffold.
13. Preparation of *Solution C* is better done at least 4 h before coating, overnight ideally. Ensure it is done at RT. The pH in this step is critical to the success of the coating. At this stage, the solution can prematurely precipitate due to pH ≥6.25 or taking too long. If this happens, start *Solution C* over with a fresh aliquot from SBF 10×.
14. Filtering is essential to remove any particles, which could potentially act as an undesirable nucleation site.
15. This incubation step may take longer than 30 min. The solution should turn turbid after 20 min (sign that the coating has worked). Early precipitation needs to be avoided, as it means that the solution concentration is different than expected, and the control over the desired CaP phase deposition is lost.
16. The second incubation does not require more than 30 min. The solution will already be turbid.
17. On the long term, it is critical to keep the scaffolds under vacuum-assisted desiccator. Ambient humidity would affect the quality of the coating over time. After coating, use the scaffolds for cell culture as soon as possible (within 2 weeks maximum).
18. If possible, proceed to cell isolation on the same day of the surgery. If this is not possible, add sterile PBS-P/S to completely immerse the bone tissues and store at 4 °C for 48 h maximum, post-surgery.
19. If using normal 175 $cm^2$ flasks, do not try to remove bone fragments prior to the Trypsin step. Proceed to the washing and trypsinization steps with the bone fragments present in the flasks. When adding the media and collecting the cell suspension, take precaution to not aspirate any bone fragments.
20. Preosteoblasts may be used up to passage 6, for optimal osteogenic differentiation. Osteogenic capacity is however donor-dependent and lower passages may need to be used for some donors.

21. If thawing frozen primary preosteoblasts, allow one passage before experimental use.
22. The use of low-attachment plates is recommended so that cells do not grow out of the scaffolds on the bottom of the wells, requiring more media changes. However, traditional plates can also be used. If so, transfer the scaffolds to new plates when switching to osteogenic media and every 2 weeks from then on.
23. Lower cell amounts can be used, but it will lead to delayed 3D confluence. No less than $2 \times 10^5$ cells/cm$^2$ should be used.
24. Cell seeding depends on the scaffold's dimensions. An average of $8 \times 10^3$ cell/mm$^3$ is recommended for seeding. Volume will vary as well. Generally, scaffold should be just wet during attachment, and medium should not overflow from the scaffolds.
25. Using disposable biopsy 5 mm punches, some hOBMT can be collected at specific time points. This allows monitoring of ECM production using SEM or IF after fixation in appropriate fixative (Glut 3% for SEM and PFA 4% for IF) [19]. Micro-computer tomography also allows to check mineralization over time.
26. Samples in 3% glutaraldehyde can be stored indefinitely provided they do not dry out.
27. Start with a working distance of 10 mm. If a large depth of field is needed, which may be needed for scaffold imaging, consider working with higher distances such as 20 mm (Fig. 2a). When higher magnification is needed, with little depth of field needed, consider reducing the voltage to 5 kV and the distance to 7 mm (Fig. 4c).
28. Parameters such as cancer cell concentration and duration of the attachment step (no more than 24 h) need to be optimized for each type of new cell to maximize and ensure reproducible seeding and optimal attachment.
29. Oven drying is critical to efficient attachment of the tissue construct sections onto the slides. Shorter times lead to inefficient section attachment to the slide and the sections may be washed off during the subsequent processing steps. Longer times in oven will lead to tissue damage with loss of micro- and macrostructures.
30. It is very important to avoid the antibody and control solutions being mixed up, in the case that both are present on the same slide.
31. If air bubbles are entrapped, quickly dip the mounted slide in Xylene and let it stand for 10 min, swirling from time to time. This will dissolve the mounting agent, allowing to recover cover glass and repeating **step 31**.

32. The intensity of protein staining between various conditions can be assessed quantitatively using standard imaging software (Image J, Visiopharm) or bone histomorphometric software (Osteomeasure). Due to variability between tissue thin sections, use at least six sections per condition, with three fields of view per section, to ensure statistical significance between conditions (Fig. 4c).
33. Using 5 mm biopsy punches, one scaffold can be divided in up to four punches. This allows studying various proteins simultaneously without compromising the fluorescent signals, which may leak in between fluorescent channels with the neighboring wavelengths. For the hOBMT, as they contain polymeric fibers and a high degree of mineralization, it is recommended not to work with multiple targets on the same sample to reduce background. Furthermore, having several samples of the same tissue construct, allows having a secondary antibody control (no primary), which is essential to validate the signals obtained from the proteins of interest.
34. Antibody concentrations need to be tested. Manufacturer instructions often suggest working concentrations for IF. If not available, start with 2–5 μg/mL.
35. Adjust the volume according to the size of the tissue sample being tested, making sure that the sample is fully covered, but do not add excessive liquid to save antibody.
36. Using phalloidin is useful to delineate the cytoskeleton of cells, however with the bioengineered tissue microtissues, phalloidin will stain both the cancer cells and the osteoblasts and may generate leakage to other channels due to strong signal. Consider working with DAPI and the protein of interest only and having a separate control with DAPI and phalloidin only. Another alternative is to work with fluorescently labeled cancer cells (red probe ideally) so that phalloidin can be avoided, yet still have cancer cell morphology information.
37. Green fluorescence probes are usually stronger that most red/-far red fluorescence probes, hence it is suggested to use green for the proteins of interest.
38. Mounting medium is not recommended for scaffolds, since mounting medium is highly viscous and will not fully penetrate the scaffolds, reducing the quality of the imaging.
39. Proceed to imaging on same/next day if possible. Samples can be imaged up to 2 weeks post staining. If longer times are required, stabilize stained samples by fixing another time for 10 min in 4% PFA/PBS.

## Acknowledgements

This work was supported by an IHBI ECR grant, Advance Queensland (AQ) Maternity Fund Award from the Queensland Government (DSITI), Young Researcher Award (2017-YR-RoW-9) from Lush (UK), supporting non-animal testing alternatives, National Health and Medical Research Council (NHMRC) Peter Doherty Early Career Research Fellowship (RF) (APP1091734), and John Mills Young Investigator Award (YI0715) from the Prostate Cancer Foundation of Australia (PCFA). The author acknowledges the TRI for the excellent core facilities that enabled this research, and thank Adler Ju from the TRI microscopy facility. The author also thanks D/Prof. Dietmar W. Hutmacher, D/Prof. Judith A. Clements AC, Dr. Elena De-Juan Pardo, Mr. Jose Malayil, Dr. Asha Mathew, Ms. Teja Yarlagadda, Dr. Lipsa Mohanty, Dr. Ali Shokoohmand, Dr. Thomas Kryza, Dr. Joan Röhl, Dr. Felicity Lawrence, Dr. Jacqui McGovern, Dr. Christoph Meinert, Dr. Aurélien Forget, Dr. Christina Theodoropoulos, and Mr. Colin Rohner.

### References

1. Gartrell BA, Saad F (2014) Managing bone metastases and reducing skeletal related events in prostate cancer. Nat Rev Clin Oncol 11 (6):335–345. https://doi.org/10.1038/nrclinonc.2014.70
2. Correia AL, Bissell MJ (2012) The tumor microenvironment is a dominant force in multidrug resistance. Drug Resist Updat 15 (1–2):39–49. https://doi.org/10.1016/j.drup.2012.01.006
3. Salamanna F, Contartese D, Maglio M, Fini M (2016) A systematic review on in vitro 3d bone metastases models. A new horizon to recapitulate the native clinical scenario? Oncotarget 7 (28):44803–44820. https://doi.org/10.18632/oncotarget.8394
4. Sung SY, Hsieh CL, Law A, Zhau HE, Pathak S, Multani AS et al (2008) Coevolution of prostate cancer and bone stroma in three-dimensional coculture: implications for cancer growth and metastasis. Cancer Res 68 (23):9996–10003. https://doi.org/10.1158/0008-5472.can-08-2492
5. Bersini S, Jeon JS, Dubini G, Arrigoni C, Chung S, Charest JL et al (2014) A microfluidic 3D in vitro model for specificity of breast cancer metastasis to bone. Biomaterials 35 (8):2454–2461. https://doi.org/10.1016/j.biomaterials.2013.11.050
6. Krishnan V, Vogler EA, Sosnoski DM, Mastro AM (2014) In vitro mimics of bone remodeling and the vicious cycle of cancer in bone. J Cell Physiol 229(4):453–462. https://doi.org/10.1002/jcp.24464
7. Hutmacher DW (2010) Biomaterials offer cancer research the third dimension. Nat Mater 9(2): 90–93. https://doi.org/10.1038/nmat2619
8. Hutmacher DW, Cukierman E (2014) Engineering of tumor microenvironments. Adv Drug Deliv Rev 79:1–2. https://doi.org/10.1016/j.addr.2014.11.001
9. Hutmacher DW, Holzapfel BM, De-Juan-Pardo EM, Pereira BA, Ellem SJ, Loessner D et al (2015) Convergence of regenerative medicine and synthetic biology to develop standardized and validated models of human diseases with clinical relevance. Curr Opin Biotechnol 35:127–132. https://doi.org/10.1016/j.copbio.2015.06.001
10. Hutmacher DW, Horch RE, Loessner D, Rizzi S, Sieh S, Reichert JC et al (2009) Translating tissue engineering technology platforms into cancer research. J Cell Mol Med 13 (8A):1417–1427. https://doi.org/10.1111/j.1582-4934.2009.00853.x
11. Sieh S, Lubik AA, Clements JA, Nelson CC, Hutmacher DW (2010) Interactions between human osteoblasts and prostate cancer cells in a

novel 3D in vitro model. Organogenesis 6 (3):181–188. https://doi.org/10.4161/org.6.3.12041

12. Sieh S, Taubenberger AV, Lehman ML, Clements J, Nelson CC, Hutmacher DW (2014) Paracrine interactions between LNCaP prostate cancer cells and bioengineered bone in 3D in vitro reflect molecular changes during metastasis. Bone 63:121–131. https://doi.org/10.1016/j.bone.2014.02.001
13. Pathi SP, Kowalczewski C, Tadipatri R, Fischbach C (2010) A novel 3-D mineralized tumor model to study breast cancer bone metastasis. PLoS One 5(1). https://doi.org/10.1371/journal.pone.0008849
14. Bock N, Loessner D, Hutmacher D, Clements J (2015) A 3D in vitro model reflecting the androgen deprivation state in prostate cancer bone metastasis. Paper presented at the 2nd prostate cancer world congress, Cairns, Australia, 17–21 Aug 2015
15. Bock N, Loessner D, Clements JA, Hutmacher DW (2015) A tissue engineered bone construct as a 3D metastasis colonization model. Paper presented at the 4th TERMIS world congress, Boston, MA, USA, 8–11 Sept 2015
16. Shokoohmand A, Bock N, Baldwin J, Kryza T, Nelson CC, Clements JA et al (2016) Development and validation of a modular 3D microenvironment platform technology for studying prostate cancer- induced bone metastasis. Paper presented at the 17th Asia-Pacific prostate cancer conference, Melbourne, Australia, 31 Aug–03 Sept 2016
17. Tevz G, Hossain AM, Bock N, Jeet V, Bathgate RA, Williams ED et al. (2015) Impeding the adaptive response of prostate cancer to androgen targeted therapies with relaxin receptor antagonist. Paper presented at the 2nd prostate cancer world congress (PCWC), Cairns, Australia, 17–21 Aug 2015
18. Kryza T, Bock N, Shokoohmand A, Tevz G, McPherson S, Lehman M et al. (2016) Modulation of protease expression in prostate cancer cells after androgen deprivation. Paper presented at the 17th Asia-Pacific prostate cancer conference, Melbourne, Australia, 31 Aug–03 Sep 2016
19. Martine LC, Holzapfel BM, McGovern JA, Wagner F, Quent VM, Hesami P et al (2017) Engineering a humanized bone organ model in mice to study bone metastases. Nat Protoc 12 (4):639–663. https://doi.org/10.1038/nprot.2017.002
20. Yang F, Wolke JGC, Jansen JA (2008) Biomimetic calcium phosphate coating on electrospun poly (epsilon-caprolactone) scaffolds for bone tissue engineering. Chem Eng J 137 (1):154–161. https://doi.org/10.1016/j.cej.2007.07.076
21. Vaquette C, Ivanovski S, Hamlet SM, Hutmacher DW (2013) Effect of culture conditions and calcium phosphate coating on ectopic bone formation. Biomaterials 34 (22):5538–5551. https://doi.org/10.1016/j.biomaterials.2013.03.088
22. Brown TD, Dalton PD, Hutmacher DW (2016) Melt electrospinning today: an opportune time for an emerging polymer process. Prog Polym Sci 56:116–166. https://doi.org/10.1016/j.progpolymsci.2016.01.001

# Chapter 3

# Real-Time and 3D Quantification of Cancer Cell Dynamics: Exploiting a Bioengineered Human Bone Metastatic Microtissue

**Nathalie Bock and Joan Röhl**

## Abstract

The study of dynamic processes in the bone metastatic compartment has been challenged by the restrictive access and limited live imaging capabilities that in vivo bone models provide. In this protocol, we show the use of a human bone metastatic bioengineered microtissue for the quantitative investigation of cancer cells in an in vitro bone-like microenvironment. Using live cell epifluorescence microscopy, traditional- and spinning disc-confocal laser scanning microscopy, we demonstrate how to obtain multidimensional real-time data of fluorescently labeled cancer cells in the metastatic microenvironment. Using 4D imaging data processing software such as ImageJ and Imaris, we show how to transform qualitative images and videos into quantitative data of cancer cell attachment, morphology, proliferation, and migration in vitro in the human bone metastatic microtissue.

**Key words** Cancer cell, Tumor microenvironment, Bone, Metastasis, Cancer model, Morphometry, Cell migration, Microtissue, Live cell imaging, Cell dynamics

## 1 Introduction

Recent advances in imaging technologies have allowed the study of cell–cell and cell–matrix interactions in different microenvironmental contexts [1]. With the development of fluorescent imaging probes combined with high-speed and -resolution microscopy and powerful imaging data processing software, it is now possible to witness intracellular and extracellular processes in cells and tissues [2]. Intravital multi-photon microscopy (IVPM) has been a key development in the field of in vivo imaging, allowing the real-time and space observation of pathologies such as metastasis [3]. IVPM enables to study dynamic processes at the single cell level, generating information of cancer cell interactions with tumor stroma at the primary and metastatic sites [4]. Yet, IVPM still faces

Jyotsna Batra and Srilakshmi Srinivasan (eds.), *Theranostics: Methods and Protocols*, Methods in Molecular Biology, vol. 2054,
https://doi.org/10.1007/978-1-4939-9769-5_3, 

technical challenges, which has limited its use. Only superficial tissues can be imaged, which is incompatible with most tumor types and orthotopic models [5]. Bone metastases, in particular, are found deep in the tissue, which heavily limits the study of any dynamic processes. Finally, while a lot of work is being done in vivo, animal models do not reproduce the biological programs specific to human disease [6], which has led to numerous clinical trial failures [7]. Altogether, working in animals is limiting our understanding of the pathophysiology of *human* bone metastasis, partly resulting in no curative treatments to date.

The current issues may be addressed by using sophisticated three-dimensional (3D) in vitro models that can bridge the gap between unrepresentative 2D models and complicated animal models [8, 9]. Tissue-engineered microtissues provide a more physiological context than traditional monolayer cultures [10], by recreating specific native 3D microenvironments that are consistent in mimicking various stages of the disease [11, 12]. They provide tailorable and reproducible platforms to study human cancer progression, metastasis, and resistance to therapeutics [13].

In the previous chapter, we have presented the fabrication of a human bone metastatic bioengineered microtissue and its use in vitro for protein characterization by immunostaining techniques and static imaging. In this chapter, by using fluorescently labeled cancer cells with the aid of confocal microscopy, we are able to further delineate the morphometric features of cancer cells in the recreated human bone-like microenvironment. Using dynamic epifluorescence microscopy and the 4D imaging data software Imaris, we can visualize and quantify the effects of drug treatments on the proliferation and migration of cancer cells on the microtissue. The methodologies proposed here open the door to a novel dimension of spatiotemporal quantitative data collection. This novel preclinical platform will contribute to better understanding of the nature of the dynamic interactions between bone and cancer, as well as serve to measure the efficiency of anti-cancer drugs in a human context.

## 2 Materials

### *2.1 Human Osteoblast-Derived Bone-Like Microtissue (hOBMT)*

Use the materials, consumables, equipment and reagents listed in the previous chapter (Subheadings 2.1–2.5), for preparation of the bioengineered human osteoblast-derived microtissues (hOBMT).

### *2.2 Cancer Cells*

Use metastatic cancer cell lines of interest, for co-culture with the hOBMT. Cells need to be transfected to express a red, fluorescent probe prior to use for the methods described hereafter (*see* **Note 1**). This protocol was tested and validated for standard metastatic prostate cancer cell lines (LNCaP, C4-2B, PC3), transfected with either mCherry or mKO2. Cells were used in passages 30–40. This protocol was also validated for mKO2-KLK14-overexpressing LNCaP models and mKO2-PSA and mKO2-PSA SNP variant-overexpressing PC3 models.

### 2.3 Cancer Cell Co-culture with hOBMT

#### 2.3.1 Materials and Consumables

1. Specimen plastic tubes (3 mL).
2. Reaction tubes (15 mL).
3. Biopsy punches 5 mm.
4. Disposable scalpels.
5. Sterile pipette tips (no filter needed), 1000 μL.
6. Optional: ultra-low attachment 24 well-plates.
7. Cell culture-treated 24 well-plates.
8. Cell counting slides for automated cell counter TC10.

#### 2.3.2 Equipment

1. Biological safety cabinet class II (laminar flow).
2. Humidified cell culture incubator set to 37 °C and 95% air, 5% $CO_2$.
3. Rocking platform mixer.
4. Pipettor (P1000).
5. Reusable PTFE board (10 × 10 cm).
6. Sterilized metal tweezers.
7. Swing-bucket centrifuge suitable for 15 mL reaction tubes.
8. 500 mL glass bottle.
9. Automated cell counter.

#### 2.3.3 Reagents

1. Ethanol: 70% v/v solution in water.
2. Sterile phosphate buffer saline (PBS): 100% v/v divalent ion-free solution.
3. Co-culture media (CCM): Roswell Park Memorial Institute 1640 medium, 10% v/v FBS, 1% v/v P/S (*see* **Note 2**).
4. 1% agar: Dissolve 1 g of agar in 100 mL serum-free RPMI in a 500 mL glass bottle. Autoclave. Store at 4 °C (*see* **Note 3**).

### 2.4 Morphometric Analysis

#### 2.4.1 Materials and Consumables

1. Paraffin film.
2. Aluminum foil.
3. Pipette tips (1000 μL).
4. Rectangular cover glasses (22 × 50 mm).
5. 24-well plates.
6. Transfer pipettes.

*2.4.2 Equipment*

1. Sterilized metal tweezers.
2. Pipettor (P1000).
3. Nikon spectral spinning disc confocal microscope fitted with a 10× PlanApo objective, using a red filter (excitation at 561 nm) (*see* **Notes 4** and **5**).
4. Imaris imaging analysis software (version 9.1.0, Bitplane AG, Zurich, Switzerland).
5. ImageJ software (version 1.51j8, NIH, USA [14]).

*2.4.3 Reagents*

1. Sterile PBS: 100% v/v divalent ion-free solution.
2. PFA: 4% v/v solution in PBS.

### 2.5 Cancer Cell Live Imaging and Migration Analysis

*2.5.1 Materials and Consumables*

1. Cell culture-treated 24-well plates.
2. Sterile pipette tips (1000 μL).

*2.5.2 Equipment*

1. Sterilized metal tweezers.
2. Pipettor (P1000).
3. Inverted epifluorescence microscope.
4. Sterilized custom-made PTFE ring inserts for 24-well plates, 16 mm diameter, 1 mm thickness, 2 mm ring size.
5. Live cell inverted epifluorescence microscope fitted with a humidified chamber, 95% air, 5% $CO_2$, set at 37 °C and cellSens Dimension software (version 1.14, Olympus, Australia).
6. OlyVIA virtual slide software (version 2.6, Olympus Soft Imaging Solutions GmbH, Germany).

*2.5.3 Reagents*

1. Co-culture media (CCM): Roswell Park Memorial Institute 1640 medium, 10% v/v FBS, 1% v/v P/S—warmed (37 °C) co-culture media (CCM).

## 3 Methods

### 3.1 Human Osteoblast-Derived Bone-Like Microtissue (hOBMT)

Prepare the human osteoblast-derived bone-like microtissues (hOBMT) as per previous chapter, Subheadings 3.1–3.5. Specifically, prepare a three-dimensional (3D) scaffold using medical-grade polycaprolactone (mPCL) by melt electrowriting (Subheading 3.1). Proceed to calcium phosphate (CaP) treatment (Subheading 3.2), leading to CaP-mPCL scaffolds. Collect human bone tissue under informed donor consent and isolate human preosteoblasts (hPOB) (Subheading 3.3). Seed and culture hPOB onto the

CaP-mPCL scaffolds for 6–10 weeks (Subheading 3.4). Proceed to quality control using scanning electron microscopy (SEM) (Subheading 3.5). In total, these procedures require 1.5–3 months.

### 3.2 Cancer Cell Attachment Assay

Cancer cell homing to bone is a critical step of the metastatic cascade. Variations between cancer cell types and the nature of the bone microenvironment according to patient age, medical history, drug treatments, etc., may lead to different degrees of cancer cell homing. When finally reaching the bone compartment, cancer cells first require attachment to ECM for survival [15] and hence attachment is a critical step to homing success [16]. While it is difficult to exactly reproduce homing in an in vitro model, we have developed a simple dynamic assay to assess cancer cell attachment to the human bone-like microenvironment, using hOBMT. This assay can be performed using preconditioned cancer cells or using preconditioned hOBMT, and in the presence of drug treatments.

The following steps must be performed in a biological safety cabinet class II (laminar flow), under aseptic conditions. Spray all tools and equipment with 70% v/v ethanol.

1. Culture the authenticated and mycoplasma-tested cancer cells transfected with a red probe, below passage 40 (LNCaP, C4-2B, PC3), for at least 2 weeks prior to co-culture with the hOBMT. If assessing the effects of drug treatments on cancer cell attachment to the hOBMT, culture the cancer cells in the treatments for the appropriate amount of time prior to experiments.
2. Select the hOBMT which will be used for the attachment assay. Using sterilized metal tweezers, place one hOBMT onto the sterilized PTFE board. Using biopsy punches, punch 5 mm tissue samples and place in the prewarmed co-culture media (CCM) media, for 10 min.
3. Aspirate the media and wash the hOBMT samples two more times with CCM, for 10 min each rinse (place the samples back in the incubator during washes).
4. If not using ultra-low attachment 24-well plates, prepare normal cell culture-treated 24-well plates as follows: use a microwave to heat the 1% sterile agar solution until it becomes liquid. Add 300 μL of the agar solution per well. Allow 5 min for gel setting.
5. Prepare the cancer cell suspensions (preconditioned or not) in CCM, containing drug treatments or not ($2 \times 10^5$ cell/mL) (*see* **Note 6**).
6. Place one single hOBMT sample in one single well of the 24-well plate. Use at least six replicates per condition, coming from different hOBMT grown from the cells obtained from the same donor.

7. Dispense 500 μL of the cancer cell suspension to each well containing hOBMT samples (leading to $10^5$ cancer cells per well). Also add 500 μL of the cancer cell suspension to six wells (control wells), which do not contain any hOBMT samples but are agar-coated (the cells collected later from these wells will be the controls).
8. Place the well plate on the rocking platform inside the incubator and monitor time.
9. At desired time points (*see* **Note 7**), such as after 3 or 24 h of co-cultures, aspirate the supernatant from all the wells and transfer it to individual 15 mL reaction tubes (one tube per hOBMT sample).
10. Proceed carefully and consistently to three more washes in CCM (1 mL each) by gently flushing CCM onto the surface of hOBMT sample, to mechanically remove unattached cells. Collect the washing CCM supernatants and add to the reaction tubes (do not mix up samples). The total volume collected should be 3.5 mL per hOBMT sample.
11. Centrifuge the reaction tubes at 300 × *g* for 3 min at room temperature.
12. Using an inverted epifluorescence microscope, double check that the cancer cells have attached reproducibly and homogeneously onto hOBMT within each condition tested.
13. Discard the supernatants and resuspend the cell pellets in 500 μL of CCM.
14. Using the automated cell counter, record the cell counts for all cell suspensions (use four technical replicates).
15. Determine the average value for control wells. To obtain the cancer cell attachment values on hOBMT per condition $\left(\overline{\mathrm{Att}_{\mathrm{hOBM}}}\right)$, use the formula below, with $\overline{C_{\mathrm{ctrl}}}$ as the average cell concentration value from control wells, $\overline{C_{\mathrm{hOBM}}}$ as the average cell concentration value collected per well, and $n$ the number of wells.

$$\overline{\mathrm{Att}_{\mathrm{hOBM}}}(\%) = \frac{\sum_{n} \frac{\overline{C_{\mathrm{ctrl}}} - \overline{C_{\mathrm{hOBM}}}}{\overline{C_{\mathrm{ctrl}}}}}{n} \times 100$$

### 3.3 Cancer Cell Morphometry on hOBMT

Morphometry, or the quantitative measurement of cellular properties, has emerged as a more powerful method than traditional techniques [17]. While histopathology is routinely used for lesion/tumor grading and takes an important part in the clinical management of diseases, it relies on a trained pathologist manually assessing sections. Conversely, morphometry can be automated by computers that apply specific algorithms to 2D or 3D fluorescent images. This strategy offers more quantitative and unbiased information, which can result in more accurate diagnostics. Specifically, when studying cancer, the morphometric features of cancer cells, such as shape factor, volume, and straightness are useful in assessing the degree of aggressiveness of a cancer cell. By submitting cancer cells to drug treatments and assessing morphometry between groups, quantitative comparison can be made. This strategy accurately informs on treatment effect, providing a more powerful output than traditional histopathology. Here, morphometric analysis can be performed with the proposed in vitro model too. The 3D morphometric properties of prostate cancer cells after initial co-culture with hOBMT are obtained by combining confocal microscopy technology with 3D imaging analysis, as per the steps below. The section is divided into three key stages; hOBMT co-culture with cancer cells, image acquisition, and image analysis.

#### 3.3.1 hOBMT Co-culture with Cancer Cells

1. Perform **steps 1–10** from Subheading 3.2, with lower cell concentrations. For LNCaP and C4-2B, use $10^5$ cells/mL. For PC3, use $0.5 \times 10^4$ cells/mL. This number needs to be optimized according to the cell type being used, especially if the cells have been re-transfected with a target of interest, after initial transfection with the red probe. As for the co-culture times, chose times ranging between 3 and 24 h (*see* **Note 7**). Use at least three hOBMT sample replicates.
2. After washing, transfer the co-culture microtissues into new 24-well plates.
3. Wash once with 1 mL of PBS.
4. Fix the co-culture microtissues using 300 μL of PFA 4% for 40 min at room temperature.
5. Rinse the co-culture microtissues with 1 mL PBS twice.
6. Add 1 mL of fresh PBS.
7. If not imaging immediately, seal the well plate with paraffin film, cover with aluminum foil and store at 4 °C.

#### 3.3.2 Imaging Acquisition

1. Prepare the spinning disc confocal microscope.
2. Set a rectangular glass cover onto the microscope stage.
3. Using tweezers, place a co-culture construct in the center of the glass cover.
4. Ensure the construct remains wet by adding a few PBS drops using a transfer pipette and sandwich the construct with a second glass cover.
5. Looking through the eyepiece with the 561 nm laser on, make sure to be imaging the correct side of the co-culture construct, presenting with attached cancer cells.

6. Using the 561 nm laser and 10× objective, select a field of view and set the limits for a *z*-stack, making sure to encompass all the cancer cells present in the view. Add a few micrometers for safety margins. Depth should be around 50 μm or less but kept consistent throughout samples.
7. Set the step size to 1 μm. Adjust the laser power, exposure time, and gain according to the fluorescence intensity of the sample (*see* **Note 4**).
8. Proceed to the acquisition. When using a 50 μm depth and a 1 μm step size, this will result in a *z*-stack of 51 images. Save the *z*-stack.
9. Repeat **steps 6–8** with five more fields of views on the same construct. Do not overlap any fields of view.
10. Proceed to the imaging of the other technical replicates and other treatment conditions, if applicable, following **steps 1–9**, with the same acquisition settings.

#### 3.3.3 Image Analysis

Morphometric parameters, such as cancer cell volume and sphericity on hOBMT can be obtained by processing the *z*-stacks with the Imaris software, while cell orientation can be obtained using ImageJ. While ImageJ is a freely available software, a license is required for Imaris, as well as powerful hardware.

#### 3.3.4 Cancer Cell Volume and Sphericity on hOBM

1. Open the Imaris imaging analysis software.
2. Go in the *Surpass* mode in the top bar and drag in a *z*-stack file. Only the cancer cells will appear (hOBMT will remain invisible, as only the red channel was recorded) (Fig. 1a).
3. Click on the *Surface* icon in the left panel, to create a new surface protocol.
4. In step 1/5 (Algorithm), leave *Favorite Creation Parameters* onto 'Default'.
5. Go to next (step 2/5): Source Channel. Ensure that the correct channel is being analyzed (561 nm). Set the *Surfaces Detail* as 1 μm and the *Thresholding* on 'Absolute Intensity'.
6. Go to next (step 3/5): Threshold. In the *Seed Points Diameter* box, enter the approximate minimum diameter value that a cell could be. This allows the software, downstream, to split any potential cells that are aggregated together (here we used 11 μm). In the *Split Touching Objects (Region Growing)*, ensure that *Enable* is ticked. This will allow to split any touching cells.
7. In the *Threshold (Absolute Intensity)*, adjust the cursor so that all cells are precisely covered by the grey mask. *See* Fig. 1b.
8. Go to next (step 4/5): Classify Seed Points. At this point, Imaris represents each single cell by a dot. By default, Imaris may suggest a 'Quality' filter based on cell intensity, which

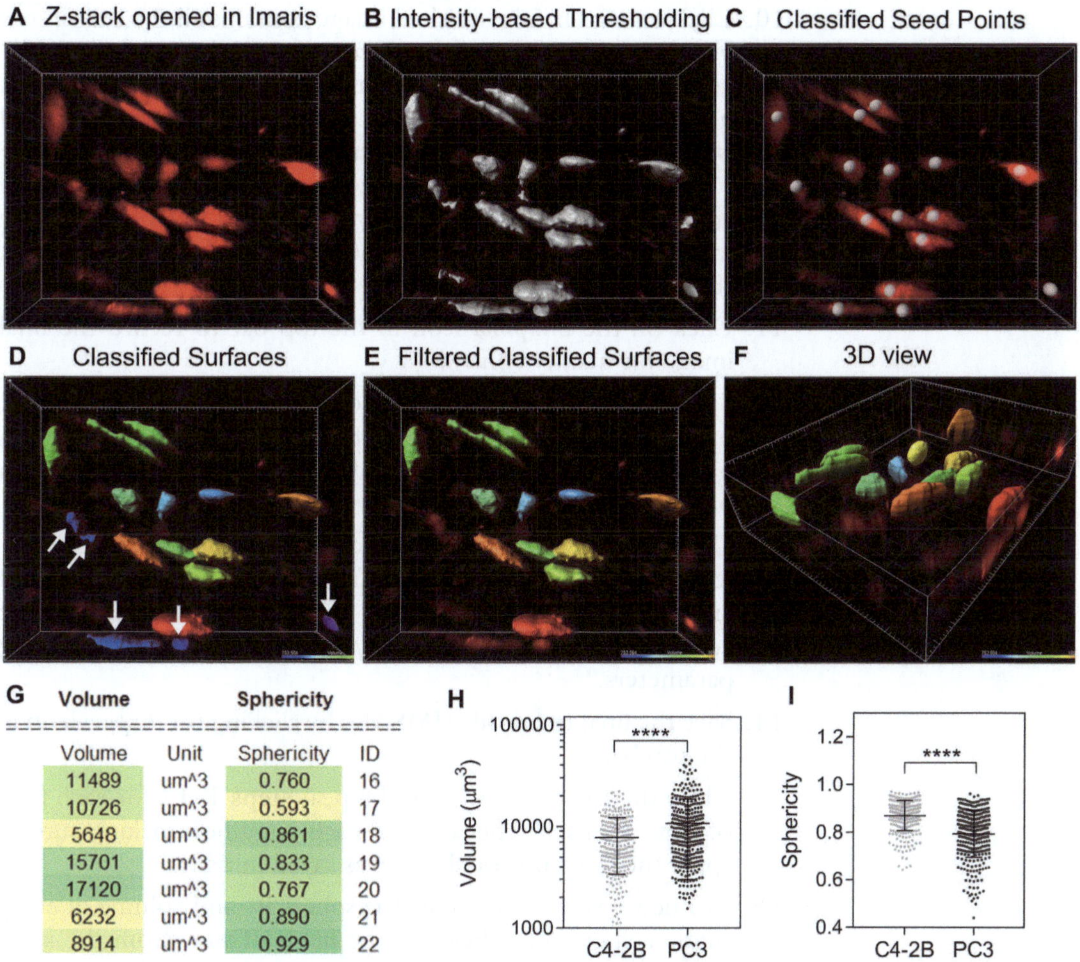


**Fig. 1** (**a–g**) Image analysis steps for the measurement of the morphometric features of cancer cells co-cultured on hOBMT, using Imaris. The white arrows show the elements to be excluded from the analysis. (**h**, **i**) Cell volume and sphericity of C4-2B and PC3 cells on hOBMT after 24 h of co-culture in the presence of 10 nM dihydrotestosterone (>9 fields of view analyzed, >280 cells total/condition). $****$ refers to *p* value <0.0001, as per the independent samples Student's *t*-test

allows to select/deselect cells, based on intensity. Adjust the cursor so that each cell is represented by an individual dot (Fig. 1c). Other filters may be added from the *Filter Type* drop-down menu.

9. Go to next (step 5/5): Classify Surfaces. At this point, Imaris highlights each individual cell with a different color, which allows assessing whether the protocol has appropriately identified all the cells. By default, Imaris may suggest a 'Number of Voxel' filter, based on cell volume. By adjusting the cursor, include any cells that were missed, remove any cells that are not single (aggregates), remove any debris and remove any cells that are touching any edges as they may be split and so are unacceptable for the analysis (Fig. 1d).

10. Click on Finish. The finished image contains all the cells in red but the ones which were analyzed are highlighted with different colors (Fig. 1e, f). At this stage, only individual cells should have been selected. If any object comprising more than one cell is selected, click 'Select' in *Pointer* (right tab). Then select the object to be deleted. From the left panel, select the *Pencil* icon and click on 'Delete'. Similarly, if one cell has been split by mistake, select both cells by holding the shift key and click 'Unify' (*see* **Note 8**).
11. Click on the *Snapshot* icon in the top bar to record the final image for quality control.
12. Save the protocol by clicking on the *Wand* icon in the left panel, then *Store Parameters for Batch*. Give the protocol a name and click 'Arena' and 'Favorite Creation Parameters' in *Store Location*. Click 'OK'.
13. Click the *Statistics* tab, and click the *Export all Statistics to File* icon, making sure to extract *Cell Volume*, *Cell Sphericity* and any other relevant parameters (Fig. 1g). Parameters can be edited by clicking the *Preferences* icon and ticking/unticking parameters.
14. Save the final Imaris file (IMS file) by clicking the *Export* icon in the top bar.
15. Close the Imaris software and repeat **steps 1–14** for each field of view, for each replicate, and each condition making sure to apply the same protocol that was saved initially.
16. In a new Excel file, compile the sphericity and volume of every single cell collected from all the fields of view from the same condition. Determine the average cell volume and average cell sphericity (Fig. 1h, i). A minimum of 200 cells per condition should be obtained.

#### *3.3.5 Cell Orientation on hOBMT*

1. In order to obtain cell orientation on hOBMT, open the ImageJ software.
2. Drag in the first *z*-stack file (*see* **Note 9**).
3. A 'Bio-Formats Import Options' window will appear. Ensure that *Stack Viewing* is on 'Hyperstack' and that all boxes are unticked except for 'Split channels'. In *Color Options*, select 'Grayscale'.
4. If using a stack that contains multiple color channels (they will appear in grayscale), close the unnecessary ones and keep only the channel with cancer cells.
5. Proceed to a maximum intensity projection of the *z*-stack, by going to *Image*, *Stacks*, *Z project*. Check that *Projection Type* is set to 'Max Intensity' and click 'OK'.

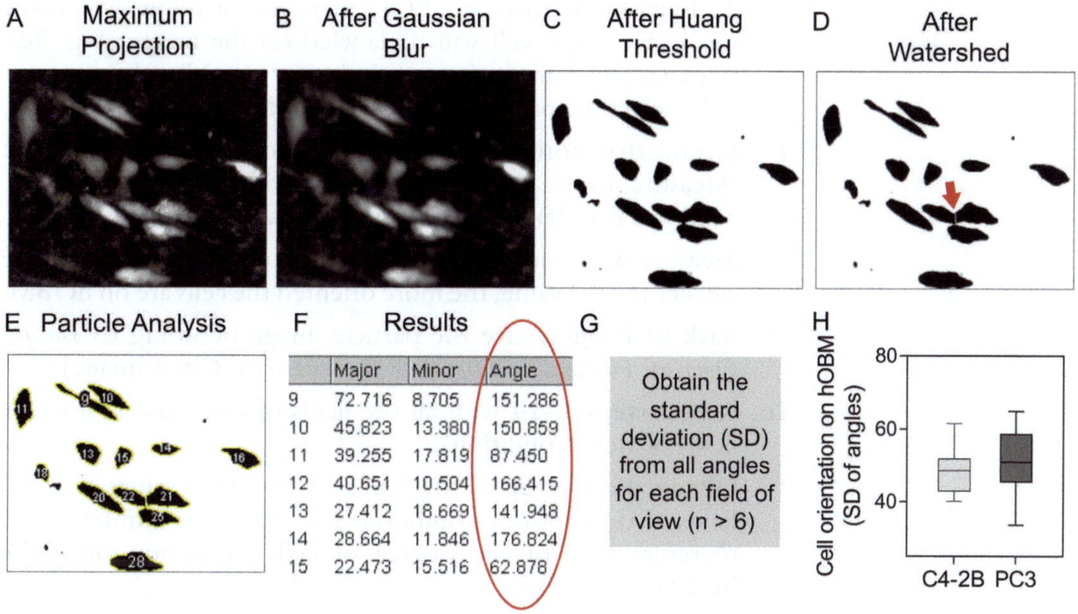


**Fig. 2** (**a–f**) Image analysis steps for the measurement of cancer cellorientation on hOBMT, using ImageJ. The red arrow shows two cells being appropriately split using the adjustable watershed plugin. (**g**) Data post-processing. (**h**) Orientation of C4-2B and PC3 cells on hOBMT after 24 h of co-culture in the presence of 10 nM dihydrotestosterone, obtained from the standard deviations of cellular angles measured in ImageJ. Represented as box plots of the standard deviations of 12 fields of view per condition

6. Close the ND2 file and keep working with the maximum intensity projection image (Fig. 2a).
7. In *Process*, go to *Filters*, *Gaussian blur*. In the window, set the *Radius* to 2.0 and click 'OK'. This step serves to blur out any speckles and helps setting a threshold in the next step (Fig. 2b).
8. In *Image*, go to *Adjust*, *Threshold*. In the window, ensure 'Dark background' is ticked. Use the 'Huang' threshold mode, to obtain the outlines of cells only and click 'Apply' (Fig. 2c).
9. For the measurement of cancer cell orientation, it is important to select single cells only. However, if two cells are in contact and orientation has not been affected, cell separation can be performed by going to *Plugins*, *Adjustable_watershed*, *Adjustable Watershed*. In the new window select a Tolerance of 1. This will split any cell aggregates into single cells (Fig. 2d).
10. Go to *Analyze*, *Set Measurements*. In the window, click 'Fit Elipse'. In *Redirect to*, select the name of the image. Click 'OK'.
11. Go to *Analyze*, *Analyze particles*. In *Size*, select '100-Infinity' (*see* **Note 10**). In *Show*, select 'nothing'. Then tick 'Clear Results', 'Exclude on edges', 'Add to Manager'. Click 'OK'.

12. A 'Region Of Interest (ROI) manager' window will come up. Every single cell will be labeled on the image (Fig. 2e). With visual confirmation, exclude any particles that are not cells (select the particle in the ROI window and click 'Delete').
13. Ensure that nothing is selected by clicking 'Deselect'. Click 'Measure' in the ROI manager. Data will appear in a 'Results' window (Fig. 2f). Copy all values into an excel file.
14. Measure the standard deviation (SD) from all angle values. The smaller the SD value, the more oriented the cells are on hOBM.
15. Back to ImageJ, save the particle image by going to *Image*, *Overlay*, *Flatten*. In *File*, click *Save as*, *Tiff*. Close ImageJ.
16. Repeat **steps 1–14** with all the fields of view (use at least six fields of view/condition).
17. Obtain the average of the standard deviation values from all fields of view for one condition (Fig. 2g). This number is the representation of the degree of cell orientation onto the hOBMT (Fig. 2h).

### 3.4 Cancer Cell Live Imaging and Migration on hOBMT

Similar to morphometry, cancer cell motility is a critical step in the metastatic process. Therapeutically speaking, the ability to reduce cancer cell migration may be a key way to reduce cancer spread to and growth within distant organs. Importantly, during metastatic processes in vivo, cancer cells migrate in confined spaces by gradually degrading their surrounding extracellular matrix (ECM), or moving around on preexisting tracks created by anatomical structures. Matrix remodeling takes an important part during the development of the metastatic niche and collagen fibers are a key substrate that fuels metastatic spread [18]. By providing a biomimetically relevant microenvironmental platform that mimics the structure and compositional platforms encountered in vivo, we can better study how therapeutic interventions can efficiently halt cancer cell migration. The 3D biomimetic bone-like model proposed here represents an appropriate and reproducible platform to study cancer cell migration. Quantitative analysis can be obtained by using live epifluorescence microscopy, to capture cell motility onto the hOBMT, combined with 3D imaging analysis software, as per the steps below. The section is divided into three key stages; hOBMT co-culture with cancer cells, image acquisition and image analysis.

#### 3.4.1 hOBMT Co-culture with Cancer Cells

1. Perform **steps 1–10** from Subheading 3.2, with lower cell concentrations. For LNCaP and C4-2B, use $10^5$ cells/mL. For PC3, use $0.5 \times 10^4$ cells/mL. This number needs optimization according to the cell type being used, especially if the cells have been re-transfected with a target of interest (*see* **Note 6**). Instead of using 5 mm biopsy punches, cut each hOBMT

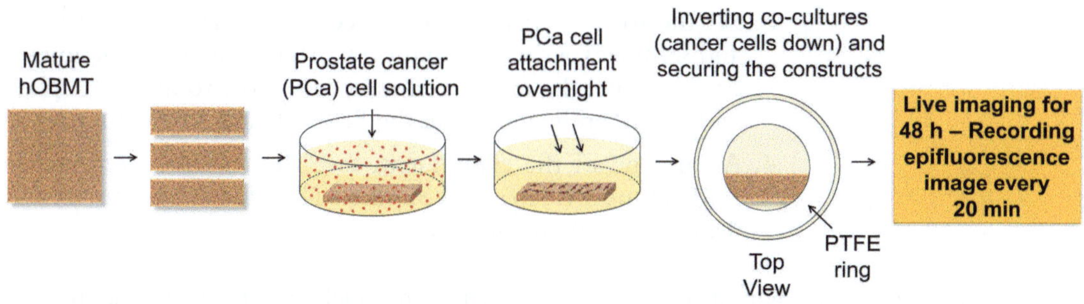


**Fig. 3** Schematic of cancer cells/hOBMT co-culture preparation for live epifluorescence imaging

construct in three identical hOBMT stripes using a sterile scalpel (Fig. 3), *see* **Note 11**. For one condition, have at least two technical replicates from different scaffolds, i.e. two hOBMT stripes from different hOBMT large scaffolds. Do the cancer cell seeding overnight (record time and be consistent in future experiments).

2. After washing, transfer the co-culture microtissues into new (uncoated) 24-well plates that contain 0.5 mL of CCM (and containing drug treatments, if applicable). Place the cancer cell side facing the plastic. Check using an epifluorescence microscope.
3. Use the PTFE ring inserts to secure down the co-cultures by gently pressing down with the tweezers (Fig. 3). This step ensures that the microtissues do not move during image acquisition.
4. Gently aspirate 90% of the media and add fresh media (containing drug treatments, if applicable).

#### 3.4.2 Imaging Acquisition

1. Place the plate in the preheated live cell chamber set a 37 °C, with 95% air, 5% $CO_2$, connected to the live cell microscope.
2. Using the 4× objective and the 561 nm laser, set an appropriate exposure time (use the same time within each experiment and remember that cell proliferation will lead to a stronger signal by the end of the experiment). Starting from one corner of the scaffold, set the focus of the first field of view, click 'add'. Manually move to an adjacent area to set the second field of view, making sure not to overlap with the first field of view. Set the focus and click 'add' again to add the second field of view to the experimental protocol. Repeat this step and add as many fields of view as needed to cover the whole construct (this should be around 8–10 fields of view per sample). Set the focus and brightness for simultaneous bright field recording, for general topography.

3. Set the software parameters to record an image every 20 min and set the experiment duration to 48 h. Start the experiment. At this stage, the motorized stage will allow to move the plate and record an image for all the fields of view, every 20 min.
4. At the end of the experiment, save the reconstructed videos (145 frames/video) as VSI files. The videos can be viewed using the OlyVIA software.

### 3.4.3 Image Analysis

Migration parameters, such as the track length, track straightness, cellular mean speed, and mean square displacement can be obtained by processing the image series obtained from the live epifluorescence microscope with the Imaris software.

1. Open the Imaris Imaging analysis software.
2. Go in the *Surpass* mode and drag in the reconstructed video. Both the red channel and bright field channel will appear (Fig. 4a). In the 'Display Adjustment' window, deselect the bright field channel. By moving the time cursor at the bottom of the window, you can see all frames from $t = 0$ h (Fig. 4b) to $t = 48$ h (Fig. 4c).
3. Click on the *Spots* icon in the left panel, to create a new spots protocol.
4. In step 1/6 (Algorithm), leave *Favorite Creation Parameters* onto 'Default'.
5. Go to next (step 2/6): Source Channel. Ensure that the correct channel is being analyzed (561 nm).

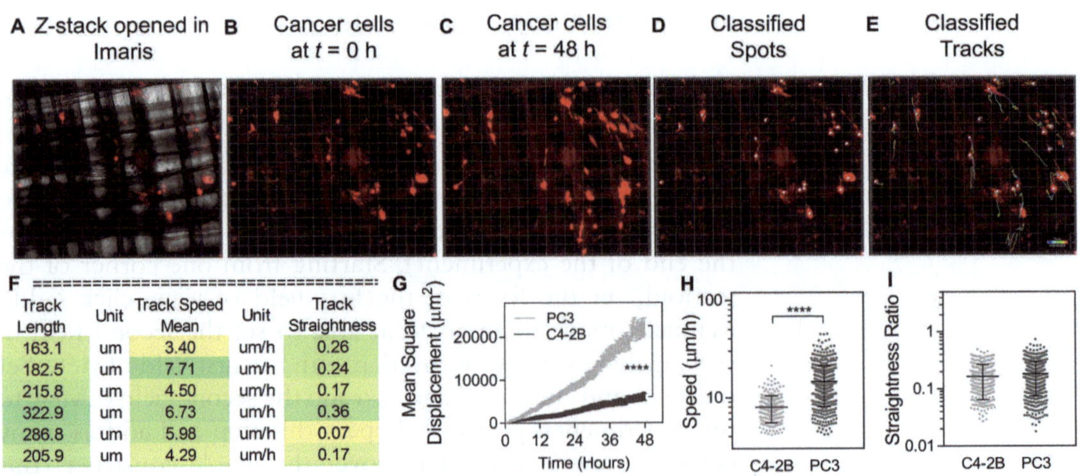


| Track Length | Unit | Track Speed Mean | Unit | Track Straightness |
|---|---|---|---|---|
| 163.1 | um | 3.40 | um/h | 0.26 |
| 182.5 | um | 7.71 | um/h | 0.24 |
| 215.8 | um | 4.50 | um/h | 0.17 |
| 322.9 | um | 6.73 | um/h | 0.36 |
| 286.8 | um | 5.98 | um/h | 0.07 |
| 205.9 | um | 4.29 | um/h | 0.17 |

**Fig. 4** (**a–f**) Analysis steps for the measurement of migration properties of cancer cells co-cultured on the hOBMT, up to 48 h, using Imaris. (**g–i**) Mean square displacement, cellular speed, and track straightness ratio of C4-2B and PC3 cells onto hOBMT for 48 h in the presence of 10 nM dihydrotestosterone ($n > 245$ tracks analyzed/condition)

6. Go to the *Slice View* icon in the top bar and drag the cursor across several cells to read out their approximate diameters (found on the right panel, as *Distance* in μm).
7. Go back to the protocol by clicking the *3D view* icon in the top bar and enter the average diameter per cell + approximately 20% in the *Estimated XY Diameter* box (here we used 30 μm) (*see* **Note 12**).
8. Go to next (step 3/6): Classify Spots. At this point, Imaris represents each single cell by a dot (Fig. 4d). By default, Imaris may suggest a 'Quality' filter, which allows to select/deselect cells, based on fluorescence intensity. Move the cursor until all cells are represented by an individual dot. More filters may be added from the *Filter Type* drop-down menu.
9. Go to next (step 4/6): Edit Spots. Do not edit any parameter.
10. Go to next (step 5/6): Tracking. Select 'Autoregressive Motion' from the *Algorithm* drop down menu. In *Parameters*, set the *Max Distance* as 20 μm. This will eliminate any track where a cell would have travelled more than 20 μm between two frames. Set the *Max Gap Size* as 5. This will allow the tracked cell to be unfocused (track momentarily lost) for no more than five frames (*see* **Note 13**).
11. Go to next (step 6/6): Classify Tracks. Add a 'Track Duration' filter, and set it at 145 s. Here 's' refers to the number of frames (1 frame = 20 min). Selecting the maximum number of frames (145) enables to consider only tracks which were successfully identified throughout the 48-h duration of the experiment (Fig. 4e) (*see* **Note 14**). More filters can be added based on speed, track length, etc. from the *Filter Type* drop down menu.
12. Click on Finish. The image displayed on the screen contains all the tracks that were analyzed. At this stage, only individual cells should have been tracked. If a cell is touching any other cell during the duration of the experiment, affecting their migration track, click on 'Select' in *Pointer* (right tab). Select the track to be deleted. From the left panel, select the *Pencil* icon. Click 'Delete'.
13. Click on the *Snapshot* icon in the top bar to record a visual representation of all the tracks analyzed.
14. Save the protocol by clicking on the *Wand* icon in the left panel, then *Store Parameters for Batch*. Give the protocol a name and click 'Arena' and 'Favorite Creation Parameters' in *Store Location*. Click 'OK'.
15. Click the *Statistics* tab, and click the *Export all statistics to File* icon, making sure to extract *Track Speed Mean*, *Track Length*, *Track Duration*, *Track Straightness*, *Displacement*$^2$ and any other relevant parameters (this can be edited by clicking the *Preferences* icon and ticking/unticking parameters).

16. Save the final Imaris file by clicking on the *Export* icon in the top bar.
17. Close the Imaris software and repeat **steps 1–16** for each field of view, for each replicate, and each condition, making sure to apply the same protocol that was saved initially.
18. In a new Excel file, compile the desired parameters collected from all the fields of view from the same condition (Fig. 4f) and determine the overall average parameters (Fig. 4g–i) (*see* **Note 15**). A minimum of 200 tracks per condition should be obtained.

### 3.5 Cancer Cell Proliferation on hOBMT

The image series obtained by live cell imaging can further be used to estimate cancer cell proliferation on hOBM, based on the increase of fluorescence area.

#### 3.5.1 Measure Area Intensity Using ImageJ

1. Open the VSI file in ImageJ. Opening of VSI files requires a plugin such as Bio-Formats (*see* **Note 9**). Go to *Plugins, Bio-Formats, Bio-Formats Importer* and then select the VSI file.
2. A 'Bio-Formats Import Options' window will appear. Ensure that *Stack Viewing* is on 'Hyperstack' and that all boxes are unticked except for 'Split channels'. In *Color Options*, select 'Grayscale'.
3. Close the bright field channel and keep the channel showing the fluorescently labeled cancer cells (red channel).
4. Go to *Process, Filters, Gaussian Blur*. Leave the default systems (*Sigma (Radius)* as 2.00 and no box ticked).
5. Click *yes* to the processing of all the images from the series.
6. Move the slider to the last image.
7. Go to *Image, Adjust, Threshold*. In the window, ensure 'Dark background' is selected. Choose 'Huang' from the drop-down menu. Do not click 'Apply'.
8. Go to *Analyze, Set Measurements*. In the window, tick 'Area' and 'Limit to threshold'. In *Redirect to*, select the name of the image. Click 'OK'.
9. Go to *Image, Stacks*, Measure Stack. Data will appear in a 'Results' window. Copy all values into an excel file.
10. In Excel, normalize to the image at $t = 0$ h (first image).

## 4 Notes

1. The CaP-mCPL scaffolds used for generating the hOBMT microtissues can generate autofluorescence, which is accentuated when excited at 488 nm. It is thus recommended to use a

red probe for cancer cells transfection, such as mCherry or mKO2. This will allow using the 561 nm channel for data collection, facilitating live imaging and morphometric analysis.

2. Co-culture media can be different, according to the cells being tested. Serum-free media can be used as well. When assessing the effects of drugs on cancer cell attachment, add the drugs or respective controls to the coculture media.
3. The agar solution is used to pre-coat normal 24-well plates, in order to prohibit cell attachment in the plates, and to maximize cancer cell attachment to hOBM. This step is not necessary if ultra-low attachment plates are used.
4. Laser power, exposure time and gain need to be adjusted for any different probes and cell lines, but will need to be kept constant within one type of sample.
5. The use of spinning disc confocal microscopy allows acquiring images much faster than a traditional confocal microscope. While using a traditional confocal microscope is possible, it is not recommended, due to the large amounts of images needed to reach statistical significance.
6. The cancer cell concentration to be used for the attachment assay, for the morphometry analysis, and for the migration assay, need to be optimized for each type of cell to ensure the success of the assays. It is not recommended to use less than $10^5$ cells per well for the attachment assay, due to increasing counting errors associated with using lower cell numbers.
7. A minimum of 1 h is suggested for aggressive metastatic cell lines and a minimum of 3 h for other metastatic cell lines. It is not recommended to do the attachment assay past 24 h of co-culture for two main reasons. Cells may have started to divide, leading to unrepresentative numbers compared to cell numbers used initially. Some cell types also tend to aggregate rapidly overtime, which can result in less accurate counting and can impair morphometric analysis, which requires cells to be single.
8. It is important to set a relevant algorithm with appropriate filters, so that minimum post-processing is done. For example, a filter that automatically removes all the cells that are in contact with any edge can be added upfront.
9. ImageJ requires a plugin such as the Bio-Formats plugin, so that ND2 files and VSI files can be opened with ImageJ. The Bio-Formats plugin [19] can be downloaded at https://www.openmicroscopy.org/bio-formats/downloads/. The file named 'bioformats_package.jar' has to be saved into the Plugins directory within ImageJ. Restart ImageJ. Bio-Formats will be found in the Bio-Formats option under the Plugins menu.

10. The Size step depends on the sample (cell size) and image acquisition (magnification), and may need to optimization for different samples.
11. Technically speaking, any geometry can be used. We however suggest having one scaffold cut at least in two, in order to use one part for the control condition and the other one for the test condition. We suggest not to use less than one-third of a scaffold per condition. With one-third of a scaffold and the seeding parameters suggested here, this will lead to at least ten fields of view being analyzed, which should give at least 100 tracks for analysis. Lower scaffold areas may not provide enough tracks, in turn reducing statistical significance.
12. It is recommended to overestimate the *Estimated XY Diameter*. This assumption will lead to more stable tracks being analyzed, by refraining unrepresentative track splitting when a cell is just branching in and out (i.e. elongating past their average diameters).
13. According to the type of cell being studied (more or less motile) these parameters may be edited to ensure the most representative tracking.
14. It is important to consider tracks that have the same duration when working with track length, mean square displacement, or any other cumulative parameters, for comparison purposes. However, it may be difficult to reach statistical significance as only a few tracks will have been tracked for the full 48 h. Hence a lower duration can be selected, but this needs to be consistent throughout all replicates and conditions for comparison.
15. Any parameter that refers to time such as Track Speed is expressed in 's' in Imaris. However, 's' refers to one frame (which was set here as 20 min). Hence it is critical to translate any parameter that used 's' into the corrected unit. For example, 1 μm/s in Imaris means 1 μm/frame, i.e. 3 μm/h.

## Acknowledgements

This work was supported by an IHBI ECR grant, Advance Queensland (AQ) Maternity Fund Award from the Queensland Government (DSITI), Young Researcher Award (2017-YR-RoW-9) from Lush (UK), supporting non-animal testing alternatives, National Health and Medical Research Council (NHMRC) Peter Doherty Early Career Research Fellowship (RF) (APP1091734), and John Mills Young Investigator Award (YI0715) from the Prostate Cancer Foundation of Australia (PCFA). The author acknowledges the TRI for the excellent core facilities that enabled this research, and thank Adler Ju from the TRI microscopy facility. The author also thanks D/Prof. Dietmar W. Hutmacher, D/Prof. Judith A. Clements AC, Dr. Elena De-Juan-Pardo, Mr. Jose Malayil, Dr. Asha Mathew, Ms. Teja Yarlagadda, Dr. Sally Stephenson, Dr. Brett Hollier.

## References


1. Eils R, Athale C (2003) Computational imaging in cell biology. J Cell Biol 161(3):477–481. https://doi.org/10.1083/jcb.200302097
2. Dallas SL, Veno PA (2012) Live imaging of bone cell and organ cultures. In: Helfrich MH, Ralston SH (eds) Bone research protocols, Methods in molecular biology, vol 816, 2nd edn. Humana Press, Totowa, pp 425–457. https://doi.org/10.1007/978-1-61779-415-5_26
3. Jain RK, Munn LL, Fukumura D (2002) Dissecting tumour pathophysiology using intravital microscopy. Nat Rev Cancer 2(4):266–276. https://doi.org/10.1038/nrc778
4. Bonapace L, Wyckoff J, Oertner T, Van Rheenen J, Junt T, Bentires-Alj M (2012) If you don't look, you won't see: intravital multiphoton imaging of primary and metastatic breast cancer. J Mammary Gland Biol Neoplasia 17(2):125–129. https://doi.org/10.1007/s10911-012-9250-8
5. Dondossola E, Alexander S, Holzapfel B, Logothetis C, Hutmacher D, Friedl P (2015) Preclinical monitoring of prostate cancer lesions in bone by intravital multiphoton microscopy. Cancer Res 75:2. https://doi.org/10.1158/1538-7445.am2015-5175
6. Landgraf M, McGovern JA, Friedl P, Hutmacher DW (2018) Rational design of mouse models for cancer research. Trends Biotechnol. https://doi.org/10.1016/j.tibtech.2017.12.001
7. Harrison RK (2016) Phase II and phase III failures: 2013–2015. Nat Rev Drug Discov 15(12):817–818. https://doi.org/10.1038/nrd.2016.184
8. Hutmacher DW (2010) Biomaterials offer cancer research the third dimension. Nat Mater 9(2): 90–93. https://doi.org/10.1038/nmat2619
9. Hutmacher DW, Cukierman E (2014) Engineering of tumor microenvironments. Adv Drug Deliv Rev 79(2):1. https://doi.org/10.1016/j.addr.2014.11.001
10. Fischbach C, Chen R, Matsumoto T, Schmelzle T, Brugge JS, Polverini PJ et al (2007) Engineering tumors with 3D scaffolds. Nat Methods 4(10):855–860
11. Salamanna F, Contartese D, Maglio M, Fini M (2016) A systematic review on in vitro 3d bone metastases models. A new horizon to recapitulate the native clinical scenario? Oncotarget 7 (28):44803–44820. https://doi.org/10.18632/oncotarget.8394
12. Taubenberger AV (2014) In vitro microenvironments to study breast cancer bone colonisation. Adv Drug Deliv Rev 79:135–144. https://doi.org/10.1016/j.addr.2014.10.014
13. Loessner D, Holzapfel BM, Clements JA (2014) Engineered microenvironments provide new insights into ovarian and prostate cancer progression and drug responses. Adv Drug Deliv Rev 79:193–213. https://doi.org/10.1016/j.addr.2014.06.001
14. Schneider CA, Rasband WS, Eliceiri KW (2012) NIH image to ImageJ: 25 years of image analysis. Nat Methods 9(7):671–675. https://doi.org/10.1038/nmeth.2089
15. Guan X (2015) Cancer metastases: challenges and opportunities. Acta Pharm Sin B 5 (5):402–418. https://doi.org/10.1016/j.apsb.2015.07.005
16. Mishra A, Shiozawa Y, Pienta KJ, Taichman RS (2011) Homing of cancer cells to the bone. Cancer Microenviron 4(3):221–235. https://doi.org/10.1007/s12307-011-0083-6
17. Collan Y, Kosma VM (1989) Morphometry in cancer diagnostics. In: Goldson AL (ed) Cancer management in man: detection, diagnosis, surgery, radiology, chronobiology, endocrine therapy. Springer Netherlands, Dordrecht, pp 134–144. https://doi.org/10.1007/978-94-009-2536-6_10
18. Paul CD, Mistriotis P, Konstantopoulos K (2017) Cancer cell motility: lessons from migration in confined spaces. Nat Rev Cancer 17(2):131–140. https://doi.org/10.1038/nrc.2016.123
19. Linkert M, Rueden CT, Allan C, Burel JM, Moore W, Patterson A et al (2010) Metadata matters: access to image data in the real world. J Cell Biol 189(5):777–782. https://doi.org/10.1083/jcb.201004104

# Part II

## Molecular Diagnostics

# Chapter 4

# Exosomes Extraction and Identification

**Xiaoxin Wu, Salah Ali A. Showiheen, Antonia Rujia Sun, Ross Crawford, Yin Xiao, Xinzhan Mao, and Indira Prasadam**

## Abstract

Exosomes are small endosome-derived lipid nanoparticles (50–120 nm in diameter), actively secreted by exocytosis in most living cells. Recently, there is a growing interest of research focused on studying the exosome functions and to understand ways to use them for therapeutic applications in a wide variety of disorders, such as cancer, cardiovascular, neurodegenerative, and musculoskeletal diseases. Recently, a number of techniques have been developed for the isolation of exosomes such as ultracentrifugation, micro-filtration centrifugation, gradient centrifugation, and size-exclusion chromatography. In this chapter, we reveal the protocol and key insights into the isolation, purification, and characterization of exosomes using ultracentrifugation method.

**Key words** Exosome, Ultracentrifugation, Size distribution, TEM, Western blot, Labeling

## 1 Introduction

Exosomes are a type of bi-lipid membrane, nano-sized extracellular vesicles with a diameter of about 30–100 nm, which are mainly between the size of microvesicles (50–1000 nm) and exosome-like vesicles (20–50 nm) [1, 2]. First discovered in the maturing mammalian reticulocyte in 1987, exosomes are now reported to be present in all kinds of eukaryotic cells and biological fluids including blood, saliva, urine, cerebrospinal fluid, and cell culture medium [3, 4]. Exosomes are multivesicular body formed by the invagination of intracellular lysosomal microparticle and released into the extracellular matrix after the vesicle membrane fusing with the cell membrane or directly from the plasma membrane [5]. They were first considered as waste products from cells; however, it is now reported that they have specialized functions and play a key role in processes such as coagulation, intercellular signaling, and cell communication as a genetic container of proteins, RNA, and DNA [6]. As one of the important pathways of extracellular signaling,

Jyotsna Batra and Srilakshmi Srinivasan (eds.), *Theranostics: Methods and Protocols*, Methods in Molecular Biology, vol. 2054, https://doi.org/10.1007/978-1-4939-9769-5_4, 

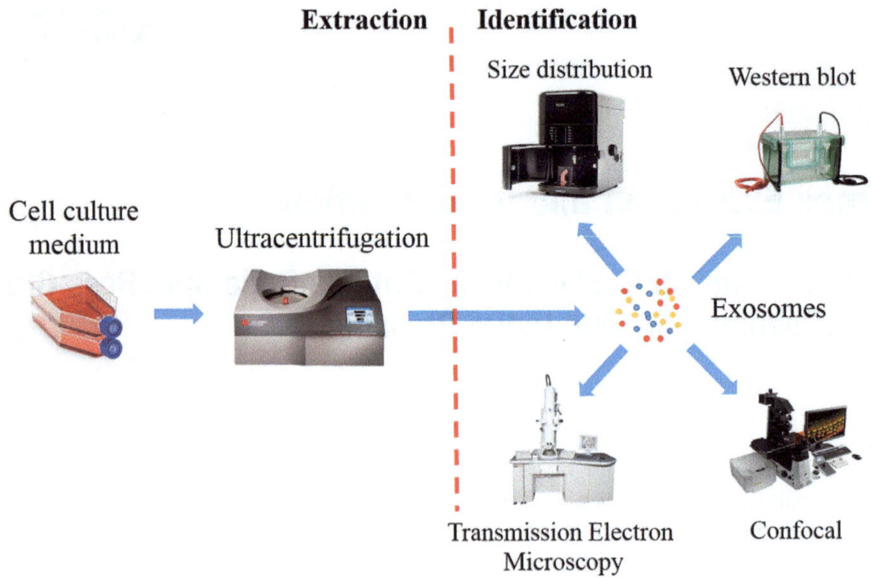


**Fig. 1** Exosome extraction and identification

exosomes are able to transmit their contents to different cells and participate in many physiological processes [7].

In recent years extensive research has been focused on exosomes for clinical applications including the use for prognosis, therapy, and as biomarkers for health and disease [8]. The purpose of this chapter is to give simple and reliable methods for purifying and characterizing exosomes. The extraction methods of exosomes mainly include ultracentrifugation, micro-filtration centrifugation, gradient centrifugation, and size-exclusion chromatography [9, 10]. And the characterization includes the size, shape, and biomarkers. Here we introduce how to use ultracentrifuge, the most common way, to isolate exosomes from primary cell culture medium and the way to characterize them using size distribution, transmission electron microscopy (TEM), and Western blot (Fig. 1). At last, we share one convenient and fast way to label exosomes using PKH67 labeling technique.

## 2 Materials

Prepare all solutions using ultrapure water (prepared by purifying deionized water, to attain a sensitivity of 18 MΩ-cm at 25 °C) and analytical-grade reagents. Prepare and store all reagents at room temperature (unless indicated otherwise). Diligently follow all waste disposal regulations when disposing waste materials. Do not add sodium azide to reagents.

### 2.1 Exosome Isolation and Size Distribution

1. Dulbecco's Modified Eagle Medium (DMEM): containing 1 g/L D-Glucose, L-Glutamine, and 110 mg/L Sodium Pyruvate, store at 2–8 °C.
2. 1×PBS: The composition of 1× PBS solution is: 137 mM NaCl; 2.7 mM KCl; 4.3 mM $Na_2HPO_4$; 1.47 mM $KH_2PO_4$. Adjust to a final pH of 7.4. Store at room temperature. Cool before use (4 °C).
3. Beckman Coulter Microfuge 18 Centrifuge.
4. Beckman Coulter OptimaTM MAX-80XP Ultracentrifuge (Type TLA 110 rotor).
5. NS300 NanoSight (ATA Scientific, Australia) fitted with a NS300 flow-cell top plate and a 405 nm laser.

### 2.2 Transmission Electron Microscopy Preparation

1. Formvar carbon-coated grid.
2. Solution 1% Uranyl Acetate (UA): Mix 2 g of UA power with 30 ml distilled water until the power is fully dissolved. Adjust pH to 3.5 by adding 1 M HCl. Add double distilled water to a final volume (5% solution) of 40 ml. Filter through a 0.45 μm filter. Dilute the solution to 1% UA. The solution should be kept at 4 °C in the dark.
3. Transmission electron microscopy: TEM, JEOL JEM-1400.

### 2.3 Western-Blot

1. 10× RIPA buffer with protease inhibitor (lysis buffer): Mix per 1 ml 10× RIPA buffer with 1 tablet PI (Protease Inhibitor Cocktail Tablets). Store at 4 °C.
2. BCA reagent A&B: BCA Protein Assay Reagent A and BCA Protein Assay Reagent B.
3. BCA protein standard: Albumin Standard.
4. Acrylamide: 40% Acrylamide/Bis Solution.
5. Tris pH 8.8/6.8 solution: Mix 121 g Tris Base with ultrapure water and adjust pH to 8.8 for resolving gel or 6.8 for stacking gel. Make up to 1 L with ultrapure water.
6. 10% APS: Mix 0.1 g Ammonium Persulfate with 1 ml ultrapure water.
7. TEMED (*see* **Note 1**).
8. Loading buffer: Mix 4× Laemmli Sample Buffer with 2-Mercaptoethanol as 9:1.
9. 10× SDS-PAGE: Mix with 60.4 g Tris base, 20 g Sodium Dodecyl Sulfate (use a mask) and 288 g Glycine. Make up to 2 L with ultrapure water.
10. Transfer buffer: Mix 800 ml 1× SDS-PAGE with 200 ml methanol (*see* **Note 2**).
11. Odyssey Blocking Buffer.

12. 10× TBS: Mix 160 g NaCl, 60 g Tris base and water. Adjust pH to 7.4 with HCl. Make up to 2 L with water.
13. TBST: Mix 1 L 1× TBS with 0.25–1 ml Tween 20 for washing.
14. Antibody: Exosomal marker antibody sampler kit.
15. Odyssey Blot Imager.

#### 2.4 Inflorescences Examination

1. PKH67: PKH67 Green Fluorescent Cell Linker Mini Kit.
2. Cell culture medium: Mix 445 ml DMEM solution with 50 ml FBS (no exosome, *see* **Note 3**) and 5 ml Penicillin–Streptomycin Solution.
3. 1%FBS: Mix 1 ml cell culture medium above with 9 ml DMEM solution.
4. 4% paraformaldehyde solution: PFA, add 40 g of PFA powder to 800 ml PBS. Increase the pH to allow the PFA power to dissolve. Make up to 1 L using 200 ml PBS. Freeze aliquots upright at −20 °C for long term storage.
5. Confocal laser scanning microscope.

## 3 Methods

Carry out all procedures at room temperature unless otherwise specified.

#### 3.1 Exosome Isolation

1. Change the Osteoblasts (OBs) cell culture medium (CM) to DMEM solution when they are 80–90% confluence and collect the medium after 24 h.
2. Centrifuge the OBs CM using a Beckman Coulter Microfuge 18 Centrifuge at 300 × *g* at 4 °C for 10 min to remove detached cells.
3. Collect the supernatant and filter through 0.22 μm filters to remove contaminating apoptotic bodies, microvesicles, and cell debris.
4. Centrifuge the clarified CM in a Beckman Coulter OptimaTM MAX-80XP Ultracentrifuge at 100,000 × *g* at 4 °C for 90 min with a Type TLA 110 rotor to pellet exosome.
5. Collect the supernatant in case for further study.
6. The exosome-containing pellets were resuspended in 1–2 ml of ice-cold PBS.
7. A second round of ultracentrifugation (same as above) is carried out, and the finial exosome pellets are resuspended in 100 μl PBS after carefully removing the supernatant.
8. Store the exosome solution in −80 °C for further use.

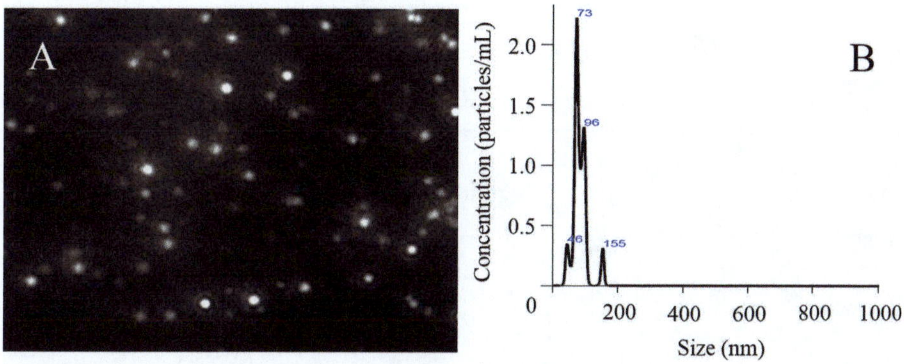


**Fig. 2** Exosome size distribution: (**a**) diluted five times, demonstrates a size distribution of particles consistent with the size range of exosomes; (**b**) diluted 50 times

### 3.2 Size Distribution

1. Dilute the exosome solution (1:10, 20 or 50) and analyze at camera levels of 16 at 25 °C.
2. Set the detection threshold for all samples as 8. Utilize the automatic settings for the maximum jump distance and blur settings. Set the sample pump speed to 50.
3. Load PBS for cleaning.
4. Load 500 μl samples into the sample chamber for recordings including video.
5. Analyze the data on the NTA software 3.0 (ATA Scientific, Australia) (Fig. 2).

### 3.3 Transmission Electron Microscopy

1. Suspend the exosome solution and load onto a Formvar carbon-coated grid with a drop of 5 μl.
2. Wait 10 min for air-drying, remove the excess exosomes with filter paper.
3. Turn the sample side upside down on one drop of 1% UA for 10–60 s depending on the concentration of exosomes.
4. Wash the sample twice by putting the grid on a drop of water and dry it by using filter paper at room temperature.
5. Image the exosomes under TEM (Fig. 3).

### 3.4 Western-Blot

Protein samples: Mix 45 μl exosome solution with 5 μl 10× RIPA (PI) for each sample.

#### 3.4.1 BCA

1. Mix 60 μl BCA protein standard with 540 μl ultrapure water (1:10 dilute).
2. Make eight tubes as the table showing below:
3. For each tube, add three wells using 50 μl/well to a 96-well plate.

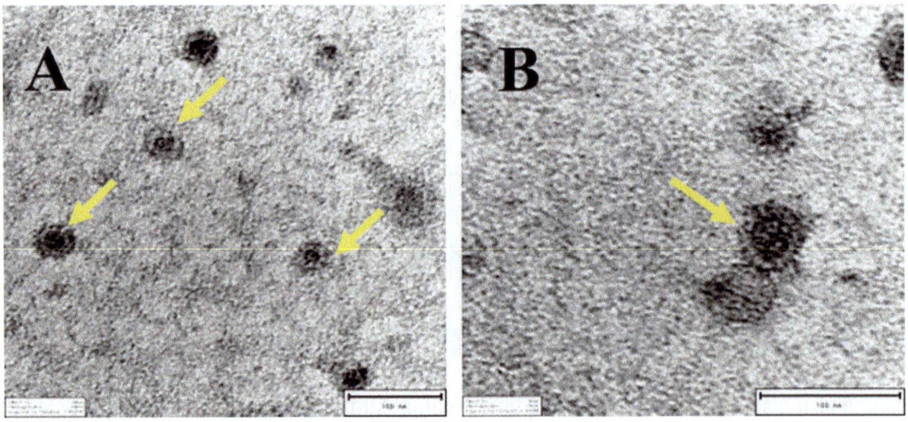


**Fig. 3** Transmission electron microscopy results. Yellow arrows: exosomes. EM image of exosomes demonstrates cup-shape morphology, scale bar = 100 nm

**Table 1**
**BCA protein standard**

| | 1 | 2 | 3 | 4 | 5 | 6 | 7 | 8 |
|---|---|---|---|---|---|---|---|---|
| Ultrapure water | 200 μl | 190 μl | 180 μl | 160 μl | 140 μl | 120 μl | 100 μl | 0 |
| Diluted standard | 0 | 10 μl | 20 μl | 40 μl | 60 μl | 80 μl | 100 μl | 200 μl |

4. Add 5 μl of each protein sample to two new wells at the same 96-well plate. After that, add 45 μl ultrapure water to these protein sample wells.
5. Reagent mix: Mix with BCA reagent A&B as

$$B = 150\ \mu l * \text{wells}/51$$
$$A = 50 * B$$

6. Add 150 μl reagent mix to each well in **steps 3** and **4**.
7. Incubate for 30 min at 37 °C.
8. Scan the plate using Benchmark Plus (BIO-RAD, Microplate spectrophotometer) at 560 nm.
9. Analyze and calculate the protein concentration (Table 1).

*3.4.2 Gels Preparation*

1. Prepare the 10% separating gel by mixing as below (*see* **Notes 4** and **5**) (Table 2):
2. Cast gel within a 7.25 cm * 10 cm * 1.5 mm gel cassette. Allow space for stacking gel and gently overlay with isobutanol or water and wait for 30 min.
3. Pour out the isobutanol or water.
4. Prepare the 4% stacking gel by mixing as below (Table 3).

**Table 2**
**10% separating gel**

| | Acrylamide | Water | Tris pH 8.8 | 10% SDS | 10% APS | TEMED |
|---|---|---|---|---|---|---|
| Each 1.5 mm gel | 1.83 ml | 2.71 ml | 2.71 ml | 0.073 ml | 0.037 ml | 0.0037 ml |

**Table 3**
**4% stacking gel**

| | Acrylamide | Water | Tris pH 6.8 | 10% SDS | 10% APS | TEMED |
|---|---|---|---|---|---|---|
| Each 1.5 mm gel | 0.33 ml | 2.51 ml | 0.41 ml | 0.033 ml | 0.017 ml | 0.0033 ml |

5. Fill in the space above the first gel and insert a 10-well gel comb immediately without introducing air bubbles. Wait for 30 min.

*3.4.3 Protein Denaturation*

1. Dilute the protein samples depending on the protein concentration results using lysis buffer.
2. Mix diluted protein samples with loading buffer as 3:1 and mix well to disperse gently.
3. Heat the mix in **step 2** at 95 °C for 5–10 min.
4. Centrifuge at 12,000 × *g* for 1 min. Wait until the temperature comes back to room temperature.

*3.4.4 Electrophoresis and Transfer*

1. Add 15 μg of the protein samples and marker in every lane (*see* **Note 6**). Electrophorese at 80 V for 30 min and switch to 120 V for 60 min after that in room temperature until the dye front reached the bottom of the gel.
2. Open the gel plates with the use of a spatula. The gel remains on one of the glass plates. Cut the gel and get rid of the parts where there is no sample on. Rinse the gel with water and transfer carefully to a container with western blot transfer buffer.
3. Cut a nitrocellulose membrane and six pieces of filter paper to the size of the gel and immerse in transfer buffer.
4. Put them in an order in the transfer buffer as below (*see* **Note 7**) (Fig. 4).
5. Gently put the "sandwich" to the transfer slot and set 100 V, 400 mA for 100 min on ice.

*3.4.5 Blocking and Antibody Incubation*

1. Open the "sandwich" and gently take out the membrane. Block the membrane with blocking buffer for 1 h at room temperature.

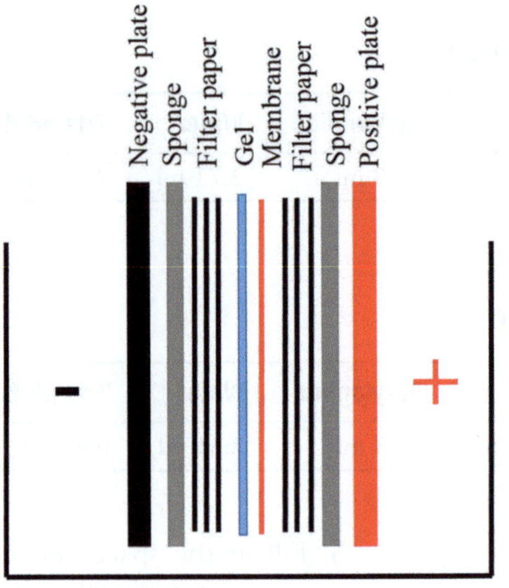


**Fig. 4** The order of plastic plate, sponges, filter paper, gel, and membrane

2. Cut the membrane and incubate them in different antibodies (1:1000 dilution, diluted in blocking buffer) at 4 °C overnight.
3. Wash 3× with TBST, 5 min each time.
4. Incubated with IRDye® 800CW Goat anti-Rabbit IgG (H + L) (1: 5000; LI-COR Biotechnology, USA) for 30–60 min avoiding light, at room temperature.
5. Wash as in **step 3**.
6. Scan the membrane using Odyssey Blot Imager.

### 3.5 Inflorescences Examination

1. Mix 25 μl exosome solution with 225 μl of Diluent C (*see* **Note 8**) and gently pipet to insure complete dispersion (*see* **Note 9**). Do not vortex and do not let exosomes stand in Diluent C for long periods of time.
2. Add 1 μl of the PKH67 ethanolic dye solution to 250 μl of the Diluent C in a polypropylene centrifuge tube and mix well to disperse.
3. Rapidly add the 250 μl exosome suspension (**step 1**) to 250 μl Dye solution (**step 2**) and immediately mix the sample by pipetting (*see* **Note 10**).
4. Incubate the exosome/dye suspension from **step 3** for 1–5 min. Because staining is so rapid, longer times provide no advantage.
5. Stop the staining by adding an equal volume (1 ml) of serum or other suitable protein solution (e.g., 1%FBS) and incubate for 1 min to allow binding of excess dye.

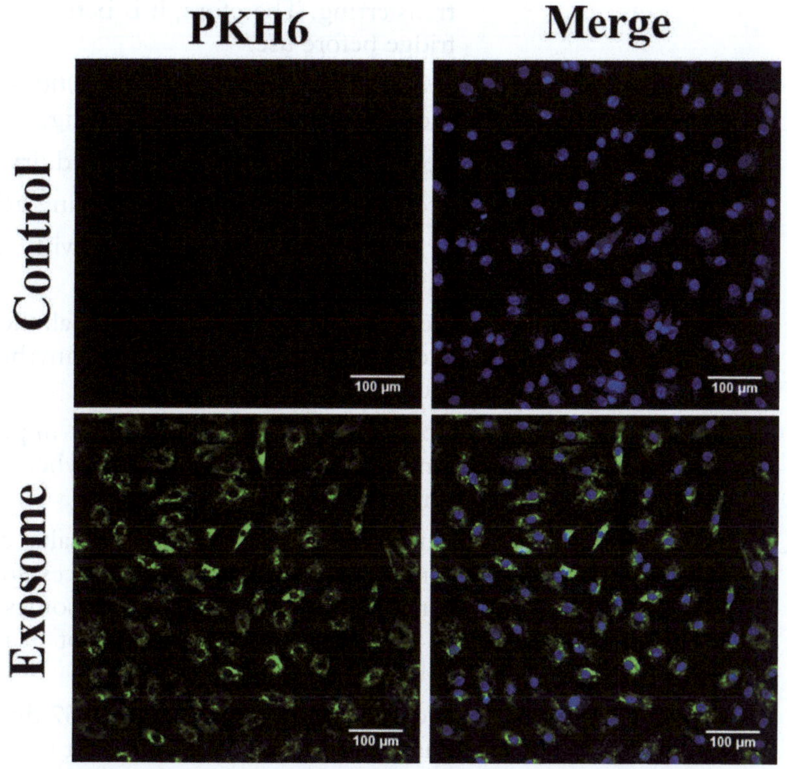


**Fig. 5** Inflorescences examination: SBOs-derived exosomes were labeled with PKH67 and cocultured with chondrocytes for 24 h; fluorescence signals were examined by confocal microscope. Scale bars: 100 μm

6. Resuspend the labeled exosome in cell culture medium (no exosome). Use this medium for cell culturing for 24 h.
7. Wash the cell with PBS 2–3 times. Fix by 4% PFA for 20–30 min.
8. Wash the fixed cell with PBS 2–3 times. Stain the cell with DAPI for 10 min then wash with PBS three times.
9. Confocal images are taken using a confocal laser scanning microscope (Fig. 5).

## 4 Notes

1. TEMED is a highly flammable liquid and vapor, which is harmful if swallowed or inhaled. It can cause severe skin burns and eye damage as well. Should be used in the hood.
2. When making the transfer buffer, the solution will release the heat and increase the temperature which is not good for

transferring. Therefore, it is better to cool the solution in the fridge before use.

3. Pre-centrifuge the FBS at 100,000 × *g* overnight and collect the supernatant. Store in the fridge.
4. Before making the gels, wash and dry all the glass plates first.
5. Prepare separating gel and stacking gel in the hood.
6. It is better to load all the lanes with samples or marker, otherwise the brands will be unflat.
7. Gently and completely get rid of all the bubbles using spatula in the transfer buffer. Do not use anything with color to cut the membranes and filter paper.
8. For reproducible results, it is important to minimize the amount of exosome solution when they are resuspended in Diluent C.
9. The presence of physiologic salts causes the dye to form micelles and substantially reduces staining efficiency. Therefore, it is important that the exosomes be suspended in Diluent C at the time dye is added, not in medium or buffered salt solutions.
10. Do not add the ethanolic PKH67 dye directly to the exosome suspension in Diluent C. Adjust exosome and dye concentrations to avoid staining in very small (<100 μl) or very large (>5 ml) volumes.

## References


1. Edgar JR (2016) Q&A: what are exosomes, exactly? BMC Biol 14:46. https://doi.org/10.1186/s12915-016-0268-z
2. Taylor DD, Zacharias W, Gercel-Taylor C (2011) Exosome isolation for proteomic analyses and RNA profiling. Methods Mol Biol 728:235–246. https://doi.org/10.1007/978-1-61779-068-3_15
3. Johnstone RM, Adam M, Hammond JR, Orr L, Turbide C (1987) Vesicle formation during reticulocyte maturation. Association of plasma membrane activities with released vesicles (exosomes). J Biol Chem 262 (19):9412–9420
4. van der Pol E, Boing AN, Harrison P, Sturk A, Nieuwland R (2012) Classification, functions, and clinical relevance of extracellular vesicles. Pharmacol Rev 64(3):676–705. https://doi.org/10.1124/pr.112.005983
5. Booth AM, Fang Y, Fallon JK, Yang JM, Hildreth JE, Gould SJ (2006) Exosomes and HIV Gag bud from endosome-like domains of the T cell plasma membrane. J Cell Biol 172 (6):923–935. https://doi.org/10.1083/jcb.200508014
6. Osier N, Motamedi V, Edwards K, Puccio A, Diaz-Arrastia R, Kenney K et al (2018) Exosomes in acquired neurological disorders: new insights into pathophysiology and treatment. Mol Neurobiol. https://doi.org/10.1007/s12035-018-1054-4
7. Wang Y, Yu D, Liu Z, Zhou F, Dai J, Wu B et al (2017) Exosomes from embryonic mesenchymal stem cells alleviate osteoarthritis through balancing synthesis and degradation of cartilage extracellular matrix. Stem Cell Res Ther 8(1):189. https://doi.org/10.1186/s13287-017-0632-0

8. Wang J, Sun X, Zhao J, Yang Y, Cai X, Xu J et al (2017) Exosomes: a novel strategy for treatment and prevention of diseases. Front Pharmacol 8:300. https://doi.org/10.3389/fphar.2017.00300
9. Corso G, Mager I, Lee Y, Gorgens A, Bultema J, Giebel B et al (2017) Reproducible and scalable purification of extracellular vesicles using combined bind-elute and size exclusion chromatography. Sci Rep 7(1):11561. https://doi.org/10.1038/s41598-017-10646-x
10. Wang J, Yao Y, Wu J, Li G (2015) Identification and analysis of exosomes secreted from macrophages extracted by different methods. Int J Clin Exp Pathol 8(6):6135–6142

# Chapter 5

# Profiling MicroRNA Markers in Plasma: Looking into Better Approaches and Recommendations

**Farhana Matin and Jyotsna Batra**

## Abstract

The revelation of stable microRNA (miRNA) species in body fluids has led to the speculation of disease-related alterations in miRNA expression levels as indicative of disease state making them attractive minimally invasive biomarkers for the diagnosis and prognosis of cancer and other diseases. Although miRNA expression profiling in body fluids holds great promise, working with low amounts of RNA in plasma and serum represents several challenges during purification, relative quantification, normalization, and data analysis. Here, we present an experimental protocol for miRNA profiling in plasma using plasma/serum-specific miRNA purification and RT-qPCR to identify potential miRNA biomarkers. We also discuss the challenges encountered during the miRNA profiling process and provide recommendations for robust purification and relative quantification of miRNAs in patient plasma samples.

**Key words** MicroRNAs, Plasma, Biomarker, Profiling, Theranostics, Personalized medicine

## 1 Introduction

MicroRNAs (miRNAs) are evolutionarily conserved ~22 nucleotides single stranded RNA molecules that inhibit messenger RNA (mRNA) transcription or translation to protein through preferential binding to the 3′ untranslated region (3′ UTR) of the mRNA transcript [1]. This may lead to direct/indirect modulation of biological processes such as proliferation, migration, apoptosis, differentiation, and development. miRNA biogenesis is a tightly regulated process, and any transcriptional/genomic alterations during miRNA processing or in disease conditions may lead to aberrant miRNA expression [1]. Therefore, pathological conditions may be detectable using miRNA expression profiles from tumor tissue and body fluid samples. Stratification of patients based on differential expression of miRNAs and the therapeutic targeting of such miRNAs would enable patient-specific tailored intervention, referred to as theranostics, a commonly used term in the field of personalized medicine nowadays [1].

Jyotsna Batra and Srilakshmi Srinivasan (eds.), *Theranostics: Methods and Protocols*, Methods in Molecular Biology, vol. 2054, https://doi.org/10.1007/978-1-4939-9769-5_5, 

Deregulated miRNA expression in tumor tissues is associated with tumor stage, grade, and aggressiveness as reported in the Cancer Genome Atlas (TCGA) dataset for many urological cancers [2–4]. These miRNA expression changes may be mirrored and easily detected in body fluids [5] providing a rationale to use miRNAs as potential minimally invasive biomarkers. Circulating miRNAs are detectable in body fluids in a free form, bound to ribonucleoprotein complexes [6, 7] or packaged into extracellular vesicles [8]. Although the precise function of secreted vesicles is not completely understood, results suggest that it may be a means by which miRNAs enable cells to communicate with each other [9]. Hence, miRNAs function as intracellular mediators that can be isolated from plasma, serum, urine, and other body fluids and their deregulation in human diseases is evident from the rising number of published studies [10].

The phenol-chloroform extraction method that relies on phase separation of molecules based on their differential solubilities is widely used to isolate total RNA from a variety of samples, including plasma and serum. Combining this method to column-based purification ensures RNA elution in a highly purified form and several commercially available kits following this principle have been developed for downstream applications [11]. However, miRNA detection in body fluids represents several challenges which need to be taken into consideration. Plasma and serum contain a small percentage of RNA since the buffy coat containing white blood cells and platelets is the main source of RNA. Therefore, a carrier RNA, such as bacterial ribosomal RNA, which mimics nucleic acids, may be used to improve miRNA yield and recovery during RNA extraction [12]. The carrier RNA adds concentration and proximity to the small amount of target miRNAs present in the sample allowing their efficient elution and preventing them from being irretrievably bound to the column membrane during the isolation process. To obtain an accurate miRNA expression profile, it is important to make sure that the plasma/serum fraction of blood is correctly processed and does not contain RNA from blood cells [13]. This also minimizes the chances of confounding variations that may occur between circulating miRNA studies due to inconsistent miRNA processing. Additionally, it is important to ensure that the plasma/serum samples are not hemolysed as even minute traces of red blood cells may greatly affect the miRNA expression profile [14]. Therefore, it is necessary to screen the samples for hemolysis by determining the absorbance of hemoglobin measured by spectrophotometry at a wavelength of 414 nm [12, 15]. Although serum samples are cleaner compared to plasma samples due to the absence of clotting factors, plasma contains a more diverse miRNA population as many miRNAs are lost during the coagulation process [16]. In addition processing of plasma samples minimizes procedural variation caused by differences in clotting efficiency during serum collection.

Data normalization is another challenging aspect of miRNA profiling studies because commonly used RNA reference genes, for example RNU6, are not expressed in plasma and serum. Alternatively, miRNAs identified through statistical algorithms such as NormFinder, geNorm, and BestKeeper are reported to be stably expressed in the literature [17, 18]. Examples include miR-16-5p, miR-1228-3p, let-7a, miR-25-3p, miR-93-5p, and miR-106b-5p, which may be used as reference genes for normalization [17, 18]. However, limitations such as inconsistent expression under various experimental conditions, small sample size, source of miRNAs, and the lack of other disease groups in studies are yet to be overcome. Therefore, exogenous spike-in RNAs derived from *Caenorhabditis elegans*, for example cel-miR-39, are more commonly used for normalization purposes and monitoring RNA extraction efficiency [12, 19]. A fixed amount and copy number of the exogenous control is added to an equal volume of plasma during RNA extraction resulting in consistent cel-miR expression between different samples, which in turn can be used for normalization of RT-qPCR data. Another normalization approach known as "global mean normalization", which takes into account the mean expression of all miRNAs detected in a sample as the normalization factor, can be employed during analysis of a high number of miRNAs instead of using a single reference control for normalization [12].

In this chapter, we describe an experimental methodology for the detection of miRNAs in plasma samples by RT-qPCR. We have divided the procedure into two major steps consisting of (1) miRNA isolation and (2) miRNA relative quantification by RT-qPCR.

## 2 Materials

### 2.1 miRNA Isolation from Plasma

1. Silica membrane-based miRNA purification kit.
2. Spike-in control (Cel-miR-39).
3. Carrier RNA/Ribosomal RNA.
4. Chloroform.
5. Ethanol (80% and 100%).
6. Tris-EDTA (TE) buffer (containing 10 mM Tris, 1 mM EDTA, pH ~8.0) for reconstitution of Cel-miR-39 primer.
7. Sterile RNase-free pipet tips.
8. 1.5 and 2 mL microcentrifuge tubes.

9. Microcentrifuge (with rotor for 2 mL tubes) for centrifugation at 4 °C and at room temperature (15–25 °C).
10. NanoDrop spectrophotometer.

### 2.2 miRNA Relative Quantification by RT-qPCR

1. miRNA reverse transcription kit.
2. miRNA-specific forward primer.
3. SYBR green PCR master mix.
4. miRNA universal primer (reverse primer for all miRNAs).
5. miRNA reverse transcription control (miRTC) primer.
6. 384-well reaction plates.
7. Optical adhesive film.
8. Multichannel pipette/tips.
9. Real-time PCR system.

## 3 Methods

### 3.1 miRNA Isolation from Plasma

The methodology described here combines phenol/guanidine-based lysis of plasma samples and silica membrane-based purification of total RNA (*see* **Note 1**). The lysis reagent and chloroform are initially added to plasma samples to facilitate lysis, denature protein complexes/RNases, and remove residual DNA/proteins, followed by phase separation by centrifugation (*see* **Note 2**). Exogenous cel-miR-39 mimic and bacterial ribosomal RNA are added after the addition of denaturant and prior to addition of chloroform to permit normalization between samples, and control for varying RNA purification yields and amplification efficiency (*see* **Notes 3** and **4**). Following centrifugation the upper aqueous phase is separated and ethanol is added to provide appropriate binding conditions to the purification column for all RNA molecules, including miRNAs, from approximately greater than 18 nucleotides in size.

1. Thaw frozen plasma sample on ice before starting or use fresh plasma sample. Cool the benchtop centrifuge to 4 °C.
2. Dilute 2 μL of plasma sample 10× to check for the presence of hemolysis on a Nanodrop spectrophotometer.
3. Go to Protein Labels in Nanodrop settings and select Other Proteins.
4. Set λ3 at 414 nm and measure 2 μL of diluted plasma sample on the Nanodrop spectrophotometer (*see* **Note 5**).
5. The appearance of a peak at this wavelength indicates the presence of hemolysis in the sample.

6. Reconstitute lyophilized cel-miR-39 miRNA mimic by adding 300 μL RNase free-water resulting in a $2 \times 10^{10}$ copies/μL stock which is further diluted to $4 \times 10^{9}$ copies/μL by adding 4 μL of $2 \times 10^{10}$ copies/μL to 16 μL RNase free water, followed by dilution to $1.6 \times 10^{8}$ copies/μL working solution by adding 2 μL of $4 \times 10^{9}$ copies/μL to 48 μL RNase free water.
7. Place 150 μL of plasma sample in a 1.5 mL tube and centrifuge for 10 min at 16,000 × *g* and 4 °C to remove additional cellular nucleic acids attached to cell debris (*see* **Notes 6–8**).
8. Add 5 volumes (i.e. 750 μL) of lysis reagent and mix by vortexing or pipetting up and down.
9. Place the tube containing the lysate on the benchtop at room temperature (15–25 °C) for 5 min.
10. Add 3.5 μL of spike-in control ($1.6 \times 10^{8}$ copies/μL working solution prepared as described above) to the sample and mix thoroughly. After RT-qPCR, the $C_T$ value obtained with the spike-in control primer targeting the synthetic control miRNA permits normalization between samples, which can control for varying RNA purification yields and amplification efficiency.
11. Add 0.5 μg of carrier RNA/ribosomal RNA to the sample for improved RNA recovery.
12. Add chloroform of an equal volume to the starting sample (i.e. 150 μL) and shake vigorously by vortexing for 15 s.
13. Place the tube containing the lysate on the benchtop at room temperature (15–25 °C) for 3 min.
14. Centrifuge for 15 min at 12,000 × *g* at 4 °C.
15. After centrifugation, heat the centrifuge up to room temperature if the same centrifuge will be used for the next centrifugation steps.
16. Transfer the upper aqueous phase (~450 μL) to a new collection tube without transferring any interphase material.
17. Add 1.5 volumes (i.e. 675 μL) of 100% ethanol and mix thoroughly by pipetting up and down several times.
18. Pipet 700 μL of the sample, including any precipitate that may have formed, into a RNA purification column in a supplied 2 mL collection tube.
19. Close the lid gently and centrifuge at 10000 × *g* for 30 s at room temperature (15–25 °C).
20. Discard the flow-through and reuse the collection tube.
21. Repeat **steps 14–16** for the remainder of the sample.
22. Add 700 μL of washing buffer containing guanidine salt and ethanol onto the RNA purification column. Small RNAs that

are <200 nucleotides in size remain attached to the RNA purification column. This washing step is necessary after pre-clearing the sample with the lysis reagent, especially for the isolation of small RNAs, such as miRNAs.

23. Close the lid gently and centrifuge for 30 s at 10,000 × *g* to wash the column.
24. Discard the flow-through and reuse the collection tube.
25. Add 500 μL of mild washing buffer containing ethanol onto the RNA purification column to remove impurities, for example, salts that are still on the column due to buffers used earlier during the purification process.
26. Close the lid gently and centrifuge for 30 s at 10,000 × *g* to wash the column.
27. Discard the flow-through and reuse the collection tube.
28. Add 500 μL of 80% ethanol (prepared by diluting 100% ethanol in RNase free water) onto the RNA purification column.
29. Close the lid gently and centrifuge for 2 min at 10,000 × *g* to wash the column membrane.
30. Carefully remove the column from the collection tube to avoid any contact with the flow-through and discard the flow-through and the collection tube.
31. Place the column into a new 2 mL collection tube.
32. Open the lid of the column and centrifuge at full speed for 10 min to dry the membrane. To avoid damage to the lids, place the columns in the centrifuge with one empty position in between.
33. Discard the collection tube containing the flow-through.
34. Place the column in a new 1.5 mL collection tube.
35. Add 14 μL of RNase-free water directly to the center of the RNA purification column membrane. Elution with 14 μL of RNase-free water results in a 12 μL eluate as the dead volume of the column is 2 μL.
36. Close the lid gently and centrifuge for 1 min at full speed to elute the RNA.
37. Measure the eluted RNA concentration using a Nanodrop spectrophotometer (*see* **Note 9**).

### 3.2 miRNA Profiling in Plasma

For miRNA profiling in plasma samples, we used a miRNA reverse transcription kit, miRNA-specific forward primers, universal primer, and SYBR green PCR master mix for relative quantification of specific miRNAs. The total RNA that contains miRNAs isolated as described in Subheading 3.1 was used as the starting material for cDNA synthesis, enabling detection of multiple miRNAs from a single cDNA reaction by RT-qPCR. In our experience and as reported previously, the RNA yield obtained from a small volume of plasma is below the quantification limit by spectrophotometry, hence, it is impractical to start with a fixed concentration of RNA for reverse transcription. Therefore, using a fixed volume of starting RNA is reasonable provided statistical adjustments are made during data analysis [12, 20, 21]. We have also used a miRNA reverse transcription control (miRTC) to assess reverse transcription efficiency during miRNA profiling experiments.

#### 3.2.1 cDNA Synthesis

1. Thaw and mix by flicking the tubes nucleics mix, HiSpec buffer, and RNase-free water at room temperature. The HiSpec buffer is used for subsequent quantification of mature miRNAs only using miRNA-specific forward primers and universal primer during RT-qPCR.
2. Prepare the reverse transcription master mix on ice by adding 2 μL nucleics mix, 4 μL HiSpec buffer, 7 μL RNase-free water, and 2 μL reverse transcriptase mix (removed from −20 °C right before master mix preparation and mixed gently).
3. Add 5 μL of template RNA (i.e. 10 pg–2 μg RNA as starting material) to the master mix to get a total volume of 20 μL, gently mix and store on ice.
4. Incubate for 60 min at 37 °C.
5. Incubate for 5 min at 95 °C to inactivate the reverse transcriptase and place on ice.
6. If proceeding to RT-qPCR immediately, dilute the cDNA 22× by adding 420 μL of RNase-free water to ensure 50 pg–3 ng cDNA/PCR reaction when 2 μL of cDNA is added per reaction.

#### 3.2.2 RT-qPCR

1. Thaw SYBR green PCR master mix, miRNA-specific forward primers, universal primer, template cDNA (diluted 22× to enable pipetting of 2 μL cDNA using a multichannel pipette resulting in 11× cDNA dilution per reaction), and RNase-free water at room temperature (15–25 °C).
2. Prepare a reaction mix by adding 2.5 μL SYBR green PCR master mix, 0.5 μL universal primer, and 0.5 μL miRNA-specific forward primer.
3. Add 2 μL of template cDNA into each well of a 384-well PCR plate to get a total volume of 5.5 μL per reaction. The reaction volume was halved keeping the cDNA input consistent with the recommendations from the manufacturer to minimize cost without compromising the quality of the PCR run.

4. Seal the plate with the optical adhesive film and briefly centrifuge for a few seconds to remove air bubbles and collect residual volume from the sides of the well.
5. Program the real-time PCR cycler for a standard run at 95 °C for 15 min to activate HotStarTaq DNA polymerase followed by a three-step cycling condition including denaturation, annealing, and extension at 94 °C for 15 s, 55 °C for 30 s, and 70 °C for 30 s, respectively, with a total number of 40 cycles and 5 μL reaction volume.

#### 3.2.3 Data Analysis

There are various means of normalization of data from plasma miRNA profiling experiments as discussed previously (*see* **Note 10**). However, there is no general agreement on the normalization method used in studies to date. Here we used the spike-in method of normalization because the amount of the exogenous control added to each sample can be maintained consistently throughout samples. This overcomes the need for using an internal control which cannot be quantified accurately due to its low concentration in the isolated RNA. Therefore, we normalized the data by using the formula $\Delta C_T$ = mean $C_T$ miRNA of interest − mean $C_T$ cel-miR-39 for every sample assuming $C_T = 40$ for undetermined miRNA expression to account for lowly expressed miRNAs. We represented the $\Delta C_T$ values as negative $\Delta C_T$ for the miRNAs of interest to keep the directionality same for miRNA expression and $C_T$ values (i.e. high $C_T$ means high miRNA expression). Data analysis can be performed using Excel, SPSS, and GraphPad Prism or other statistical packages.

## 4 Notes

1. This protocol can also be used for isolation of cell-free total RNA, including miRNAs, from small volumes (up to 200 μL) of other body fluids such as serum and urine.
2. Since body fluids do not typically contain significant amounts of DNA, DNase 1 digestion is not necessary as the lysis reagent coupled with column purification removes trace amounts of DNA if present.
3. Since plasma has a low RNA content, the use of bacterial ribosomal RNA as discussed in this protocol, or a carrier RNA, such as MS2 RNA is recommended to improve the RNA yield during the extraction process.
4. RNA extraction efficiency can be monitored using exogenous spike-in controls, such as cel-miR-39, used in this protocol which also provides a means of data normalization. Similarly, reverse transcription performance can be monitored by using

an internal synthetic RNA control, miRNA reverse transcription control (miRTC), already present in the nucleics mix during cDNA synthesis.

5. The quantification of miRNAs exclusively present in red blood cells, such as miR-16-5p and miR-451, and the detection of hemoglobin at 414 nm on a Nanodrop spectrophotometer are ways of confirming that the plasma samples are hemolysis free.
6. The RNA purification column used in this protocol has a maximum binding capacity of 45 μg of RNA and loading volume of 700 μL. Therefore, it is ideal to use up to 200 μL of starting sample.
7. Plasma samples should be centrifuged at 16,000 × *g* for 10 min at 4 °C to pellet any insoluble components/debris present in plasma.
8. The presence of enzyme inhibitors in plasma makes it difficult to obtain a better sensitivity even after increasing the sample input as the quantity of inhibitors also increases and they may vary between methods of extraction and from one sample to another.
9. Plasma has a low RNA content; hence a 260/280 ratio of ~2 indicating pure RNA may not be achievable in all samples using the Nanodrop spectrophotometer. Therefore, the use of a bioanalyzer is required if precise measurements are desired; however, this technique is expensive compared to the other techniques.
10. It is important to investigate the normalization method used for data analysis for each study. Although the use of exogenous spike-in controls is preferred in most cases, identification of a miRNA endogenous control that is consistently expressed across all samples using statistical algorithms such as NormFinder, geNorm, and BestKeeper is beneficial for data analysis. Global mean normalization is a method of choice for miRNA screening assays, such as miRNA PCR arrays, which involve the detection of a large number of miRNAs.

## Acknowledgments

This work is supported by the NHMRC CDF to A/Prof Jyotsna Batra and QUT Postgraduate Research Award (QUTPRA) and QUT HDR Tuition Fee Sponsorship to Farhana Matin.

## References


1. Matin F, Jeet V, Clements JA, Yousef GM, Batra J (2016) MicroRNA theranostics in prostate cancer precision medicine. Clin Chem 62 (10):1318–1333. https://doi.org/10.1373/clinchem.2015.242800
2. Cancer Genome Atlas Research N (2015) The molecular taxonomy of primary prostate cancer. Cell 163(4):1011–1025. https://doi.org/10.1016/j.cell.2015.10.025
3. The Cancer Genome Atlas Research N, Creighton CJ, Morgan M, Gunaratne PH, Wheeler DA, Gibbs RA et al (2013) Comprehensive molecular characterization of clear cell renal cell carcinoma. Nature 499:43. https://doi.org/10.1038/nature12222. https://www.nature.com/articles/nature12222#supplementary-information
4. The Cancer Genome Atlas Research N, Weinstein JN, Akbani R, Broom BM, Wang W, Verhaak RGW et al (2014) Comprehensive molecular characterization of urothelial bladder carcinoma. Nature 507:315. https://doi.org/10.1038/nature12965. https://www.nature.com/articles/nature12965#supplementary-information
5. Weber JA, Baxter DH, Zhang S, Huang DY, Huang KH, Lee MJ et al (2010) The microRNA spectrum in 12 body fluids. Clin Chem 56(11):1733–1741. https://doi.org/10.1373/clinchem.2010.147405
6. Arroyo JD, Chevillet JR, Kroh EM, Ruf IK, Pritchard CC, Gibson DF et al (2011) Argonaute2 complexes carry a population of circulating microRNAs independent of vesicles in human plasma. Proc Natl Acad Sci U S A 108 (12):5003–5008. https://doi.org/10.1073/pnas.1019055108
7. Vickers KC, Palmisano BT, Shoucri BM, Shamburek RD, Remaley AT (2011) MicroRNAs are transported in plasma and delivered to recipient cells by high-density lipoproteins. Nat Cell Biol 13(4):423–433. https://doi.org/10.1038/ncb2210
8. Franzen CA, Blackwell RH, Foreman KE, Kuo PC, Flanigan RC, Gupta GN (2016) Urinary exosomes: the potential for biomarker utility, intercellular signaling and therapeutics in urological malignancy. J Urol 195(5):1331–1339. https://doi.org/10.1016/j.juro.2015.08.115
9. Junker K, Heinzelmann J, Beckham C, Ochiya T, Jenster G (2016) Extracellular vesicles and their role in urologic malignancies. Eur Urol 70(2):323–331. https://doi.org/10.1016/j.eururo.2016.02.046
10. Fendler A, Stephan C, Yousef GM, Kristiansen G, Jung K (2016) The translational potential of microRNAs as biofluid markers of urological tumours. Nat Rev Urol 13 (12):734–752. https://doi.org/10.1038/nrurol.2016.193
11. Meerson A, Ploug T (2016) Assessment of six commercial plasma small RNA isolation kits using qRT-PCR and electrophoretic separation: higher recovery of microRNA following ultracentrifugation. Biol Methods Protoc 1(1): bpw003
12. Matin F, Jeet V, Moya L, Selth LA, Chambers S, Australian Prostate Cancer B et al (2018) A plasma biomarker panel of four MicroRNAs for the diagnosis of prostate cancer. Sci Rep 8(1):6653. https://doi.org/10.1038/s41598-018-24424-w
13. Blondal T, Jensby NS, Baker A, Andreasen D, Mouritzen P, Wrang Teilum M et al (2013) Assessing sample and miRNA profile quality in serum and plasma or other biofluids. Methods 59(1):S1–S6. https://doi.org/10.1016/j.ymeth.2012.09.015
14. Kirschner MB, Kao SC, Edelman JJ, Armstrong NJ, Vallely MP, van Zandwijk N et al (2011) Haemolysis during sample preparation alters microRNA content of plasma. PLoS One 6(9):e24145. https://doi.org/10.1371/journal.pone.0024145
15. Shah JS, Soon PS, Marsh DJ (2016) Comparison of methodologies to detect low levels of hemolysis in serum for accurate assessment of serum microRNAs. PLoS One 11(4): e0153200. https://doi.org/10.1371/journal.pone.0153200
16. Wang K, Yuan Y, Cho JH, McClarty S, Baxter D, Galas DJ (2012) Comparing the microRNA spectrum between serum and plasma. PLoS One 7(7):e41561. https://doi.org/10.1371/journal.pone.0041561
17. Niu Y, Wu Y, Huang J, Li Q, Kang K, Qu J et al (2016) Identification of reference genes for circulating microRNA analysis in colorectal cancer. Sci Rep 6:35611. https://doi.org/10.1038/srep35611
18. Zhao H, Ma TF, Lin J, Liu LL, Sun WJ, Guo LX et al (2018) Identification of valid reference genes for mRNA and microRNA normalisation in prostate cancer cell lines. Sci Rep 8(1):1949. https://doi.org/10.1038/s41598-018-19458-z
19. Sharova E, Grassi A, Marcer A, Ruggero K, Pinto F, Bassi P et al (2016) A circulating

miRNA assay as a first-line test for prostate cancer screening. Br J Cancer 114 (12):1362–1366. https://doi.org/10.1038/bjc.2016.151

20. Kroh EM, Parkin RK, Mitchell PS, Tewari M (2010) Analysis of circulating microRNA biomarkers in plasma and serum using quantitative reverse transcription-PCR (qRT-PCR). Methods 50(4):298–301. https://doi.org/10.1016/j.ymeth.2010.01.032

21. Marabita F, de Candia P, Torri A, Tegner J, Abrignani S, Rossi RL (2016) Normalization of circulating microRNA expression data obtained by quantitative real-time RT-PCR. Brief Bioinform 17(2):204–212. https://doi.org/10.1093/bib/bbv056

# Chapter 6

# Isolation and Quantification of MicroRNAs from Human Saliva

**Sri Ram Arunachalam, Kai Dun Tang, and Chamindie Punyadeera**


## Abstract

Oral premalignant disorders (OPMD) have relatively high malignant transformation rates to Oral Cancers (OC). Oral carcinogenesis is a multistep process that originates as epithelial hyperplasia followed by epithelial dysplasia, leading to fully malignant phenotypes. Early detection can be lifesaving but is currently not possible due to the lack of early diagnostic tools. The current diagnostic methods such as biopsy sampling, tumor tissue staining, and imaging techniques require skilled personnel and are invasive, painful, time-consuming, and expensive. Saliva has gained momentum as the diagnostic fluid of the future due to its noninvasive nature, ease of sampling, multiple samples can be collected with ease and more importantly does not require skilled personnel. The use of saliva in cancer diagnostics is an emerging and an expanding field. MicroRNA (miRNA) play a role in cancer initiation and progression and the expression changes of miRNA have been investigated as a potential biomarker in cancer studies. In this chapter, we describe a robust and cost-effective protocol to isolate and enrich miRNA from saliva samples. Profiling miRNAs in saliva samples can form part of the clinical management of OPMD and OC patients in the future.


**Key words** Oral premalignant disorder (OPMD), MicroRNA (miRNA), Early detection, Saliva, Noninvasive, miRNA, Diagnosis, Prognosis, Biomarker

## 1 Introduction

Oral premalignant disorder (OPMD) is a disease or syndrome, which, if untreated, may lead to oral cancer (OC). Approximately 90% of OCs are derived from squamous cell type, and often known as oral squamous cell carcinoma (OSCC) [1]. The majority of OSCC are expected to arise from already existing OPMD, which includes leukoplakia, erythroplakia, actinic cheilosis, oral submucous fibrosis (OMSF), oral lichen planus, and snuff patch [1–6]. If OSCC are diagnosed at an early stage, the survival rates are high. Unfortunately, only 31% of OSCC are diagnosed at an early stage [7]. As a consequence, the 5-years survival rate drops down to 20%, in contrast to 80% survival rate when OSCC patients are diagnosed at an early stage of the disease [8].

Jyotsna Batra and Srilakshmi Srinivasan (eds.), *Theranostics: Methods and Protocols*, Methods in Molecular Biology, vol. 2054,
https://doi.org/10.1007/978-1-4939-9769-5_6, 

Clinical examination, imaging, and histological analysis using tumor biopsy samples are the current standard diagnostic tools. However, these techniques require trained and highly skilled personnel as well as invasive, painful, and time-consuming techniques. Tolonium chloride or toluidine blue dye, Oral CDx brush biopsy, fluorescence imaging, optical spectroscopy, vital tissue staining, clinical examination, photodynamic detection, and chemiluminescent kits are some of the recently used clinical diagnostic tools [9, 10]. Therefore, there is an unmet clinical need to develop noninvasive, easy-to-use tools to identify OPMD patients before clinical manifestation.

There is emerging evidence linking oral health to systemic disease [11]. As such attention has tuned to saliva as a biological fluid. Whole saliva consists of several salivary and non-salivary components. It contains secretions from the salivary glands (parotid, submandibular, and sublingual glands, and minor salivary glands) and non-salivary components such as the gingival crevicular fluid, nasal and bronchial secretions, serum and blood derivatives from wounds, desquamated epithelial linings, food components, and microorganisms that reside in the oral cavity [12, 13]. Human saliva is gaining momentum as the diagnostic biofluid of the future because of its ease of collection, noninvasiveness, multiple samples can be collected from a person, no requirement for trained personnel, and as such an ideal biofluid for screening studies [14]. We and others have shown that saliva can be used as a diagnostic fluid to detect heart diseases [15–17] and head and neck cancers [18–21]. The use of saliva as a prognostic tool in managing cancer patients is an upcoming field [22, 23].

Tumor recurrence and mortality of a cancer patient depends on the time of detection. Scientists across the globe are investigating molecular changes that can be used as biomarkers to facilitate early diagnosis. Pathogenesis of OPMD involves genetic and epigenetic aberrations leading to either gain of oncogenes or a loss of tumor suppressor genes [24]. MicroRNAs (miRNAs) are small noncoding RNAs, approximately 19–25 nucleotides long. A number of studies have suggested that cancer initiation, progression, and development are associated with alteration in miRNA expression levels [25–28]. It is believed that one-third of all protein-coding genes in the human genome are controlled by miRNAs [29]. miRNAs are reported to play a vital role in posttranscriptional regulation of gene expression [30–32]. The discovery of their presence not only in tissues but also in body fluids, including blood, urine, and cerebrospinal fluid, together with their changes in expression under a number of pathological conditions, has implicated miRNAs as informative biomarkers of disease.

Unlike isolating total RNA from body fluids, miRNAs are very stable and as such do not undergo hydrolysis. There is also accumulating scientific evidence to demonstrate that miRNA is more stable than RNA in the saliva, making it a great tool for the gene expression and profiling studies. Various studies have demonstrated the clinical utility of miRNAs when making therapeutic decisions [33]. However, exploiting miRNAs in this capacity requires methodological rigor. In this work, we describe a detailed protocol to isolate and quantify miRNA from human saliva samples.

## 2 Materials

### 2.1 Saliva Sample Collection

1. 50 mL Falcon tubes—for sample collection.
2. Dry ice—for transportation of saliva samples from the hospital to the laboratory.

### 2.2 Saliva Sample Processing

1. 1.5 mL Eppendorf tubes.
2. QIAzol reagent (Qiagen, Valencia, CA, USA).

### 2.3 miRNA Isolation and Quantification

1. QIAzol reagent (Qiagen).
2. 100% ethanol.
3. NucleoSpin® miRNA kit (Macherey-Nagel, Düren, Germany).
4. Binding buffer.
5. Wash buffer.
6. RNase-free water/Elution buffer.
7. Nandrop ND-1000 spectrometer (Thermo Fisher Scientific, Waltham, USA).
8. Qubit® 3.0 Fluorometer (Thermo Fisher Scientific).

### 2.4 cDNA Synthesis

1. miRNA template.
2. miScript II RT Kit (Qiagen).
3. miScript HiSpec buffer (Qiagen).
4. Reverse transcriptase mix.
5. RNase-free water.
6. BioRadT100™ Thermal Cycler (Bio-Rad, Hercules, CA, USA).

### 2.5 miRNA Primer Assay

1. cDNA template.
2. miScript miRNA primer assays (Qiagen).
3. SYBR Green master mix.
4. Universal primer.
5. RNase-free water.
6. QuantStudio 7 Flex Real-Time PCR (Applied Biosystems, Carlsbad, CA, USA).

### 2.6 miRNA PCR Array

1. cDNA template.
2. miScript™ miRNA HC PCR arrays (Qiagen, catalogue number MIHS-3102Z).
3. SYBR Green master mix.
4. Universal primer.
5. RNase-free water.
6. QuantStudio 7 Flex Real-Time PCR (Applied Biosystems).

## 3 Methods

### 3.1 Saliva Sample Collection from Participants

The unstimulated saliva (drool) samples from OPMD patients and healthy controls are collected in 50 mL sterile Falcon tubes (*see* **Note 1**) across three clinical sites in Brisbane, Australia. In brief, the volunteers were asked to sit in a comfortable upright position and were asked to rinse their mouth with water (to remove food debris) and were asked to tilt their heads down and to pool saliva in the mouth for about 2–5 min [34]. They were then asked to expectorate saliva onto the Falcon tubes kept on ice. Saliva samples were transported on dry ice to the laboratory (*see* **Note 2**).

### 3.2 Saliva Sample Processing

The optimized in house protocol is followed for sample processing. Aliquot the unprocessed saliva into 1.5 mL Eppendorf tubes and store in −80 °C as one fraction (*see* **Note 3**). 200 μL of drool is taken as another fraction and 800 μL of QIAzol reagent is added and stored in −80 °C until further use (Fig. 1).

### 3.3 miRNA Isolation

There are several optimized commercially available kits that can be used when extracting miRNA from body fluids. Assay kits such as miRNAeasy (Qiagen), mirVana miRNA Isolation Kit (Ambion), NucleoSpin miRNA kit (Macherey-Nagel), and traditional trizol/TRI-Reagent followed by precipitation are some of the extraction methods.

In this study, we describe a method using NucleoSpin miRNA kit (Macherey-Nagel) in conjugation with QIAzol reagent (Qiagen) to isolate high yield miRNA from saliva as per our previous publications [20, 35]. Total RNA isolation (*see* **Note 4**), extraction (*see* **Note 5**), and miRNA isolation (*see* **Note 6**) are the three different options for extraction. We are describing a method to isolate and enrich for miRNA alone. Add 800 μL QIAzol reagent to 200 μL unprocessed whole mouth saliva. Vortex briefly and incubate at room temperature (RT) for 5 min. Add 160 μL chloroform, vortex vigorously and incubate at RT for 3 min. Centrifuge the samples at full speed for 15 min at 4 °C. Transfer approximately 600–800 μL of the upper clear aqueous layer (*see* **Note 7**) to a new 2 mL collection tube and add 200 μL 100% ethanol. Mix briefly and

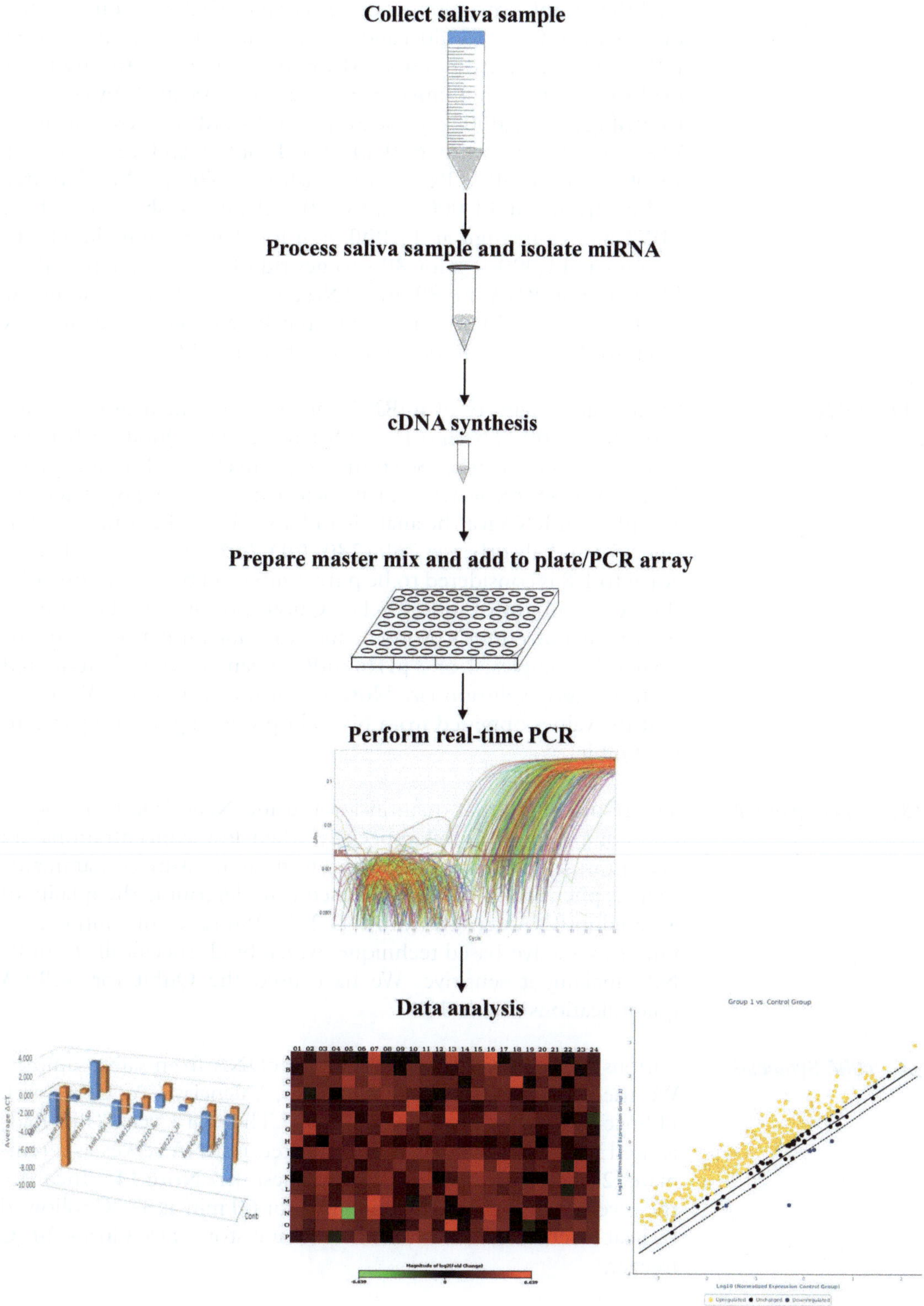


**Fig. 1** Workflow for saliva collection and processing

load the sample into blue ring NucleoSpin RNA column. Centrifuge at 11,000 × *g* for 30 s and use the blue ring column for large RNA enrichment. Add 800 μL MX buffer to the flow through and load the solution mix onto green ring NucleoSpin RNA column. Centrifuge at 11,000 × *g* for 30 s and discard the flow through. Wash the column with 600 μL MW1 buffer and centrifuge at 11,000 × *g* for 30 s. Repeat the wash with 700 μL MW2 buffer and centrifuge at 11,000 × *g* for 30 s. A final wash with 250 μL MW2 and centrifuge at 11,000 × *g* for 2 min. Spin the empty column at 11,000 × *g* for 30 s to get rid of any remaining buffer. Elute the miRNA in 30 μL RNase-free water by spinning at 11,000 × *g* for 30 s. Repeat the spin by adding eluted miRNA again to the column in order to get a better yield.

#### *3.3.1 miRNA Quantification*

Quality and quantity of miRNA can be determined using Nanodrop ND-1000 (Thermo Fisher Scientific) and Qubit® 3.0 Fluorometer (Thermo Fisher Scientific) respectively. For Nanodrop, use 1 μL of RNase-free water to blank before use (*see* **Note 8**). Then use 1–2 μL of miRNA for the analysis and note down the concentration (ng/μL) and absorbance 260/280. OD 260/280 greater than or equal to 1.8 is considered to be pure. Qubit can be used to quantify the exact amount of miRNA. For Qubit, prepare the standards as per the manufacturer's protocol and read/measure it before quantifying the samples. 1 or 2 μL of miRNA can be used when diluted with the assay solution (*see* **Note 9**) for quantification. We found that the values obtained from Nanodrop were higher compared to the Qubit.

#### *3.3.2 Nanodrop vs. Qubit*

The results show that quantification using Nanodrop is not accurate compared to Qubit especially when low concentrations are present in samples. Since the Nanodrop can measure contaminants such as phenol and ethanol, we used it to determine the quality of isolated miRNA derived from OD 260/280 ratio. In contrast, the Qubit uses a dye-based technique, which binds specifically to miRNAs making it sensitive. We have used the Qubit for miRNA quantifications (Table 1).

### *3.4 cDNA Synthesis*

Various kits are available to synthesize cDNA from saliva samples. We used miScript II RT Kit (Qiagen, Valencia, CA, USA) with HiSpec buffer (*see* **Note 10**). T100™ Thermal Cycler (Bio-Rad) is used for cDNA synthesis. The isolated miRNA serves as a template (250 ng to 1 μg) for cDNA synthesis (*see* **Note 11**). Incubate the reverse transcription reaction mix for 60 min at 37 °C followed by heat inactivation for 5 min at 95 °C and store cDNA in a −20 °C freezer.

**Table 1**
**Quantification of isolated miRNA on a Nanodrop and a Qubit**

| Sample | Sample Id | Nanodrop | 260/280 | Qubit |
|---|---|---|---|---|
| Group 1 | 1 | 640.37 | 2.21 | 377.6 |
| Group 1 | 2 | 258.25 | 2.01 | 120 |
| Group 1 | 3 | 159.09 | 1.99 | 79.8 |
| Group 1 | 4 | 163.22 | 1.93 | 76.2 |
| Group 1 | 5 | 155.06 | 1.99 | 68.6 |
| Group 2 | 6 | 637.46 | 1.96 | 136 |
| Group 2 | 7 | 217.68 | 1.77 | 79 |
| Group 2 | 8 | 183.74 | 1.91 | 75.6 |
| Group 2 | 9 | 123.02 | 1.8 | 74.4 |
| Group 2 | 10 | 100.52 | 2.01 | 50.8 |

### 3.5 miRNA Primer Assay

In order to test the efficiency of converted cDNA, we used specific primers. We used nine primer pairs based on previously published work. Quantitative real-time PCR is used for primer assay. Dilute the cDNA and use 3 ng per reaction. QuantiTect SYBR Green PCR master mix is used with universal primer and target-specific primer and cDNA. QuantStudio 7 Flex Real-Time PCR (Applied Biosystems) is used to carry out the primer assay. Reaction conditions are 95 °C for 15 min to activate the hot start Taq DNA polymerase, denaturation at 94 °C for 15 s, annealing at 55 °C for 30 s and extension at 70 °C for 30 s for 40 cycles. Perform the analysis by the ΔΔCT method. The samples that are consistent among the groups and have a significant difference between OPMD and controls are selected to perform the PCR array.

### 3.6 miRNA PCR Array

Various options are available for interrogating miRNAs expression profiles. GeneChip, miRNA microarray, and probe are some of the commercially available options for miRNA expression profiling. We used miScript™ miRNA HC PCR arrays (Qiagen, catalogue number MIHS-3102Z) as per the manufacturer's protocol. This allows profiling of 372 miRNA targets. 1.5 ng of cDNA per well is used for the array. QuantStudio 7 Flex Real-Time PCR (Applied Biosystems) is used to carry out the reaction. Prepare the master mix and aliquot in PCR tubes strips in order to use the multichannel pipette (*see* **Note 12**). Reaction conditions are 95 °C for 15 min to activate the hot start Taq DNA polymerase, denaturation at 94 °C for 15 s, annealing at 55 °C for 30 s, and extension at 70 °C for 30 s for 40 cycles.

### 3.7 Analysis

Use the miScript miRNA PCR Array Data Analysis web portal for the data analysis (https://www.qiagen.com/au/shop/genes-and-pathways/data-analysis-center-overview-page?akamai-feo=off). This web portal is the straightforward easiest way for the data analysis. Once the raw data is uploaded and parameters are set, the analysis is done within a fraction of a second.

Manual analysis can be performed using the ΔΔCT method, by calculating ΔCT from the CT values after normalizing with SNORDs the endogenous normalizer. Graphs can be plotted using ΔCT or Fold change. Graph pad prism is used to make a graph and statistical analysis.

## 4 Notes

1. All the OPMD patients and healthy controls should sign an informed consent.
2. Storage and handling will impact the quality and quantity of biomolecules. So the samples should be stored in dry ice immediately to minimize the degradation and transport to the processing site.
3. Aliquot 200 μL of unprocessed sample for miRNA isolation and store remaining fraction for further use. If there is no unprocessed sample, the QIAzol mixed fraction can be used directly for isolation of miRNA.
4. Use total RNA isolation to extract small RNA and large RNA in one single fraction.
5. Use separated isolation to extract small and large RNA in two separate fractions.
6. Use miRNA isolation to extract only small RNA.
7. Carefully pipette out the clear aqueous phase without disturbing the bottom layer. If the phase is mixed, repeat the spin and transfer the aqueous layer. Store the bottom phase for DNA isolation.
8. For the blank, use the same water or buffer in which miRNA is eluted. Also, wipe the samples each time after measuring the concentration in order to avoid contamination and false result.
9. Make sure to select the correct volume of miRNA used (1 or 2 μL) for the assay to get the accurate concentration. As Qubit is so sensitive, failure to selecting the correct volume will lead to incorrect quantification.
10. miScript II RT Kit comes with HiSpec buffer and HiFlex buffer. Use 5× HiSpec buffer to prepare cDNA if profiling miRNA using miScript miRNA HC PCR Array.

11. Use minimum 125–250 ng miRNA as a template as it is the minimum recommended RNA input as per the manufacturer's protocol.
12. Multichannel pipette loads sample in alternative wells. Carefully load the master mix into all the wells and change the tips each time to avoid contamination, as each well has different primers.

## Acknowledgments

Our work was supported by the Janssen: Pharmaceutical Companies of Johnson & Johnson and Translational Research Institute (TRI) Spore Grant.

### References

1. Huber MA, Tantiwongkosi B (2014) Oral and oropharyngeal cancer. Med Clin North Am 98 (6):1299. https://doi.org/10.1016/j.mcna.2014.08.005
2. Greer RO Jr (2011) Oral manifestations of smokeless tobacco use. Otolaryngol Clin North Am 44(1):31–56, v. https://doi.org/10.1016/j.otc.2010.09.002
3. Sharan RN, Mehrotra R, Choudhury Y, Asotra K (2012) Association of betel nut with carcinogenesis: revisit with a clinical perspective. PLoS One 7(8):e42759. https://doi.org/10.1371/journal.pone.0042759
4. van der Waal I (2009) Potentially malignant disorders of the oral and oropharyngeal mucosa; terminology, classification and present concepts of management. Oral Oncol 45 (4–5):317–323. https://doi.org/10.1016/j.oraloncology.2008.05.016
5. van der Waal I (2010) Potentially malignant disorders of the oral and oropharyngeal mucosa; present concepts of management. Oral Oncol 46(6):423–425. https://doi.org/10.1016/j.oraloncology.2010.02.016
6. Yardimci G, Kutlubay Z, Engin B, Tuzun Y (2014) Precancerous lesions of oral mucosa. World J Clin Cases 2(12):866–872. https://doi.org/10.12998/wjcc.v2.i12.866
7. Siegel R, Ma J, Zou Z, Jemal A (2014) Cancer statistics, 2014. CA Cancer J Clin 64(1):9–29. https://doi.org/10.3322/caac.21208
8. Salazar C, Calvopiña D, Punyadeera C (2014) miRNAs in human papilloma virus associated oral and oropharyngeal squamous cell carcinomas. Expert Rev Mol Diagn 14(8):1033–1040
9. Messadi DV (2013) Diagnostic aids for detection of oral precancerous conditions. Int J Oral Sci 5(2):59–65. https://doi.org/10.1038/ijos.2013.24
10. Gillenwater A, Papadimitrakopoulou V, Richards-Kortum R (2006) Oral premalignancy: new methods of detection and treatment. Curr Oncol Rep 8(2):146–154
11. Zhang X, Schulz B, Walsh T, Atherton J, Kostner K, Punyadeera C (2017) Saliva: linking oral health with heart failure. Eur J Heart Fail 19(S1):1. https://doi.org/10.1002/ejhf.833
12. Cheng Y-SL, Rees T, Wright J (2014) A review of research on salivary biomarkers for oral cancer detection. Clin Transl Med 3:3–3. https://doi.org/10.1186/2001-1326-3-3
13. Kaufman E, Lamster IB (2002) The diagnostic applications of saliva—a review. Crit Rev Oral Biol Med 13(2):197–212
14. Pfaffe T, Cooper-White J, Beyerlein P, Kostner K, Punyadeera C (2011) Diagnostic potential of saliva: current state and future applications. Clin Chem 57(5):675–687
15. Zhang X, Walsh T, Atherton JJ, Kostner K, Schulz B, Punyadeera C (2017) Identification and validation of a salivary protein panel to detect heart failure early. Theranostics 7 (18):4350–4358. https://doi.org/10.7150/thno.21727
16. Punyadeera C, Dimeski G, Kostner K, Beyerlein P, Cooper-White J (2011) One-step homogeneous C-reactive protein assay for saliva. J Immunol Methods 373(1–2):19–25. https://doi.org/10.1016/j.jim.2011.07.013

17. Foo JYY, Wan Y, Kostner K, Arivalagan A, Atherton J, Cooper-White J et al (2012) NT-ProBNP levels in saliva and its clinical relevance to heart failure. PLoS One 7(10): e48452. https://doi.org/10.1371/journal.pone.0048452
18. Lim Y, Wan Y, Vagenas D, Ovchinnikov DA, Perry CF, Davis MJ et al (2016) Salivary DNA methylation panel to diagnose HPV-positive and HPV-negative head and neck cancers. BMC Cancer 16(1):749. https://doi.org/10.1186/s12885-016-2785-0
19. Chai RC, Lim Y, Frazer IH, Wan Y, Perry C, Jones L et al (2016) A pilot study to compare the detection of HPV-16 biomarkers in salivary oral rinses with tumour p16(INK4a) expression in head and neck squamous cell carcinoma patients. BMC Cancer 16:178. https://doi.org/10.1186/s12885-016-2217-1
20. Wan Y, Vagenas D, Salazar C, Kenny L, Perry C, Calvopiña D et al (2017) Salivary miRNA panel to detect HPV-positive and HPV-negative head and neck cancer patients. Oncotarget 8(59):99990–100001. https://doi.org/10.18632/oncotarget.21725
21. Tang KD, Kenny L, Perry C, Frazer I, Punyadeera C (2017) The overexpression of salivary cytokeratins as potential diagnostic biomarkers in head and neck squamous cell carcinomas. Oncotarget 8(42):72272–72280. https://doi.org/10.18632/oncotarget.19731
22. Zanotti L, Paderno A, Piazza C, Pagan E, Bignotti E, Romani C et al (2017) Epidermal growth factor receptor detection in serum and saliva as a diagnostic and prognostic tool in oral cancer. Laryngoscope 127(11):E408–e414. https://doi.org/10.1002/lary.26797
23. Nagler RM (2009) Saliva as a tool for oral cancer diagnosis and prognosis. Oral Oncol 45(12):1006–1010. https://doi.org/10.1016/j.oraloncology.2009.07.005
24. Zahran F, Ghalwash D, Shaker O, Al-Johani K, Scully C (2015) Salivary microRNAs in oral cancer. Oral Dis 21(6):739–747
25. Shiiba M, Uzawa K, Tanzawa H (2010) MicroRNAs in head and neck squamous cell carcinoma (HNSCC) and oral squamous cell carcinoma (OSCC). Cancers 2(2):653–669. https://doi.org/10.3390/cancers2020653
26. Lu J, Getz G, Miska EA, Alvarez-Saavedra E, Lamb J, Peck D et al (2005) MicroRNA expression profiles classify human cancers. Nature 435:834. https://doi.org/10.1038/nature03702. https://www.nature.com/articles/nature03702#supplementary-information
27. Volinia S, Calin GA, Liu C-G, Ambs S, Cimmino A, Petrocca F et al (2006) A microRNA expression signature of human solid tumors defines cancer gene targets. Proc Natl Acad Sci U S A 103(7):2257–2261
28. Mitchell PS, Parkin RK, Kroh EM, Fritz BR, Wyman SK, Pogosova-Agadjanyan EL et al (2008) Circulating microRNAs as stable blood-based markers for cancer detection. Proc Natl Acad Sci U S A 105 (30):10513–10518
29. Nagadia R, Pandit P, Coman WB, Cooper-White J, Punyadeera C (2013) miRNAs in head and neck cancer revisited. Cell Oncol 36 (1):1–7
30. Jansson MD, Lund AH (2012) MicroRNA and cancer. Mol Oncol 6(6):590–610. https://doi.org/10.1016/j.molonc.2012.09.006
31. Lin S, Gregory RI (2015) MicroRNA biogenesis pathways in cancer. Nat Rev Cancer 15 (6):321
32. Peng Y, Croce CM (2016) The role of MicroRNAs in human cancer. Signal Transduct Target Ther 1:15004. https://doi.org/10.1038/sigtrans.2015.4
33. Sethi N, Wright A, Wood H, Rabbitts P (2014) MicroRNAs and head and neck cancer: reviewing the first decade of research. Eur J Cancer 50 (15):2619–2635. https://doi.org/10.1016/j.ejca.2014.07.012
34. Mohamed R, Campbell J-L, Cooper-White J, Dimeski G, Punyadeera C (2012) The impact of saliva collection and processing methods on CRP, IgE, and Myoglobin immunoassays. Clin Transl Med 1:19–19. https://doi.org/10.1186/2001-1326-1-19
35. Salazar C, Nagadia R, Pandit P, Cooper-White-J, Banerjee N, Dimitrova N et al (2014) A novel saliva-based microRNA biomarker panel to detect head and neck cancers. Cell Oncol 37 (5):331–338

# Chapter 7

# The Isolation and Analysis of Circulating Tumor Cells


## Tianyu Guo, Elzbieta Stankiewicz, Xueying Mao, and Yong-Jie Lu



## Abstract

Circulating tumor cells (CTCs), with their close association with cancer metastasis, the most aggressive feature of solid tumors, represent an important aspect of "liquid biopsy," which provides minimally- or noninvasive approaches for cancer detection and disease status monitoring. CTC analysis has shown the potential clinical applications in several cancer types and has been approved by FDA for clinical use in advanced breast, prostate, and colorectal cancer prognosis. In this chapter, we describe a CTC isolation method using a cell size and deformability-based system, Parsortix, and the immunofluorescence staining method to detect CTCs with both epithelial and mesenchymal features. We also describe a repeated fluorescence in situ hybridization (FISH) approach to detect alterations of multiple genomic regions on the same CTCs after immunofluorescence analysis. This approach allows the study of CTCs as a biomarker for cancer detection, prognosis, and therapeutic response monitoring, as well as the study of the heterogeneity of CTCs and cancers.



**Key words** Circulating tumor cell, Epithelial-to-mesenchymal transition, Enumeration, Immunofluorescence staining, Fluorescence in situ hybridization


## 1 Introduction

Circulating tumor cells (CTCs) are rare cancer cells disseminated from the tumor site to the bloodstream. There has been intensive research on CTCs in recent years. Enumeration and molecular characterization of CTCs has been demonstrated to be of potential clinical value in different aspects, including cancer prognosis, treatment response prediction, and resistance monitoring [1, 2]. Moreover, CTCs may also assist study of cancer heterogeneity and cancer development [3]. With the increasing sensitivity to capture CTCs, ours and others' unpublished work also shows the potential of using CTC analysis for cancer detection/diagnosis. There are a range of devices that enable CTCs isolation based on different principles. The most common CTC isolation methods can be divided into three types: antibody-based approaches, physical property-based approaches and function-based approaches. Regardless of isolation principles, the characterization of CTCs is


Jyotsna Batra and Srilakshmi Srinivasan (eds.), *Theranostics: Methods and Protocols*, Methods in Molecular Biology, vol. 2054, https://doi.org/10.1007/978-1-4939-9769-5_7, © Springer Science+Business Media, LLC, part of Springer Nature 2019

mostly based on antigen expression. Most of the current CTC analysis methods rely on the expression of epithelial-specific proteins, such as EpCAM and cytokeratin (CK), and absence of leukocyte-specific surface antigens, CD45, on CTCs. One representative of this CTC isolation and characterization strategy is CellSearch system, which has been approved by the Food and Drug Administration (FDA) for detecting and enumerating CTCs as prognostic indicator for patients with metastatic breast, prostate, and colorectal cancers. However, many studies have demonstrated that epithelial-to-mesenchymal transition (EMT) and mesenchymal-to-epithelial transition (MET) are important processes for CTC dissemination and distant metastasis formation [4, 5]. Therefore, it is crucial to bring in mesenchymal markers, such as vimentin (Vim), for CTC characterization. Other markers, such as stem cell markers, may also be of interest in different studies.

Although most CTC analysis have been concentrated on CTC enumeration, other studies, such as genetic analysis, functional studies and CTC culture have also been performed in recent years, showing great values in cancer research and potential clinical applications [6–9]. In this chapter, we describe a CTC isolation and analysis method, which we have applied successfully in prostate cancer study [10, 11]. The method consists of CTC isolation using a size-based system, Parsortix (Angel PLC Europe), CTC characterization using immunofluorescence staining, and repeated FISH analysis of genomic changes in the CTCs. This isolation method can also capture CTC clusters, whose clinical relevance have been implicated in some studies [12, 13]. The immunofluorescence staining described here enables the detection of epithelial ($CK^+Vim^-CD45^-$), EMTing ($CK^+Vim^+CD45^-$) and mesenchymal ($CK^-Vim^+CD45^-$) CTCs. We also provide an optimized approach to robustly wash off immunofluorescence signals and perform multiple rounds of FISH analysis on the same cells, which facilitates the genomic analysis of CTCs.

## 2 Materials

### 2.1 Slide Preparation

1. 0.1% Poly-L-lysine (25988-63-0, Sigma).
2. PAP pen.
3. Microscope slides.

### 2.2 CTC Isolation and Fixation onto Slides

1. Blood collection tubes: K2EDTA vacutainer (BD).
2. Virkon disinfectant.
3. Dulbecco's Phosphate Buffer Saline without calcium chloride and magnesium chloride (DPBS) (Sigma-Aldrich).
4. Isolation buffer: DPBS containing 1% Bovine Serum Albumin (BSA) and 2 mM EDTA. To make 500 ml of isolation buffer,

add 5 g BSA and 5 ml of 0.2 M EDTA into 495 ml of DPBS and mix well till the BSA is completely dissolved. Store the isolation buffer at 4 °C for maximum 1 month.

5. Parsortix™ system (Angle, UK).
6. Ficoll-Paque PLUS (GE Healthcare).
7. 50 ml LeucoSep tube (Greiner Bio-One Ltd).
8. 50 ml conical tube.
9. 1–250 μl low-binding, wide-orifice tip (VWR).
10. 1.5 ml siliconized low-retention tube (Fisher Scientific Ltd).
11. Cold acetone (keep in −20 °C).
12. 0.075 M potassium chloride (KCl).
13. Eppendorf tube.

### 2.3 Immunofluorescence Staining

1. Cover glass.
2. Phosphate Buffer Saline (PBS).
3. PBST: PBS containing 0.05% TWEEN® 20.
4. Permeabilization solution: PBS containing 0.1% Triton™ X-100.
5. Blocking solution: PBS containing 10% normal donkey serum.
6. Staining solution: PBS containing 2% BSA and 2 mM EDTA.
7. Pre-conjugated antibodies shown in Table 1 (*see* **Note 1**).
8. Mounting medium: *SlowFade*™ gold antifade mountant with DAPI (Invitrogen™).
9. Ariol image analysis system (Leica Microsystems (Gateshead) Ltd., UK), equipped with an Olympus BX61 microscope.

### 2.4 FISH

#### 2.4.1 Stripping off Immunofluorescence Signals

1. Stripping buffer: 2% SDS, 0.8% β-mercaptoethanol, 0.0625 M Tris–HCl, pH 6.8. Stripping buffer should be freshly made.
2. PBS.
3. Water bath.
4. Coplin jar.

**Table 1**
**Pre-conjugated antibodies for the detection of CTCs by immunofluorescence staining**

| Antigen | Conjugate | Host species | Clone | Clonality | Supplier |
|---|---|---|---|---|---|
| Cytokeratin (CK) | Fluorescein-5-isothiocyanate (FITC) | Mouse | CK3-6H5 | Monoclonal | Miltenyi Biotec |
| CD45 | Phycoerythrin (PE) | Mouse | 5B1 | Monoclonal | Miltenyi Biotec |
| Vimentin (Vim) | Alexa Fluor® 647 | Rabbit | EPR3776 | Monoclonal | Abcam |

#### 2.4.2 Slide Pretreatment for Probe Hybridization

1. Fixation solution: 75% methanol, 25% acetic acid.
2. 70% acetic acid.
3. 70% ethanol.
4. 90% ethanol.
5. 100% ethanol.
6. PBS.
7. Coplin jar.

#### 2.4.3 Probe Preparation

1. FISH probe.
2. BioPrime DNA labeling system (Invitrogen).
3. Teramthylrhodamin-5-D-UTP and/or Fluorescein-12-dUTP (Roche).
4. Special 10× dNTP: 1 mM dATP, 1 mM dGTP, 1 mM dCTP, 0.65 mM dTTP.
5. Microspin G50 columns.
6. TE buffer.
7. PCR thermocycler.
8. Cot-1 DNA (1 μg/μl).
9. Salmon sperm (10 μg/μl).
10. 4 M NaCl.
11. 100% Ethanol.
12. 80% ethanol.
13. Heat block.

#### 2.4.4 Probe Hybridization

1. Hybridization buffer: 60%v/v deionized formamide, 12% v/v dextran sulfate, 2.4× saline-sodium citrate (SSC) buffer, 0.14 mM EDTA pH 8.0, 400 μg/ml Salmon sperm.
2. Heat block.
3. Hot plate.
4. Humidity box (slide box lined with wet tissue on the bottom).
5. Rubber seal.
6. Cover slips.
7. 37 °C incubator.

#### 2.4.5 Post-hybridization Wash

1. 2× SSC buffer.
2. SSCT buffer: 4× SSC, 0.2% TWEEN® 20. Add 100 ml of 20× SSC and 1 ml of TWEEN® 20 to 400 ml of water.
3. 50% formamide/2× SSC buffer: mix 50 ml of formamide with 10 ml of 20× SSC and 40 ml of water. This buffer should be freshly made before use.

4. PBS.
5. Water bath.
6. 70% ethanol.
7. VECTASHIELD Antifade Mounting Medium with DAPI (Vector Labs, US).
8. Humidity box (slide box lined with wet tissue).
9. Hot plate.
10. Coplin jar.
11. Cover slips.

#### 2.4.6 Removing the Hybridized FISH Probes

1. 2× SSC solution.
2. SSCT.
3. 70% formamide/2× SSC: mix 70 ml of formamide with 10 ml of 20× SSC and 20 ml of water. This buffer should be freshly made before use.
4. Water bath.
5. Coplin jar.

## 3 Methods

If not specified, all procedures are done at the room temperature (RT).

### 3.1 Slide Preparation

1. Draw a small circle (about 5 mm in diameter) on the slide using PAP pen. Air-dry (*see* **Note 2**).
2. Add 20 μl of Poly-L-lysine in the circle and incubate at RT for 45 min (add another 10 μl to the slide every 15 min to prevent slide from drying).
3. Remove excess of Poly-lysine left on the slide using tissue and air-dry. Slides can be stored in 4 °C for maximum 1 month.

### 3.2 Standard CTC Isolation

Blood is drawn into 10 ml K2EDTA vacutainers and kept at RT before use. Normally 7.5 ml of blood is drawn into each tube and the following protocol is for a 7.5 ml sample of whole blood, but it is possible to scale for different blood volumes. Changes should be applied if starting with a different blood volume (*see* Subheading 3.5). The blood samples are processed within 2 h after collection (*see* **Note 3**). All instruments and disposables should be decontaminated with 4% Virkon after use.

#### 3.2.1 Enrichment of Mononuclear Cells from the Whole Blood

1. Fill 50 ml LeucoSep tube with 15.3 ml of Ficoll-Paque PLUS and spin at 1000 × *g* for 30 s at RT. The separation medium is now located below the frit.

2. Prepare 10 ml of DPBS in a 50 ml conical tube. Wet the pipette using DPBS and add 1 ml of DPBS to the LeucoSep tube to cover the frit.
3. Turn the blood vacutainer up and down to mix the separated plasma and cells. Gently transfer 7 ml of whole blood to the LeucoSep tube along the tube wall, leaving 0.5 ml of whole blood in the EDTA tube (*see* **Note 4**). Add 6 ml of DPBS to the LeucoSep tube to dilute sample and rinse remaining blood in the pipette and blood on the tube wall.
4. Centrifuge the Leucosep tube for 15 min at 1000 × *g* without brakes. Prepare 10 ml of DPBS in a 50 ml new conical tube.
5. After spin, a layer of cells can be seen just above the frit (Fig. 1a). Decant all the supernatant containing cells from LeucoSep tube to a new 50 ml conical tube.
6. Rinse the LeucoSep tube with 5 ml of DPBS using 10 ml pipette and transfer it to the 50 ml conical tube with cells. Avoid suction of the Ficoll-Paque PLUS through the frit and avoid pressing the pipette against the frit.
7. Repeat **step 6** using the same 10 ml pipette. The LeucoSep tube can be discarded.
8. Centrifuge the conical tube with cells at 200 × *g* for 8 min at RT with the brakes set to ON (set deceleration at 5). The cells will be pelleted at the bottom of the conical tube (Fig. 1b).

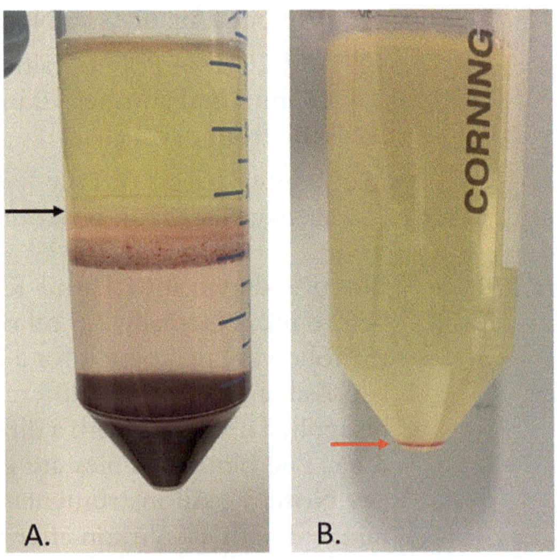


**Fig. 1** Mononuclear cells enrichment for CTC isolation. (**a**) Mononuclear cells in the whole blood are enriched and form a white layer above the frit in Leucosep tube (indicated by black arrow); (**b**) mononuclear cells are pelleted at the tube bottom and the pellet appears little red color (indicated by red arrow)

9. After centrifugation, use 25 ml pipette to gently aspirate off the supernatant as much as possible without disturbing the cell pellet. Use a P1000 pipette to remove the remaining supernatant if too much supernatant left (up to 0.5 ml can be left in the tube).
10. Prepare 4.5 ml of isolation buffer in a new conical tube. Add 200 μl of isolation buffer to the pellet and gently tap the tube on the bench clockwise for about 2 min to loosen the pellet until it is completely resuspended. Gently resuspend the cells with 200 μl low-binding tip and transfer the cells to the original EDTA vacutainer.
11. Add another 200 μl of isolation buffer to rinse the residual cells in the conical tube with the same low-binding tip and transfer the liquid to the original EDTA vacutainer.
12. Repeat **step 11**.
13. Rinse the residual cells with another 1 ml of isolation buffer using P1000 tip and transfer to the original EDTA vacutainer.
14. Repeat **step 13** twice. The final volume of liquid in the EDTA vacutainer should be around 4 ml. The sample is ready to be run on the Parsortix™ system.

*3.2.2 Isolation of CTCs Using Parsortix™ System*

1. Prime the system: select the program PX2_PF and run with isolation buffer. A new 6.5 μm cassette should be inserted to replace the cleaning cassette.
2. Isolate CTCs: select the program PX2_S99F. Follow the instruction on the machine and attach the EDTA vacutainer with cells from Subheading 3.2.1, **step 14** to the machine. During the isolation, shake the EDTA vacutainer every 10 min to avoid the cells precipitation.
3. Harvest CTCs: select the program PX2_H. Collect 200 μl of cell suspension in a 1.5 ml siliconized low-retention tube.
4. Clean the system: Insert cleaning cassette. Select the program PX2_CT or PX2_C and clean the machine by running DPBS as a sample. After the cleaning, the machine is ready for another run.

*3.2.3 Cell Transfer to Slide and Fixation*

1. Spin the harvested cell from Subheading 3.2.2, **step 3** at 1000 × *g* for 3 min at RT. Prepare 30 μl of 0.075 M KCL in an Eppendorf tube.
2. Gently take out the supernatant using P200 pipette without disturbing the pellet (*see* **Note 5**) and leave around 5–10 μl solution.
3. Add 10 μl of 0.075 M KCl into pellet using low-binding, wide-orifice tip. Gently click the bottom of tube to loosen the pellet. Pipette up and down gently using the low-binding, wide-

orifice tip to resuspend the cells and transfer the cells onto the PAP pen circled area on the prepared slide from Subheading 3.1, **step 3**.

4. Repeat **step 3** using the same tip and transfer all cell solution onto the slide.
5. Dry the slide using hair dryer with cool air settings. We use an in-house developed slide drier for this purpose.
6. Place the slide on ice. Carefully add 10 μl of ice-cold acetone to the cells on the slide. Fix the cells using ice-cold acetone for 18 min on ice. Add more ice-cold acetone as needed so the slide will not dry out during fixation. Keep acetone at −20 °C when not in use.
7. Air-dry the slide.
8. Store the slide at 4 °C. Fixed slides can be kept at 4 °C for 1 week.

### 3.3 Cell Characterization by Immunofluorescence Staining

1. Rehydrate the slide prepared as Subheading 3.2.3 with PBS for 2 min.
2. Wash the slide with PBS-T for 1 min.
3. Wash the slide with PBS for 2 min.
4. Add 30 μl of blocking solution onto the slide and incubate for 10 min at RT.
5. Rinse the slide with PBS twice and incubate with PBS for 2 min.
6. Add 40 μl of anti-CD45-PE (1:50 diluted in staining solution) onto the slide and incubate for 15 min at RT in dark.
7. Rinse with PBS three times and then further wash with PBS for 3 min three times (at least 10 min in total).
8. Add 30 μl of 0.1% Triton X-100 and incubate for 5 min at RT in dark.
9. Rinse the slide with PBS twice and incubate in PBS for 2 min in dark.
10. Incubate with 30 μl of anti-CK-FITC (1:10 diluted in staining solution) and anti-Vimentin- Alexa Fluor® 647 (1:75 diluted in staining solution) for 30 min at RT in dark.
11. After incubation, rinse the slide with PBS for three times and wash with PBS for three times, 3 min each time.
12. Air-dry the slide in dark.
13. Add 4 μl of mounting medium and cover with cover slip. The slide can be stored at 4 °C for future scan (*see* **Note 6**).
14. Scan the slides with Ariol scanning system using FITC, Cy3, and far-red channels and 20× optical zoom lens with seven Z stacks (Fig. 2).

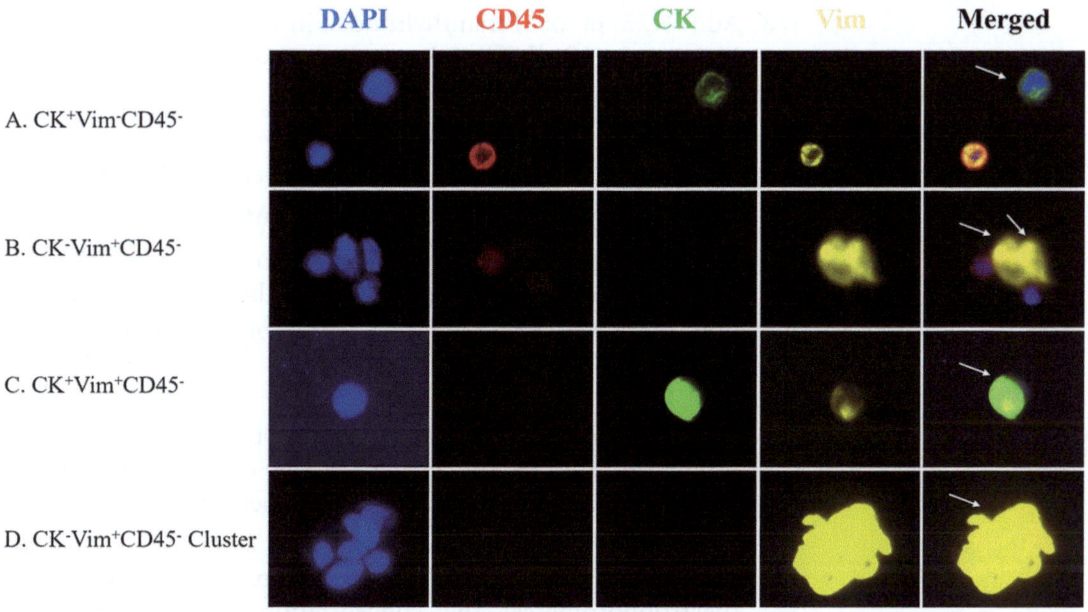


**Fig. 2** Representative immunofluorescence images for different types of CTCs in prostate cancer patients. (**a**) A representative of epithelial CTC (CK+Vim−CD45−) indicated by the arrow; (**b**) two representative mesenchymal CTCs (CK−Vim+CD45−) indicated by the arrow; (**c**) a representative of EMT CTC (CK+Vim+CD45−) indicated by the arrow; (**d**) a representative mesenchymal CTCs (CK−Vim+CD45−) cluster indicated by the arrow

### *3.4 FISH Analysis*

#### *3.4.1 Stripping off Immunofluorescence Signals*

1. Prepare 50 °C water bath in the fume hood containing a coplin jar with stripping buffer.
2. Wash the slide in stripping buffer at 50 °C for 20 min with some agitation during the wash.
3. Wash the slide in PBS at RT three times for 5 min.
4. Air-dry the slide.

#### *3.4.2 Pretreat the Slide for Probe Hybridization*

1. Incubate the slide in fixation solution for 10 min.
2. Dip the slide in 70% acetic acid for 10 min.
3. Wash the slide with PBS 3 × 5 min.
4. Dip the slide in 70% ethanol for 3 min.
5. Dip the slide in 90% ethanol for 3 min.
6. Dip the slide in 100% ethanol for 3 min.
7. Air-dry the slide.

#### *3.4.3 Probe Labeling*

1. Dilute 300 ng of probe DNA (*see* Chapter 16 for more information of probe DNA preparation) in water to the final volume of 24 μl. Add 20 μl of Random Primers Solution.
2. Denature the probes in the PCR thermocycler for 5 min at 95 °C, then cool them on ice for 10 min.

3. Add 1.75 μl of Teramthylrhodamin-5-D-UTP (Red dye)/ Fluorescein-12-dUTP (Green dye) and 5 μl of the special 10× dNTP.
4. Add 1 μl of Klenow Fragment.
5. Place the probes in the PCR thermocycler for 3 h at 37 °C.
6. After 3 h of labeling, add 5 μl of Stop buffer.
7. Prepare Microspin G50 columns: loosen the cup of the column, break off the bottom plug, place the column in a collection tube and spin at 3500 rpm (1400 × *g*) for 1 min. Wash the column twice with 200 μl of TE buffer at 3500 rpm (1400 × *g*) for 1 min.
8. Transfer the column to an Eppendorf tube, add the labeled probe solution to the center of the column and spin at 3500 rpm (1400 × *g*) for 2 min. The labeled probe is collected in the Eppendorf tube.
9. Check the concentration of the cleaned probes with Nanodrop spectrophotometer (the concentration is expected to be 90–150 ng/μl).
10. Precipitate the probe: add 30 μl of COT-1 DNA, 1 μl of salmon sperm, 4 μl of 4 M NaCl and 222 μl of ethanol to the probe. Mix well and put the probe in −80 °C for 20–30 min.
11. Centrifuge the probe at 13,000 rpm (19000 × *g*) for 20 min at 4 °C.
12. Remove the supernatant and rinse the pellet with 200 μl of fresh 80% ethanol. Spin at 13,000 rpm (19000 × *g*) for 10 min at 4 °C.
13. Remove the supernatant and air-dry the pellet.
14. Add 20 μl of nuclease-free water (final probe concentration should be about 200 ng/μl), mix well and place the samples on a heat-block at 50 °C for 5 min to completely resuspend the probe.
15. Store the probes at −20 °C.

#### 3.4.4 Probe Hybridization

1. Take out hybridization buffer and labeled FISH probes from −20 °C and keep them on ice.
2. Preheat the hybridization buffer on the heat-block for 5 min at 50 °C and vortex to ensure that all ingredients are completely dissolved. Place the humidity box in the 37 °C incubator.
3. Mix 1.5 μl (around 300 ng) of the labeled probe with 9.6 μl of hybridization buffer for each slide. Pipette the mixture up and down to ensure the probes are well resuspended in the buffer. Pipette the hybridization mix onto the slide and cover with 22 × 22 mm coverslip.

4. Seal the edges of the coverslip with rubber cement.
5. Put the slide on the hot plate at 98 °C for 10 min for denature.
6. Put the denatured slides into humidity box in 37 °C incubator for overnight hybridization. Make sure that the lid is closed properly and there is enough moisture to keep the slides humid during overnight incubation.

*3.4.5 Post-hybridization Washing*

1. In the fume hood prepare 42 °C water bath with two coplin jars of 50% formamide/2× SSC and one coplin jar of 2× SSC.
2. Take out the hybridized slide from humidity box and carefully remove the coverslip.
3. Rinse the slide in 42 °C 2× SSC for three times.
4. Wash the slide in 42 °C 50% formamide/2× SSC buffer for 4 min twice.
5. Wash the slide in SSCT for 3 min with agitation on the shaker.
6. Wash the slide in PBS for 5 min twice.
7. Dip the slide in 70% ethanol for 20 s twice to dehydrate.
8. Air-dry the slide.
9. Add 10 μl of VECTASHIELD Antifade Mounting Medium with DAPI on the slide. Cover it with cover slip. Keep in the fridge at 4 °C in dark. For long-term storage slides can be kept at −20 °C (Fig. 3).

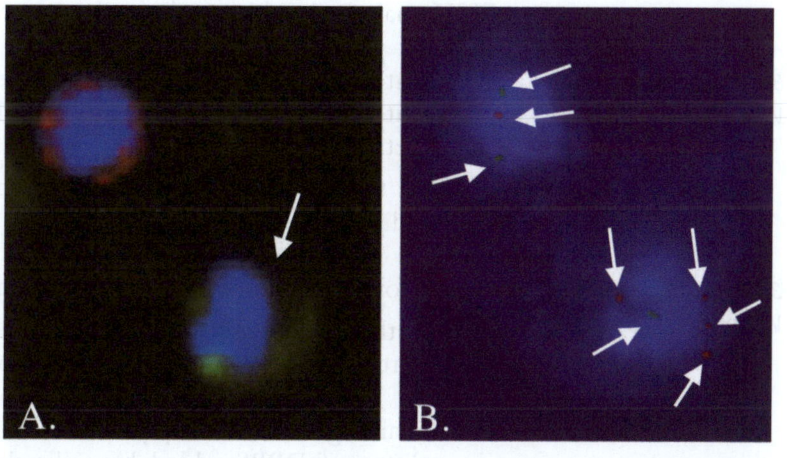


**Fig. 3** Representative FISH images of a prostate cancer epithelial (CK+Vim−CD45−) CTC post-immunofluorescence staining. (**a**) Immunofluorescence image of one epithelial (CK+Vim−CD45−) CTC (indicated by an arrow). CK is in green, Vim is in yellow, CD45 is in red, and DAPI is in blue; (**b**) FISH analysis by probes of androgen receptor (AR) (red) and FBXL4 (green) on the CTC showed in (**a**). Gain of AR (four copies) and loss of FBXL4 (one copy left) were shown on the CTC. The adjacent lymphocyte presents normal AR (one copy) and FBXL4 (two copies) status. FISH signals are indicated by arrows

#### *3.4.6 Removing the Hybridized Probes for FISH Rehybridization*

1. Prepare 68 °C water bath in the fume hood containing a coplin jar with 2× SSC solution and another coplin jar with 70% formamide/2× SSC buffer.
2. Meantime, place the slide with coverslips in SSCT buffer for 5 min at room temp to remove the cover slips.
3. Incubate the slide in 2× SSC, 2 × 10 min with agitation to remove mounting medium with DAPI.
4. To remove old FISH signal, incubate the slide at 68 °C in 2× SSC for 2 min, followed by 70% formamide/2× SSC buffer for 4 min to denature the slide and probe.
5. Wash the slide in 2× SSC for 2 min followed by two washes with water for 3 min.
6. Air-dry the slide. Slide is ready for hybridization with another set of FISH probes.

### 3.5 Scaling CTC Isolation Protocol for Different Blood Volumes

We suggest using 7.5 ml whole blood for standard CTC isolation, and the volume of blood should stay the same across different clinical cases to be comparable. However, it is possible to isolate CTCs from different volumes of blood. For situations where the starting volume is either higher or lower than 7.5 ml, the protocol for CTC isolation can be adapted using the following metrics. We do not recommend using this protocol for less than 3 ml of peripheral blood as CTCs appear at very low number in the blood. Small volume of blood may cause big variance. Regardless of starting blood volume, the process of immunofluorescence staining and FISH analysis should stay the same.

#### *3.5.1 Starting Blood Volume Smaller Than 5 ml*

1. Dilute the blood with isolation buffer at 1:1 ratio. Turn the vacutainer up and down to mix the blood and isolation buffer together.
2. Run the sample directly on the machine as indicated in Subheading 3.2.2.

#### *3.5.2 Starting Blood Volume from 5 to 10 ml*

1. Follow the protocol from Subheading 3.2.1, **steps 1** and **2**.
2. Gently transfer the whole blood to the LeucoSep tube along the tube wall, leaving 0.5 ml whole blood in the EDTA tube. Add DPBS to the LeucoSep tube to dilute sample and rinse remaining blood in the pipette and blood on the tube wall. The volume of DPBS added here should be 1 ml less than the volume of blood added to the LeucoSep tube.
3. The rest of the protocol is the same as Subheading 3.2.1 from **step 4**.

*3.5.3 Starting Blood Volume Bigger Than 10 ml*

1. When the starting blood volume is bigger than 10 ml, we suggest to equally separate the blood into two or more tubes (range from 5 to 10 ml per tube) and run the blood as independent samples as indicated in Subheading 3.5.2 to guarantee the sufficiency and quality of CTC isolation.

## 4 Notes

1. In our experience, these pre-conjugated antibodies work well on prostate cancer CTC slides. Other antibodies with different conjugates can also be used. However, antibodies should always be tested before use.
2. This size of the PAP pen circle should be modified based on the cell number put on the slide. The more cells being put on the slide, the bigger PAP pen circle is needed to avoid cell overlapping. The PAP pen circle should be totally dried as wet PAP pen can be washed off and cause dirty background after immunofluorescence staining.
3. The best recovery of CTCs and the best quality of immunofluorescence staining occurs in samples processed soon after blood collection. Storing the blood overnight before processing results in decreased quality of immunofluorescence signals on isolated cells.
4. 0.5 ml of whole blood is left in the vacutainer because we found that mixing some red blood cells with enriched monocytes would help samples to run through the system.
5. After the spin, the cells will be pelleted at the bottom of the low-retention tube. The pellet may be difficult to be observed as it is very small. Special care should be taken when taking out the supernatant.
6. The slide can be stored at 4 °C in the dark for few days, but the fluorescence will bleach overtime. To obtain best quality signal, it is advisable to scan the slides within 3 days after staining.

### References


1. Mateo J, Gerlinger M, Rodrigues DN, de Bono JS (2014) The promise of circulating tumor cell analysis in cancer management. Genome Biol 15(8):448. https://doi.org/10.1186/s13059-014-0448-5
2. Xu L, Shamash J, Lu YJ (2015) Circulating tumor cells: a window to understand cancer metastasis, monitor and fight against cancers. J Cancer Res Updates 4(1):13–29
3. Hayes D, Paoletti C (2013) Circulating tumour cells: insights into tumour heterogeneity. J Intern Med 274(2):137–143. https://doi.org/10.1111/joim.12047
4. Alix-Panabieres C, Pantel K (2014) OPINION challenges in circulating tumour cell research. Nat Rev Cancer 14(9):623–631. https://doi.org/10.1038/nrc3820
5. Yu M, Bardia A, Wittner B, Stott S, Smas M, Ting D et al (2013) Circulating breast tumor cells exhibit dynamic changes in epithelial and mesenchymal composition. Science 339

(6119):580–584. https://doi.org/10.1126/science.1228522

6. Yu M, Bardia A, Aceto N, Bersani F, Madden MW, Donaldson MC et al (2014) Cancer therapy. Ex vivo culture of circulating breast tumor cells for individualized testing of drug susceptibility. Science 345(6193):216–220. https://doi.org/10.1126/science.1253533
7. Hodgkinson CL, Morrow CJ, Li Y, Metcalf RL, Rothwell DG, Trapani F et al (2014) Tumorigenicity and genetic profiling of circulating tumor cells in small-cell lung cancer. Nat Med 20(8):897–903. https://doi.org/10.1038/nm.3600
8. Alix-Panabieres C, Bartkowiak K, Pantel K (2016) Functional studies on circulating and disseminated tumor cells in carcinoma patients. Mol Oncol 10(3):443–449. https://doi.org/10.1016/j.molonc.2016.01.004
9. Guo T, Wang CS, Wang W, Lu Y-j (2016) Culture of circulating tumor cells – holy grail and big challenge. Int J Cancer Clin Res 3:065
10. Xu L, Mao X, Imrali A, Syed F, Mutsvangwa K, Berney D et al (2015) Optimization and evaluation of a novel size based circulating tumor cell isolation system. PLoS One 10(9): e0138032. https://doi.org/10.1371/journal.pone.0138032
11. Xu L, Mao X, Guo T, Chan P, Shaw G, Hines J et al (2017) The novel association of circulating tumor cells and circulating megakaryocytes with prostate cancer prognosis. Clin Cancer Res 23(17):5112–5122. https://doi.org/10.1158/1078-0432.CCR-16-3081
12. Aceto N, Bardia A, Miyamoto D, Donaldson M, Wittner B, Spencer J et al (2014) Circulating tumor cell clusters are oligoclonal precursors of breast cancer metastasis. Cell 158(5):1110–1122. https://doi.org/10.1016/j.cell.2014.07.013
13. Fabisiewicz A, Grzybowska E (2017) CTC clusters in cancer progression and metastasis. Med Oncol 34(1). https://doi.org/10.1007/s12032-016-0875-0

Chapter 8

# The Isolation and Characterization of Circulating Tumor Cells from Head and Neck Cancer Patient Blood Samples Using Spiral Microfluidic Technology

Arutha Kulasinghe, Majid Ebrahimi Warkiani, and Chamindie Punyadeera


## Abstract

Metastasis is responsible for 90% of cancer-related deaths. The study of circulating tumor cells (CTCs) enables the study of the units of disease responsible for the process of metastasis. While the biology of the primary tissue is relatively known, little is understood about the cells en route to distant sites. Here we describe the isolation of CTCs using the spiral microfluidic technology for the efficient sorting of CTCs from head and neck cancer (HNC) patient blood samples. Furthermore, the molecular characterization of CTCs can aid in stratifying patients for targeted therapy such as immunotherapy, which is having a profound impact in the treatment of metastatic HNC.


**Key words** Circulating tumor cells, Head and neck cancers, Distant metastasis, Spiral microfluidic technology, Immunotherapy

## 1 Introduction

Circulating tumor cells (CTCs) have shown to be a good biomarker for cancer diagnosis and prognosis in recent clinical trials. CTCs remain as few cells, typically 1–10 CTCs/ml in a background of billions of normal blood cells. Therefore, extremely sensitive technologies are required for the efficient depletion of the background cells. The FDA-approved CellSearch platform (Menari Silicon Biosystems) pre-selects for epithelial tumor cells (e.g. EpCAM) by positive selection and further characterizes cells for pan-cytokeratin (epithelial marker), CD45 (common leukocyte marker), and DAPI (nuclear marker). Enumeration of EpCAM-positive CTCs have demonstrated to be of clinical significance in a number of tumor types including breast, prostate, and colorectal cancers where CTC thresholds alone have been associated with progression-free survival (PFS) and overall survival (OS) [1]. However, the field has made a shift away from epitope-dependent

Jyotsna Batra and Srilakshmi Srinivasan (eds.), *Theranostics: Methods and Protocols*, Methods in Molecular Biology, vol. 2054, https://doi.org/10.1007/978-1-4939-9769-5_8, 

platforms to capture non-epithelial tumor cells in an attempt to capture a greater proportion of CTCs [2, 3] and to further utilize CTCs for functional characterization [4]. Microfluidic technologies have come to the fore over the past 5 years, where CTC enrichment has been achieved based as a function of cell size and deformability [5–9]. Such technologies include the spiral chip (commercialized by ClearBridge Biomedics) [10], Vortex (Vortex BioSciences) [11], Pastotix (Angle PLC) [12], and the Massachusetts General Hospital developed CTC-iCHIP [13] to name a few. In this study, we describe a spiral microfluidic CTC enrichment technology, which has been used by our team for HNC clinical applications [5, 14, 15].

The ability to capture intact CTCs enables the expansion of these rare cells outside a patient's body for the testing of therapies or deriving in vivo/ex vivo cell lines for the study of transient metastatic disease [4, 16–18]. Recent reports have demonstrated the presence of immunotherapeutic targets on CTCs, which can potentially be used to identify patients for highly targeted therapy—a current unmet clinical need [15, 19, 20].

## 2 Materials

### 2.1 Slides

1. Superfrost™ Plus (ThermoFisher Scientific).
2. Super PAP Pen (ThermoFisher Scientific).

### 2.2 Blood Collection Vacutainers

1. 9 ml K2E EDTA Vacutainer (BD).
2. 10 ml Streck tubes (Streck Inc).

### 2.3 CTC Enrichment

1. Cell Fixation buffer

   16% Formaldehyde (w/v) Methanol-free (ThermoFisher Scientific) is used to prepare fresh 4% paraformaldehyde by diluting in 1× PBS. This is stored at 4 °C.

### 2.4 Sheath Buffer

1. 50 ml of Sheath buffer consisting of 1× PBS, 2 nM EDTA, 0.5% BSA should be prepared and stored at 4 °C for a maximum of 2 weeks.

### 2.5 Red Blood Cell Lysis Buffer

1. This buffer is sourced through Astral Scientifics (Catalogue number 786–849) (*see* **Note 1**).

### 2.6 Spiral CTC Chip Setup

1. Tygon tubing is used to connect the inlet of the spiral chip to a syringe pump. The outlets of the spiral chip are connected to two sterile collection tubes (CTC outlet and waste outlet). The lengths of the outlet tubing are identical (10 cm). The chip is setup such that it is in a fixed position on an upright microscope to observe the bifurcation point near the outlets (*see* **Note 2**).

### 2.7 Phenotyping of CTCs

1. Phosphate Buffered Saline (PBS).
2. CellSearch Staining Reagent (Menari Silicon Biosystems) (*see* **Note 3**).
3. CellSearch Permeabilization buffer (Menari Silicon Biosystems).
4. CellSearch fixation buffer (Menari Silicon Biosystems).
5. CellSearch nuclear dye (Menari Silicon Biosystems).
6. Mounting medium—Prolong gold antifade (Invitrogen).
7. Superfrost Plus slides (ThermoFisher Scientific) and Coverslips.
8. Zeiss Axio 2 Upright Research Microscope.

### 2.8 EGFR DNA FISH

1. 70% ethanol.
2. 85% ethanol.
3. 96% ethanol.
4. 100% ethanol.
5. Phosphate buffered Saline (PBS).
6. Coplin Jar.
7. EGFR DNA FISH Probe (DakoCytomation).
8. DAKO hybridizer for Molecular Pathology (DakoCytomation).
9. DAKO histology FISH Accessory Kit (DakoCytomation) containing pretreatment buffer, wash buffer, stringent wash buffers and mounting media containing DAPI.
10. Humidity box.
11. Rubber sealant.
12. Water Bath.

## 3 Methods

### 3.1 Preparation of the Blood Samples Prior to the Spiral Microfluidic Chip

1. The red blood cell lysis buffer (stored at 4 °C—*see* **Note 4**) is combined with the patient's whole blood in a 50 ml Falcon (Corning) tube in a 3:1 ratio respectively.
2. The Falcon tube is inverted twice and left to stand (*see* **Note 5**). After approximately 12–15 min (*see* **Note 6**), centrifuge the blood sample at 500 × *g* for 15 min at room temperature to pellet the cells.
3. After centrifugation, remove the supernatant using a vacuum line (*see* **Note 7**). Resuspend the cell pellet in the sheath buffer to a total volume of approximately 10 ml (*see* **Note 8**).

### 3.2 Priming of the Spiral Chip

1. A priming run is performed prior to running of the clinical sample to ensure that the internals of the spiral chip are working and to prevent any bubbles from disrupting the flow within the chip. The initial priming run using the sheath buffer alone at a flow rate of 1.7 ml/min for 10–15 min.
2. If sterile conditions are required such as for CTC culture, an initial sterilization can be performed by running 70% (vol/vol) ethanol in sterile deionized water into a syringe and pushing this through the chip at 1.7 ml/min followed by the sheath buffer.

### 3.3 Processing of Blood Samples

1. Once primed, the clinical sample is loaded onto a new 10 ml syringe and carefully loaded onto the syringe pump (*see* **Note 9**) (as shown in Fig. 1). The sample is pumped through the spiral chip at a rate of 1.7 ml/min until the run is complete (*see* **Note 10**).
2. The outputs are collected and spun down at 300 × *g* for 5 min. The supernatant is discarded, and the pellet resuspended in 1 ml of 4% Paraformaldehyde.

### 3.4 Transfer of Enriched Cells onto Glass Slides

1. The CTC fraction is then transferred onto coated glass slides for phenotyping. This is achieved by cyto-centrifugation by a Shandon CytoCentrifuge 4 (ThermoFisher Scientific). The samples are loaded at a maximum of 200 μl per cytospot and centrifuged at 300 × *g* for 5 min (*see* **Note 11**).
2. Once the cytospins are prepared and clips carefully removed, the slides are allowed to air-dry for approximately 20 min at room temperature.
3. After air-drying, the cytospots are fixed with freshly prepared 4% Paraformaldehyde (ThermoFisher Scientific) for 10 min at room temperature. After fixation, the slides are washed with 1× PBS (three times) and stored short term at −20 °C or long term at −80 °C in long term storage boxes.

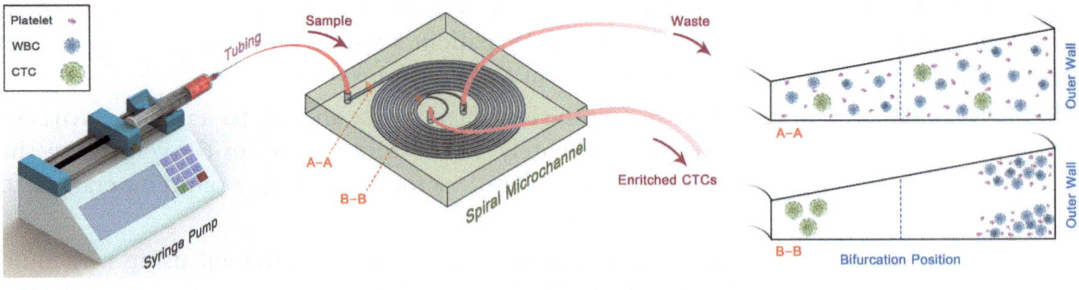


**Fig. 1** Spiral CTC chip assembly for the enrichment of circulating tumor cells from lysed blood samples. CTCs are separated as a function of cell size and deformability. (Reproduced from Kulasinghe et al., Scientific Reports 2017)

### 3.5 CTC Characterization

1. CTCs are characterized using the standard antibody staining cocktail from CellSearch (Menari Silicon Biosystems) consisting of a fixation buffer, permeabilization buffer, staining reagent, and nuclear stain.
2. The antibody cocktail stains for anti-pan-cytokeratin (CK-8,18,19) and common leukocyte marker (CD45). The cytospots are incubated in dark humid chambers with the staining cocktail at 4 °C overnight.
3. Post incubation, the cells are washed three times with 1× PBS and left to air-dry completely. Once air-dried, the cytospots are incubated with Prolong Gold (Invitrogen) to preserve the fluorescent signal, cover slipped and stored in the dark before imaging.

### 3.6 CTC Identification

1. The slides are scanned and imaged using an epifluorescent microscope (Zeiss Axio Imager 2) in the channels for PE/APC/DAPI.
2. CTCs are identified as intact, cytokeratin-positive, DAPI-positive, CD45-negaitve cells which were larger than 4 μm.
3. Cells staining positive for CD45 and DAPI and negative for pan-cytokeratin are determined as white blood cells.

### 3.7 DNA Fluorescence In Situ Hybridization

#### 3.7.1 Pretreatment of Slides

Slides containing CTC-enriched cells (*see* **Note 12**) are placed are in coplin jars containing pretreatment solution (DakoCytomation) at 98 °C for 10 min and digested with pepsin for at 37 °C for 5 min (DakoCytomation). The slides are then dipped through a graded ethanol series for 2 min intervals each (*see* **Note 13**) and allowed to air-dry.

#### 3.7.2 Probe Hybridization

Following dehydration, dual color, dual target EGFR DNA FISH assays can be performed using EGFR/CEN-7 FISH probe mix (DakoCytomation). The FISH probes should be stored at −20 °C and kept on ice prior to hybrization. The cytospots are covered with 10 μl of probe solution (*see* **Note 14**) and sealed with a coverslip and rubber sealant/cement and placed in the DAKO hybridizer for denaturation (82 °C for 5 min) followed by hybridization (45 °C for 20 h).

#### 3.7.3 Post-hybridization Washes

Stringent washes (DakoCytomation) are performed post hybridization (*see* **Note 15**) in coplin jars. The initial stringent wash is performed in a fume hood at room temperature for 2–3 min followed by 65 °C for 10 min in a water bath. This is followed by two wash steps in wash buffer (Dakocytomation) for 3 min each. The slides are then dipped in a graded ethanol series (70%, 85%, 96%) for 2 min each prior to air-drying. Mounting media containing DAPI (DakoCytomation) is finally added and cover slipped before storing in the dark.

*3.7.4 Imaging of Labeled Cells*

The slides are then visualized using an epifluorescent microscope (Zeiss Axio Imager 2). EGFR is visualized as red signal (tetramethylrhodamine isothiocyanate filter) and CEP7 as a green signal (fluorescein isothiocyanate filter) and the nuclei as blue signal with a DAPI filter. EGFR status is scored per nucleated cell as the number of EGFR signals to that of CEP7 signal.

## 4 Notes

1. This is a necessary step to reduce the cellular components passing through the microfluidic chip and has been demonstrated to have minimal effect on the enrichment process [7].
2. The spiral chip walls should be visually observed under the microscope and the bifurcation point observed continuously during the sample run to ensure that there is no cellular debris/air bubbles blocking the separation of cells. Increasing the flow rate of the sheath buffer can aid in removing residual bubbles within the chip.
3. CellSearch staining reagent (Menari Silicon Biosystems) is used for the characterization of CTCs. CTCs are defined by the CellSearch criteria as cells which are nucleated, EpCAM, Cytokeratin-8,18,19 positive and CD45 negative. CTCs are intact cells with >4 μm diameter. Further molecular confirmation of CTCs is possible using DNA fluorescence in-situ hybridization (FISH) for molecular alterations shared with the primary tumor. In head and neck cancers, EGFR is commonly amplified [21]. The staining reagent consists of 0.0006% mouse monoclonal antibodies specific to cytokeratins conjugated to phycoerythrin (PE), 0.0012% mouse anti-CD45 monoclonal antibody conjugated to allophycocyanin (APC) in buffer containing 0.5% BSA and 0.1% sodium azide. For further characterization of CTCs, EGFR DNA FISH may be used to determine molecular alterations in head and neck cancer.
4. The red blood cell lysis buffer should be brought to room temperature before processing the sample.
5. The blood sample can be placed on a gentle roller for efficient lysis.
6. The blood sample should turn “wine-red” upon complete lysis.
7. Do not disturb the cell pellet at this crucial step.
8. The resuspension procedure should be done by gently pipetting up and down to preserve the cells.

9. Remove any bubbles in the syringe by carefully loading the sample into the syringe and expelling any residual air/liquid bubbles by gently pushing the syringe in.
10. For further CTC purity, a second round of enrichment is possible by collecting the CTC output, topping up the volume to 10 ml with sheath buffer and re-running the sample. This has shown to reduce the contaminating white blood cells in the CTC outlet by a 3 log manner for downstream genetic analysis [22]. Large CTC clusters or microemboli may cause clogging issues during the running of the sample. Should this occur, stop the syringe pumps immediately, and increase the flow rate of the sheath buffer to clean the chip or alternatively use a new chip.
11. Pre-wetting of the filter paper of the cytofunnels with 1× PBS ensures the transfer of cells directly onto the slide with the least amount of absorption through the filter paper.
12. Cytospots should be demarcated using the Super PAP pen.
13. Ethanol series (70%, 85%, 96%) should be prepared fresh for each round of FISH.
14. Perform this step in a fume cabinet.
15. Gently remove the rubber sealant so as not to disturb the cytospot.

## Acknowledgments

This chapter was supported by the Translational Research Institute (TRI) Spore Grant.

### References

1. Cristofanilli M, Budd GT, Ellis MJ, Stopeck A, Matera J, Miller MC et al (2004) Circulating tumor cells, disease progression, and survival in metastatic breast cancer. N Engl J Med 351 (8):781–791. https://doi.org/10.1056/NEJMoa040766
2. Yu M, Bardia A, Wittner BS, Stott SL, Smas ME, Ting DT et al (2013) Circulating breast tumor cells exhibit dynamic changes in epithelial and mesenchymal composition. Science (New York, NY) 339(6119):580–584. https://doi.org/10.1126/science.1228522
3. Kulasinghe A, Perry C, Jovanovic L, Nelson C, Punyadeera C (2015) Circulating tumour cells in metastatic head and neck cancers. Int J Cancer 136(11):2515–2523. https://doi.org/10.1002/ijc.29108
4. Khoo BL, Grenci G, Jing T, Lim YB, Lee SC, Thiery JP et al (2016) Liquid biopsy and therapeutic response: circulating tumor cell cultures for evaluation of anticancer treatment. Sci Adv 2(7):e1600274. https://doi.org/10.1126/sciadv.1600274
5. Kulasinghe ATT, Blick T, O'Byrne K, Thompson EW, Warkiani ME, Nelson C, Kenny L, Punyadeera C (2017) Enrichment of circulating head and neck tumour cells using spiral microfluidic technology. Sci Rep 7:42517
6. Fachin F, Spuhler P, Martel-Foley JM, Edd JF, Barber TA, Walsh J et al (2017) Monolithic chip for high-throughput blood cell depletion to sort rare circulating tumor cells. Sci Rep 7 (1):10936. https://doi.org/10.1038/s41598-017-11119-x

7. Warkiani ME, Khoo BL, Wu L, Tay AKP, Bhagat AAS, Han J et al (2016) Ultra-fast, label-free isolation of circulating tumor cells from blood using spiral microfluidics. Nat Protoc 11(1):134–148. https://doi.org/10.1038/nprot.2016.003. http://www.nature.com/nprot/journal/v11/n1/abs/nprot.2016.003.html#supplementary-information
8. Warkiani ME, Guan G, Luan KB, Lee WC, Bhagat AA, Chaudhuri PK et al (2014) Slanted spiral microfluidics for the ultra-fast, label-free isolation of circulating tumor cells. Lab Chip 14(1):128–137. https://doi.org/10.1039/c3lc50617g
9. Warkiani ME, Khoo BL, Tan DS, Bhagat AA, Lim WT, Yap YS et al (2014) An ultra-high-throughput spiral microfluidic biochip for the enrichment of circulating tumor cells. Analyst 139(13):3245–3255. https://doi.org/10.1039/c4an00355a
10. Chudasama DY, Freydina DV, Freidin MB, Leung M, Montero Fernandez A, Rice A et al (2016) Inertia based microfluidic capture and characterisation of circulating tumour cells for the diagnosis of lung cancer. Ann Transl Med 4 (24):480. https://doi.org/10.21037/atm.2016.12.28
11. Sollier E, Go DE, Che J, Gossett DR, O'Byrne S, Weaver WM et al (2014) Size-selective collection of circulating tumor cells using Vortex technology. Lab Chip 14(1):63–77. https://doi.org/10.1039/c3lc50689d
12. Xu L, Mao X, Imrali A, Syed F, Mutsvangwa K, Berney D et al (2015) Optimization and evaluation of a novel size based circulating tumor cell isolation system. PLoS One 10(9):e0138032. https://doi.org/10.1371/journal.pone.0138032
13. Karabacak NM, Spuhler PS, Fachin F, Lim EJ, Pai V, Ozkumur E et al (2014) Microfluidic, marker-free isolation of circulating tumor cells from blood samples. Nat Protoc 9 (3):694–710. https://doi.org/10.1038/nprot.2014.044
14. Kulasinghe A, Schmidt H, Perry C, Whitfield B, Kenny L, Nelson C et al (2018) A collective route to head and neck cancer metastasis. Sci Rep 8(1):746. https://doi.org/10.1038/s41598-017-19117-9
15. Kulasinghe A, Kenny L, Punyadeera C (2017) Circulating tumour cell PD-L1 test for head and neck cancers. Oral Oncol 75:6–7. https://doi.org/10.1016/j.oraloncology.2017.10.011
16. Dive C, Brady G (2017) SnapShot: circulating tumor cells. Cell 168(4):742–742.e741. https://doi.org/10.1016/j.cell.2017.01.026
17. Carter L, Rothwell DG, Mesquita B, Smowton C, Leong HS, Fernandez-Gutierrez F et al (2017) Molecular analysis of circulating tumor cells identifies distinct copy-number profiles in patients with chemosensitive and chemorefractory small-cell lung cancer. Nat Med 23(1):114–119. https://doi.org/10.1038/nm.4239
18. Gao D, Vela I, Sboner A, Iaquinta PJ, Karthaus WR, Gopalan A et al (2014) Organoid cultures derived from patients with advanced prostate cancer. Cell 159(1):176–187. https://doi.org/10.1016/j.cell.2014.08.016
19. Mazel M, Jacot W, Pantel K, Bartkowiak K, Topart D, Cayrefourcq L et al (2015) Frequent expression of PD-L1 on circulating breast cancer cells. Mol Oncol 9(9):1773–1782. https://doi.org/10.1016/j.molonc.2015.05.009
20. Strati A, Koutsodontis G, Papaxoinis G, Angelidis I, Zavridou M, Economopoulou P et al (2017) Prognostic significance of PD-L1 expression on circulating tumor cells in patients with head and neck squamous cell carcinoma. Ann Oncol 28(8):1923–1933. https://doi.org/10.1093/annonc/mdx206
21. TCGA (2015) Comprehensive genomic characterization of head and neck squamous cell carcinomas. Nature 517(7536):576–582. https://doi.org/10.1038/nature14129
22. Aya-Bonilla CA, Marsavela G, Freeman JB, Lomma C, Frank MH, Khattak MA et al (2017) Isolation and detection of circulating tumour cells from metastatic melanoma patients using a slanted spiral microfluidic device. Oncotarget 8(40):67355–67368. https://doi.org/10.18632/oncotarget.18641

# Chapter 9

# Pharmacogenetics: Role of Single Nucleotide Polymorphisms

**Emrah Yucesan and Nur Ozten**


## Abstract

Genome sequencing methods have basically similar algorithms, although they show a few differences between the platforms. The human genome contains approximately three billion base pairs, and this amount is huge and therefore impossible to sequence at one step. However, this amount is not a problem for developed technology. Researchers break DNA into small random pieces and then sequence and reassemble. Library preparation, sequencing, bioinformatic approaches and reporting. High-quality library preparation is critical and the most important part of the next-generation sequencing workflow. Successful sequencing directly requires high-quality libraries. Sequencing is second step and all high-throughput sequencing approaches are generally based on conventional Sanger sequencing. After preparation of library and sequencing, later steps are completely computer-based (in silico) approaches called as bioinformatics.


**Key words** Pharmacogenetics, Genetic variants, SNPs

## 1 Introduction

Pharmacogenomics, the study of how genetic makeup determines the response to a therapeutic intervention or the effect of genetic variation on the efficacy and toxicity of drugs, is the most promising science for the pharmaceutical industry to emerge in the post-genomic era [1]. The term *pharmacogenetics* which is used more generally to describe the effect of genetic factors on drug response is often substituted by pharmacogenomics [2]. The vision of a "personalized" medicine and the development of prescriptions with a personal touch have been in the attention of the science world lately [3]. Progress by the development of cloning and high-throughput sequencing technologies, the availability of sequence of the human genome, and eventually, access to all human genes and their regulators, transcripts, and proteins as the basis for disease gene and drug target discovery. With defined reference sequences of genes and genomes, sequence comparisons within and between species became feasible and, consequently, the identification of

Jyotsna Batra and Srilakshmi Srinivasan (eds.), *Theranostics: Methods and Protocols*, Methods in Molecular Biology, vol. 2054, https://doi.org/10.1007/978-1-4939-9769-5_9, © Springer Science+Business Media, LLC, part of Springer Nature 2019

differences in DNA sequence, so-called *single nucleotide polymorphisms (SNPs)* [4]. Variations of drug transportation genes cause different responses mostly in the drug metabolizing enzymes (DMEs) or cytochrome P450 enzymes [1]. The response of drug therapy is not predictable because of single nucleotide polymorphism effects yielding beneficial to serious adverse effects [5]. Cytochrome P450 (CYP) enzymes are key players in the formation of inactive metabolites of drugs and other xenobiotics [6]. Genetic variants in the DNA of several phase I and II enzymes and transporters are involved in the activation and detoxification of drugs and other xenobiotics by predicting a part of the drug-related toxicity and efficacy of treatment, which has been shown in retrospective studies, prospective trials, and meta-analyses as well [6].

Enormous growth and development of high-throughput sequencing tools, particularly in the last two decades, has revealed the annotate of the genomic data [7]. Millions of variants have been identified throughout sequencing of entire human genome which is whole genome sequencing (WGS) [8]. Another sequencing approach is whole exome sequencing (WES). In this approach solely exons are sequenced (also exon-intron boundaries). WES is especially effective in the Mendelian diseases [9, 10]. In addition, amplicon or targeted sequencing is another approach that enables to analyze variants in the specific genomic regions [11]. Furthermore RNA sequencing enables to detect the expression levels of entire genome for different tissues, this is very useful particularly for cancer samples [12]. Recently several sequencing approaches are carried out at different labs. Such as WGS, WES, targeted sequencing, ChIP seq, RNA sequencing, small RNA-seq., single-cell DNA/RNA seq.

Variants which detect using different sequencing approaches indicate their clinical significance, predisposition, and/or drug resistance. Candidate variants are analyzed according to population data and minor allele frequency (MAF) values for specific variants. After that in silico prediction tools are used to detect variants effect (Table 1). In these analyses, which are performed entirely in silico, the pathogenicity of the resulting variation is determined by taking into account such conditions as amino acid exchange (hydrophobic, hydrophilic, acidic, basic, etc.) and evolutionary conservation of the sequence corresponding to the reference genome [13].

At present, genomic data either healthy populations or specific diseases are publicly available, open access and free (http://www.internationalgenome.org).

## 2 Materials and Methods

Genome sequencing methods have basically similar algorithms, although they show a few differences between the platforms. The human genome contains approximately three billion base pairs; this

**Table 1**
**Open access in slico prediction tools**

| | |
|---|---|
| Align-GVGD | http://agvgd.hci.utah.edu |
| SIFT | http://sift.jcvi.org |
| MutationTaster | http://www.mutationtaster.org |
| PolyPhen-2 | http://genetics.bwh.harvard.edu/pph2/ |

amount is huge and therefore impossible to sequence at one step [14]. However, this amount is not a problem for developed technology. Researchers break DNA into small random pieces and then sequence and reassemble. Furthermore, they align sequences and annotate as a final step.

### 2.1 Major Steps for NGS and Analysis

Library preparation, sequencing, bioinformatic approaches, and reporting.

1. High-quality library preparation is the critical and the most important part of the next-generation sequencing workflow (Fig. 1). Successful sequencing directly requires high-quality libraries. Recently, several library preparation kits are developed by companies such as Agilent, Bio Scientific, Kapa Biosystems, New England Biolabs, Illumina, Life Technologies, Pacific Biosciences and Roche and are available commercially [15]. Today different samples may be sequenced using kits, e.g., cell lines, fresh tissue, blood, and formalin-fixed paraffin-embedded (FFPE) samples. Library is prepared by fragmenting a genomic DNA or RNA and ligating specialized adapters followed by PCR amplification and sequencing [16, 17].
2. Sequencing is the second step, and all high-throughput sequencing approaches are generally based on conventional Sanger sequencing, which uses sequencing reagents, including florescent-labeled nucleotides. However, technologies have different methods such as paired-end, single-end, and mate pair. Same as preparation kits, there are different types of sequencing platforms developed by companies [18, 19].
3. After preparation of library and sequencing, later steps are completely computer-based (in silico) approaches called as bioinformatics. Following sequencing, genome assembly is performed in order to get entire genome fragmented where different sized fragments are aligned [20].
4. After NGS approaches, detected variants should be addressed to known clinical conditions (Fig. 1).

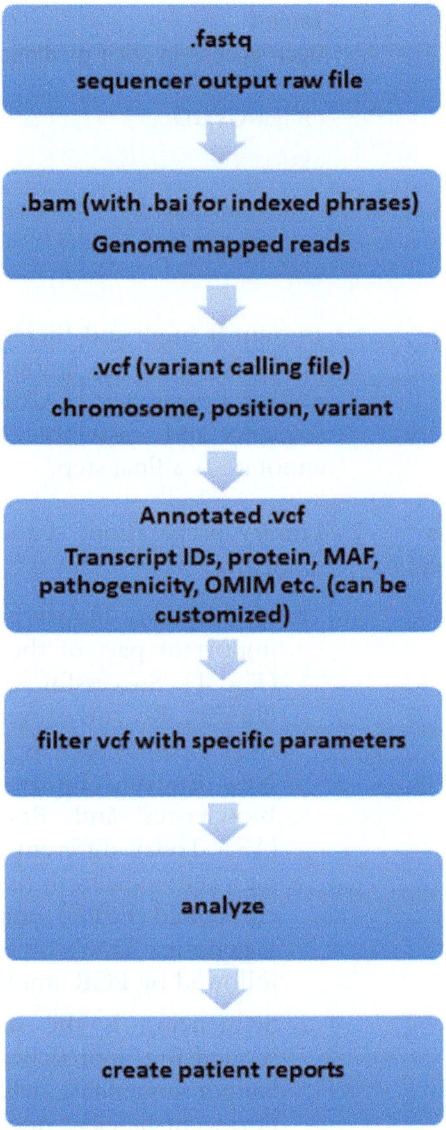


**Fig. 1** Bioinformatics workflow from .vcf file to clinical report

### 2.2 SNP, LD, and Arrays

Single nucleotide polymorphisms, frequently called SNPs, are the most common type of genetic variations. The entire human genome has almost ten million SNPs. Most SNPs have no direct effect for humans but some may be related with a specific clinical condition [21]. SNPs may be used to detect the causative genes, and follow the inheritance of disease genes within families [22]. SNPs with known specific properties that may help assume a drug response and, susceptibility to exogenic factors, e.g., toxins [1]. Some SNPs increase the potential risk for the specific disease for the specific population [23, 24]. Previously SNPs were used especially for candidate gene or linkage disequilibrium

(LD) analysis. LD allows a comprehensive genome-wide analysis; however, there are still challenges and limitations [25, 26].

The availability of numerous genetic markers, including restriction fragment length polymorphisms (RFLPs), variable number tandem repeats (VNTRs), microsatellites, and SNPs together with databases that store and distribute that information has made linkage analysis more feasible [27]. Among them SNPs are more convenient for the genotyping studies. Commonly used SNP panels varying from 10K (10,000 SNPs) to 1000K (1,000,000 SNPs) can interrogate many SNP loci in a single chip experiment and allow genome resolution as low as 10 kilo bases (kb) [28].

Various computer programs have been developed and distributed free of charge for the application of statistical tests for linkage analysis. There is also much effort in creating user-friendly software platforms, where various linkage programs are integrated with standardized input data formats and results are presented in graphic outputs. ALOHOMORA specially designed for SNP panels and easy LINKAGE applicable both to SNP and microsatellite panels are some software tools widely used for linkage analysis [29, 30].

### 2.3 Array Platforms Using SNPs

Array platforms are used for several scientific aims and include two main groups, microarray and macroarray. In this text's content, we will focus on microarray and its applications in terms of pharmacogenomics. Firstly, three microarrays are used, expression arrays, SNP arrays, methylation arrays. Expression arrays are used to detect the expression levels of large numbers of genes simultaneously or to genotype multiple regions of a genome. This is crucial particularly for cancer and drug resistance studies [31].

### 2.4 Expression Arrays

Expression arrays are valuable and widely used tools for biomedical research.

This technique provides a comprehensive view of gene activity, interaction between genes, detecting pathways in specific tissues or other samples.

Commercial arrays attempt to detect the expression level of all of the genes in the genome in one step. To evaluate the expression level, array platform uses a collection of DNA spots attached to a solid surface. Each spot contains a known and specific DNA sequence called probes. When DNA sample added into the spot, it binds highly condense with probe. So the main principle is hybridization between two DNA stands. All procedures are similar to the old blotting approach, which is called northern blot, i.e. hybridization. Although different labels feature (fluorophore, chemiluminescence, etc.), probe-target hybridization may detect and quantify the exact expression values [32]. Expression array technique can be used to find new therapeutic targets in pharmacogenomic studies [33]. Serial analysis of gene expression (SAGE) and microarray are the two important high-throughput methods for analysis of gene expression, in a single hybridization reaction. In terms of reflecting the function, phenotype, or responses to the environmental stimuli, the complete set of transcripts expressed by a cell in a population of cells is called transcriptome [34]. As single-nucleotide-polymorphism (SNP) genotyping platforms microarrays have been widely used. Although different alternative approaches have been used to detect SNPs, the most commonly used are allele discrimination by hybridization as used by Affymetrix [35], allele-specific extension, and ligation to a "bar-code" oligo which is hybridized to a universal array (the Illumina "Golden Gate Assay" [36], or in which the arrayed DNA is extended across the SNP in a single nucleotide extension reaction [37], or the Infinium Assay of Illumina [38]. DNA microarrays technique is based on the specificity of base pairing and binding of complementary strands of nucleic acids, where the homology (similarity) of two different nucleic acids was identified by Sol Spiegelman. This method was called "hybridization" of nucleic acids. Afterwards, the DNA microarray also called microarray dotted an array of DNA copies (cDNAs) corresponding to a large number of different mRNAs of known sequence onto a glass slide. The cDNAs were double-stranded and could be denatured into single strands, which could then be used to bind, or hybridize, to fluorescently labeled nucleic acid samples from cancerous or normal cells. By washing, the unbound molecules were taken away and fluorescent bound nucleic acid samples were identified by laser microscopy. Differences in microarray patterns between normal and cancerous cells could be quickly identified by fluorescent-indicated expressed genes dots [39].

### 2.5 SNP Arrays

SNP array is a type of DNA microarray which is used to detect polymorphisms within a population. As known specific polymorphisms are particularly important for drug response, SNP array application is crucial for pharmacogenomics studies. SNP array, as a hybridization method, is similar to the expression array, but remember, different genes expression levels may be different in different tissues because of gene regulation (switch on-off mechanism). However SNP arrays, which show comprehensive genomic profile using SNPs, are the same in every type of tissue in normal conditions (except mosaicism) [40].

### 2.6 Methylation Arrays

DNA methylation plays a crucial and dynamic role in the regulation gene expression, which is also known epigenetic regulation. Changes in the methylation pattern and level have been shown to contribute to cancer and also drug response. This array is used only to analyze genome-wide CpG islands. CpG islands are very important in terms of pharmacogenomics, for instance, hypermethylation at the CpG islands of a tumor suppressor gene, which leads to its silencing. Epigenetic changes play a significant role in tissue level, therefore it may affect drug response. The first step of methylation array is the same as methylation-specific PCR, which is bisulfite conversion and, then the hybridization process is the same as other array methods [41, 42].

### 2.7 Drug Sensitivity

New approaches to drug development and therapeutic efficacy are identified by microarrays. Many genes in one cell affect its response to therapeutic agents, regarding to drug exclusion, resistance, metabolism, DNA repair, etc. The expression of these genes may allow for the characterization of each cell, tissue, and tumor in regards to the mechanism of drug action, its predicted drug sensitivity, and side effects. The expression of particular gene patterns in particular tissues, as for example, cathepsin K expression in osteoclasts or inflammation-associated genes in rheumatoid tissue, may identify possible treatment strategies [43].

## References


1. Alwi ZB (2005) The use of SNPs in pharmacogenomics studies. Malays J Med Sci 12 (2):4–12
2. Peet NP, Bey P (2001) Pharmacogenomics: challenges and opportunities. Drug Discov Today 6(10):495–498
3. Roses AD (2000) Pharmacogenetics and the practice of medicine. Nature 405 (6788):857–865. https://doi.org/10.1038/35015728
4. Hoehe MR, Kroslak T (2004) Genetic variation and pharmacogenomics: concepts, facts, and challenges. Dialogues Clin Neurosci 6 (1):5–26
5. Roden DM, Altman RB, Benowitz NL, Flockhart DA, Giacomini KM, Johnson JA et al (2006) Pharmacogenomics: challenges and opportunities. Ann Intern Med 145 (10):749–757
6. de Man FM, Goey AKL, van Schaik RHN, Mathijssen RHJ, Bins S (2018) Individualization of irinotecan treatment: a review of pharmacokinetics, pharmacodynamics, and pharmacogenetics. Clin Pharmacokinet. https://doi.org/10.1007/s40262-018-0644-7
7. Muir P, Li S, Lou S, Wang D, Spakowicz DJ, Salichos L et al (2016) The real cost of sequencing: scaling computation to keep pace with data generation. Genome Biol 17:53. https://doi.org/10.1186/s13059-016-0917-0
8. Gonzaga-Jauregui C, Lupski JR, Gibbs RA (2012) Human genome sequencing in health and disease. Annu Rev Med 63:35–61. https://doi.org/10.1146/annurev-med-051010-162644
9. Meienberg J, Bruggmann R, Oexle K, Matyas G (2016) Clinical sequencing: is WGS the better WES? Hum Genet 135(3):359–362. https://doi.org/10.1007/s00439-015-1631-9
10. Belkadi A, Bolze A, Itan Y, Cobat A, Vincent QB, Antipenko A et al (2015) Whole-genome sequencing is more powerful than whole-exome sequencing for detecting exome variants. Proc Natl Acad Sci U S A 112 (17):5473–5478. https://doi.org/10.1073/pnas.1418631112
11. Stasik S, Schuster C, Ortlepp C, Platzbecker U, Bornhäuser M, Schetelig J et al (2018) An optimized targeted Next-Generation Sequencing approach for sensitive detection of single nucleotide variants. Biomol Detect Quantif 15:6–12. https://doi.org/10.1016/j.bdq.2017.12.001
12. Kukurba KR, Montgomery SB (2015) RNA sequencing and analysis. Cold Spring Harb Protoc 2015(11):951–969. https://doi.org/10.1101/pdb.top084970
13. Nalpathamkalam T, Derkach A, Paterson AD, Merico D (2014) Genetic Analysis Workshop 18 single-nucleotide variant prioritization based on protein impact, sequence conservation, and gene annotation. BMC Proc 8. (Suppl 1 Genetic Analysis Workshop 18Vanessa Olmo:S11. https://doi.org/10.1186/1753-6561-8-S1-S11
14. Marian AJ (2014) Sequencing your genome: what does it mean? Methodist Debakey Cardiovasc J 10(1):3–6

15. Head SR, Komori HK, LaMere SA, Whisenant T, Van Nieuwerburgh F, Salomon DR et al (2014) Library construction for next-generation sequencing: overviews and challenges. Biotechniques 56(2):61–64, 66, 68, passim. https://doi.org/10.2144/000114133
16. Hedegaard J, Thorsen K, Lund MK, Hein AM, Hamilton-Dutoit SJ, Vang S et al (2014) Next-generation sequencing of RNA and DNA isolated from paired fresh-frozen and formalin-fixed paraffin-embedded samples of human cancer and normal tissue. PLoS One 9 (5):e98187. https://doi.org/10.1371/journal.pone.0098187
17. Einaga N, Yoshida A, Noda H, Suemitsu M, Nakayama Y, Sakurada A et al (2017) Assessment of the quality of DNA from various formalin-fixed paraffin-embedded (FFPE) tissues and the use of this DNA for next-generation sequencing (NGS) with no artifactual mutation. PLoS One 12(5):e0176280. https://doi.org/10.1371/journal.pone.0176280
18. Liu L, Li Y, Li S, Hu N, He Y, Pong R et al (2012) Comparison of next-generation sequencing systems. J Biomed Biotechnol 2012:251364. https://doi.org/10.1155/2012/251364
19. Buermans HP, den Dunnen JT (2014) Next generation sequencing technology: advances and applications. Biochim Biophys Acta 1842 (10):1932–1941. https://doi.org/10.1016/j.bbadis.2014.06.015
20. Ekblom R, Wolf JB (2014) A field guide to whole-genome sequencing, assembly and annotation. Evol Appl 7(9):1026–1042. https://doi.org/10.1111/eva.12178
21. Auton A, Brooks LD, Durbin RM, Garrison EP, Kang HM, Korbel JO et al (2015) A global reference for human genetic variation. Nature 526(7571):68–74. https://doi.org/10.1038/nature15393
22. Hildebrandt F, Heeringa SF, Rüschendorf F, Attanasio M, Nürnberg G, Becker C et al (2009) A systematic approach to mapping recessive disease genes in individuals from outbred populations. PLoS Genet 5(1): e1000353. https://doi.org/10.1371/journal.pgen.1000353
23. Wray NR, Goddard ME, Visscher PM (2007) Prediction of individual genetic risk to disease from genome-wide association studies. Genome Res 17(10):1520–1528. https://doi.org/10.1101/gr.6665407
24. Abraham G, Inouye M (2015) Genomic risk prediction of complex human disease and its clinical application. Curr Opin Genet Dev 33:10–16. https://doi.org/10.1016/j.gde.2015.06.005
25. Wilkening S, Chen B, Bermejo JL, Canzian F (2009) Is there still a need for candidate gene approaches in the era of genome-wide association studies? Genomics 93(5):415–419. https://doi.org/10.1016/j.ygeno.2008.12.011
26. Patnala R, Clements J, Batra J (2013) Candidate gene association studies: a comprehensive guide to useful in silico tools. BMC Genet 14:39. https://doi.org/10.1186/1471-2156-14-39
27. Ball AD, Stapley J, Dawson DA, Birkhead TR, Burke T, Slate J (2010) A comparison of SNPs and microsatellites as linkage mapping markers: lessons from the zebra finch (Taeniopygia guttata). BMC Genomics 11:218. https://doi.org/10.1186/1471-2164-11-218
28. Wallace C, Dobson RJ, Munroe PB, Caulfield MJ (2007) Information capture using SNPs from HapMap and whole-genome chips differs in a sample of inflammatory and cardiovascular gene-centric regions from genome-wide estimates. Genome Res 17(11):1596–1602. https://doi.org/10.1101/gr.5996407
29. Rüschendorf F, Nürnberg P (2005) ALOHOMORA: a tool for linkage analysis using 10K SNP array data. Bioinformatics 21 (9):2123–2125. https://doi.org/10.1093/bioinformatics/bti264
30. Hoffmann K, Lindner TH (2005) easyLINKAGE-Plus—automated linkage analyses using large-scale SNP data. Bioinformatics 21 (17):3565–3567. https://doi.org/10.1093/bioinformatics/bti571
31. Inaoka K, Inokawa Y, Nomoto S (2015) Genomic-wide analysis with microarrays in human oncology. Microarrays (Basel) 4 (4):454–473. https://doi.org/10.3390/microarrays4040454
32. Ballester B, Johnson N, Proctor G, Flicek P (2010) Consistent annotation of gene expression arrays. BMC Genomics 11:294. https://doi.org/10.1186/1471-2164-11-294
33. Karczewski KJ, Daneshjou R, Altman RB (2012) Chapter 7: Pharmacogenomics. PLoS Comput Biol 8(12):e1002817. https://doi.org/10.1371/journal.pcbi.1002817
34. Wang Z, Gerstein M, Snyder M (2009) RNA-Seq: a revolutionary tool for transcriptomics. Nat Rev Genet 10(1):57–63. https://doi.org/10.1038/nrg2484
35. Wang DG, Fan JB, Siao CJ, Berno A, Young P, Sapolsky R et al (1998) Large-scale identification, mapping, and genotyping of single-nucleotide polymorphisms in the human genome. Science 280(5366):1077–1082

36. Fan JB, Oliphant A, Shen R, Kermani BG, Garcia F, Gunderson KL et al (2003) Highly parallel SNP genotyping. Cold Spring Harb Symp Quant Biol 68:69–78
37. Kurg A, Tõnisson N, Georgiou I, Shumaker J, Tollett J, Metspalu A (2000) Arrayed primer extension: solid-phase four-color DNA resequencing and mutation detection technology. Genet Test 4(1):1–7. https://doi.org/10.1089/109065700316408
38. Gunderson KL, Steemers FJ, Ren H, Ng P, Zhou L, Tsan C et al (2006) Whole-genome genotyping. Methods Enzymol 410:359–376. https://doi.org/10.1016/S0076-6879(06)10017-8
39. DeRisi J, Penland L, Brown PO, Bittner ML, Meltzer PS, Ray M et al (1996) Use of a cDNA microarray to analyse gene expression patterns in human cancer. Nat Genet 14(4):457–460. https://doi.org/10.1038/ng1296-457
40. LaFramboise T (2009) Single nucleotide polymorphism arrays: a decade of biological, computational and technological advances. Nucleic Acids Res 37(13):4181–4193. https://doi.org/10.1093/nar/gkp552
41. Bock C (2012) Analysing and interpreting DNA methylation data. Nat Rev Genet 13 (10):705–719. https://doi.org/10.1038/nrg3273
42. Kurdyukov S, Bullock M (2016) DNA methylation analysis: choosing the right method. Biology (Basel) 5(1). https://doi.org/10.3390/biology5010003
43. Rew DA (2001) DNA microarray technology in cancer research. Eur J Surg Oncol 27 (5):504–508. https://doi.org/10.1053/ejso.2001.1116

# Chapter 10

# *EGFR* Mutation Analysis in Non-small Cell Lung Carcinoma from Tissue Samples Using the Fully Automated Idylla™ qPCR System

**Simon Heeke and Paul Hofman**

## Abstract

The introduction of tyrosine-kinase inhibitors (TKI) targeting specific *EGFR* mutations for the treatment non-small cell lung cancer patients (NSCLC) dramatically increased the clinical outcome in a subset of patients harboring specific activating *EGFR* mutations. Three different generations of TKI have been developed until now, demonstrating increasing progression-free survival as well as overall survival. However, to benefit of the treatment, the analysis of the genomic content of each patient is mandatory. Additionally, resistance mutations are prevalent and occur frequently and rapidly during treatment. Therefore, tests to detect *EGFR* mutations at initial diagnosis as well as during treatment, e.g., from liquid biopsies, have been developed and implemented in clinical daily practice for theranostic purpose.

As *EGFR* mutation testing has to be highly reliable, fast, and easy to perform, the automatic qPCR system Idylla™ has been developed and implemented for clinical mutation testing from tissue samples and soon from circulating free DNA. Therefore, we here describe how the Idylla™ system can be used for the analysis of *EGFR* mutations in NSCLC patients. Importantly, as the results are massively influenced by the preanalytical steps, we also provide information on the correct sample selection to avoid nonconclusive results.

**Key words** Idylla, *EGFR*, PCR, NSCLC, Liquid biopsy, Plasma, FFPE

## 1 Introduction

With approximately 1.8 million deaths caused each year by lung cancer worldwide, it remains to be the leading cause of cancer-related death [1–4]. Incidence is higher in women and differs depending on the region and most importantly on the smoking behavior [5–9]. Generally, the histology of lung cancer can be classified into small-cell lung cancer (SCLC) and non-small cell lung cancer (NSCLC) with the latter accounting for the majority of all lung cancer cases (~85%) [10]. While chemotherapy was the standard of care in NSCLC for several years, the increasing emergence of immunotherapy in NSCLC might change the standard of

Jyotsna Batra and Srilakshmi Srinivasan (eds.), *Theranostics: Methods and Protocols*, Methods in Molecular Biology, vol. 2054, https://doi.org/10.1007/978-1-4939-9769-5_10, 

care in the future [11–13]. Interestingly, approximately 10–40% of all NSCLC patients harbor mutations in the epidermal growth factor receptor (*EGFR)*, which can be targeted by *EGFR* tyrosine kinase inhibitors (TKIs) [14]. So far, four different anti-*EGFR* TKIs have been developed and clinically approved. Gefitinib and erlotinib were approved by the Food and Drug Administration (FDA) in 2003 and 2004, respectively [15, 16]. Afatinib as first irreversible *EGFR* inhibitor was later approved in 2013 [17] followed by osimertinib, the first third-generation TKI approved by the FDA for the first-line treatment of *EGFR*-mutated NSCLC in 2015 [18]. Unfortunately, response to targeted treatment is generally short and resistance occurs usually after 7–10 month of treatment [19]. While many different resistance mechanisms have been reported, the acquisition of secondary mutations, most notably the T790M, remains to be the most relevant [20]. Therefore, the newly developed TKIs of the second and third generation (afatinib and osimertinib) have been designed aiming to show improved efficacy against the T790M mutation and are recommended in second-line treatment after failure of treatment with a first- and second-generation TKI [21].

Consequently, the assessment of the *EGFR* mutation status is mandatory in all advanced and metastatic NSCLC patients to allow the precise treatment of patients harboring *EGFR* mutations. Additionally, due to the occurrence of resistance mutations, the analysis for acquired mutations during treatment and most importantly after failed treatment is an important challenge in routine molecular pathology.

Different methods have been developed for the assessment of *EGFR* mutations derived from tissue samples as well as from plasma. As we do not want to discuss the different advantages and disadvantages here in detail, we ask the interested reader to look up the main provided resources [22–25].

In clinical routine practice assessment of targetable *EGFR* mutations, some factors play an important role and determine the most suitable method to be used:

1. Due to fast tumor progression, the treatment has to be started as early as possible to allow best treatment outcome for the patients. Thus, the turnaround time is a pivotal point and faster methods have to be preferred over time-consuming and complicated workflows.
2. The desired method has to be reproducible to avoid error as some of these results may lead to inappropriate treatment decisions. Therefore, we highly recommend relying on CE-IVD (in Europe) marked test or FDA-approved companion diagnostics (in the USA) to get most robust results. However, the use of "research use only" (RUO) tests might not be

always avoidable but should then be tightly controlled by including appropriate controls.

3. To avoid cross contamination or errors during handling of patient-derived samples we highly recommend using easy workflows over complicated and error-prone workflows.
4. To allow reliable mutation analysis, the test has to be very specific and highly sensitive. In clinical routine, sample material is often very limited and consequently tests that require less sample input should be preferred over tests which are dependent on increased sample input when the specificity is equal. Additionally, circulating free (cfDNA) concentration in blood is rather low requiring an even more sensitive test. Furthermore, as the clinical treatment is dramatically influenced by the result of the mutation analysis, highest specificity is critical.
5. The test has to be cost-effective.
6. As the analysis of the data can be very time-consuming and as it might be biased if done manually, an automatic data analysis with a reporting scheme highlighting the found mutations as well as some quality parameters is preferred.

The Idylla™ *EGFR* Mutation Test (CE-IVD) (Biocartis, Mechelen, Belgium) is compliant to those recommendations [26]. It is a fully-automated PCR system performing all necessary steps automatically. This reduces errors, hands-on time, risk of cross contamination and allows the fast and reproducible assessment of *EGFR* mutations comparable to main other techniques [27]. It allows the automatic generation of a report with the mutation status in less than one working day (~3 h). In total, 51 different *EGFR* mutations are covered by the Idylla™ system [28]. Additionally, Biocartis is currently developing a fully automated liquid biopsy version of the Idylla™ *EGFR* Mutation test, the Idylla™ ctEGFR Mutation Assay, expected to be launched as a RUO test in 2019. This assay has also a true sample-in, results-out capability where 2 ml of plasma are inserted directly into the cartridge. The assay will detect all mutations as currently present in the Idylla™ *EGFR* tissue cartridge (51 mutations), and as currently described in international guidelines, in approx. Two and a half hours. Therefore, it allows the assessment of samples derived from different sources like formalin-fixed paraffin-embedded (FFPE) samples and plasma allowing the initial *EGFR* mutation assessment as well as the analysis of resistance mutations during treatment using liquid biopsies.

We therefore, present here the workflow used for the analysis of *EGFR* mutation from FFPE tissue for the analysis of *EGFR* mutation status in NSCLC patients.

## 2 Materials

- Idylla™ EGFR Mutation Test (CE-IVD).
- Whatman™ Grade 1 Qualitative Filter Paper Circle, Diameter: 10 mm, Pore Size: 11 μm.
- Microscope slide (e.g., Thermo Scientific™ SuperFrost Plus™ Adhesion slides).
- Laboratory-grade water.
- Sterile razor blades for microtome, e.g., Accu-Edge Disposable Microtome Blades (Sakura Finetek Europe B.V., Alphen aan den Rijn, Netherlands).
- Cleaned and sterilized forceps to transfer the FFPE cut. Plastic forceps might reduce the risk of destroying the sample.
- Laboratory centrifuge with swing bucket rotor reaching at least 2000 × *g* (e.g., Heraeus™ Megafuge™ 16 with TX-200 rotor).
- Laboratory microtome to cut FFPE blocks (e.g., Thermo Fisher HM 340E).

## 3 Methods

### 3.1 General Recommendations

The quality of the results is greatly influenced by the pre-analytical steps which are performed to embed and prepare the sample.

1. Reduce the time for sample preparations to preserve genomic content. Transport of samples to the pathology laboratory as well as preparation of the sample in the laboratory has to be as short as possible (*see* **Note 1**).
2. Working in a biosafety cabinet should be imperative for biological samples even when the viral status of the patient is known. Good laboratory practices should be followed at any time.
3. Carefully prepare one sample after the other and never at the same time avoiding cross contamination and inversions of samples.
4. Waste produced during this procedure has to be placed in a biohazard bag and has to be processed accordingly.

### 3.2 EGFR Mutation Analysis from Tissue Samples

#### 3.2.1 Sample Preparation and Selection

For the analysis, use formalin-fixed and paraffin-embedded (FFPE) blocks from patient-derived material. Both, surgical specimen and biopsy samples are sufficient for the analysis (Fig. 1a, b). Furthermore, cytological specimen from, e.g., bronchioalveolar lavage can be used after being embedded in Formalin as cytoblocks [29] (*see* **Note 2**). In general, the sample must have a tumor cell content greater than 10% [28] (Fig. 1c, d). If not already performed, cut a second slide and stain it with an appropriate staining (e.g., H&E)

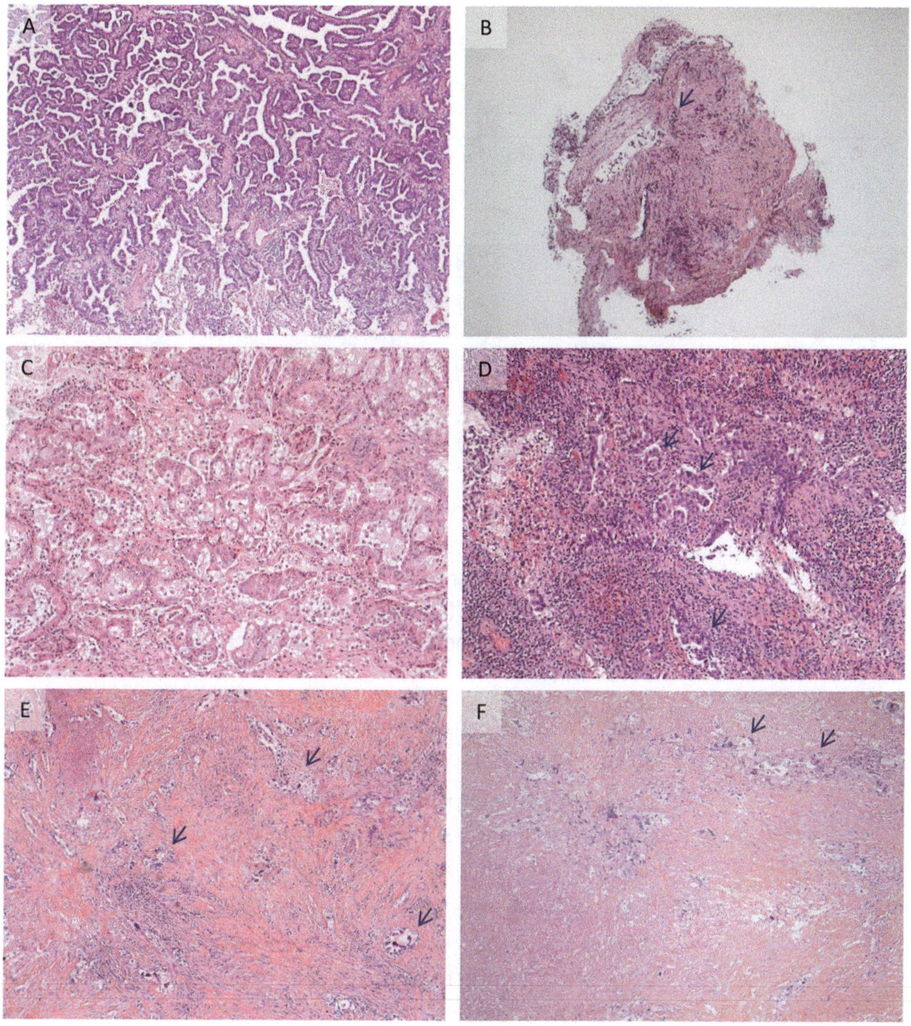


**Fig. 1** Sample selection for Idylla™ automated qPCR. For the detection of *EGFR* mutations, whole tissue slides from surgical specimen (**a**) as well as smaller biopsies (**b**) can be used. Important for the selection is the percentage of tumor cells in the sample. In (**c**), a high tumor cell percentage of >50% is shown making this sample a good candidate for the analysis, while (**d**) shows a sample with less than 10% of tumor cells (*arrow*) with a high number of inflammatory cells. Idylla™ requires at least 10% of tumor cells. A slide with a low level of tumor cells might be still suitable for the analysis. However, if possible, the starting material might be increased to allow sufficient target amplification. Additionally, performing macro or microdissection increases the tumor-cell content and decreases the number of non-tumor cells (and finally the presence of germinal DNA). Finally, the quality of the sample dramatically influences the analysis. (**e**, **f**) are tumor cell samples with high necrotic and low tumor cell content and large fibrotic areas and are therefore unsuitable for the analysis. All slides are Hematoxylin & Eosin stained from FFPE tissue samples derived from lung adenocarcinomas

for the determination of tumor cell content. Determination always has to be performed by a board certified and trained senior pathologist to avoid false assessment. Make sure to have a high-quality tumor section without necrotic areas (and if possible avoiding areas with high level of inflammatory cells) to get best results (Fig. 1e, f) (*see* **Note 3**).

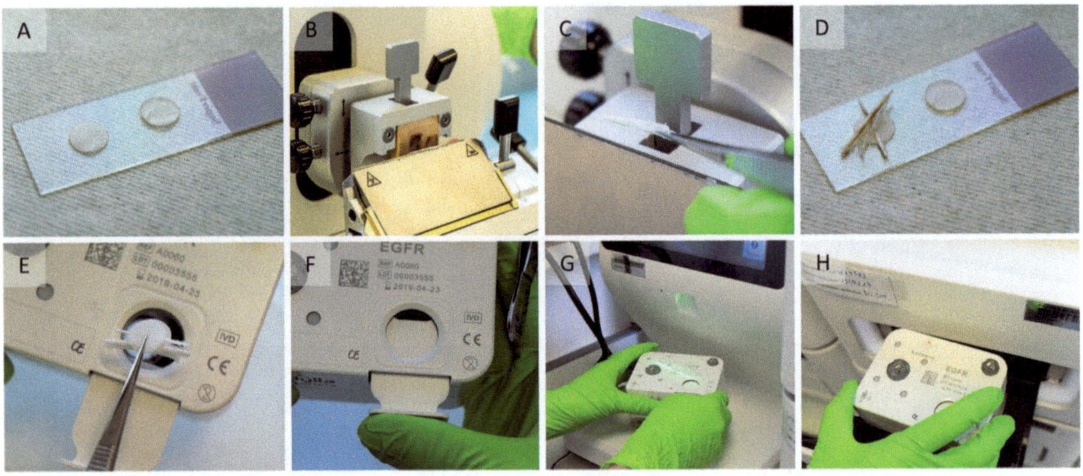

**Fig. 2** Preparation of a tissue sample for the Idylla™ automated qPCR. As described in the protocol, two waterman filters have to be wetted and placed on a microscope slide (**a**). A tissue slide has to be cut from a FFPE block (**b**) taken (**c**) and placed on a wetted waterman filter (**d**). The waterman/FFPE-sample is added in the Idylla™ cartridge (**e**) and closed (**f**). Finally, the cartridge has to be scanned on the Idylla™ system (**g**), the protocol has to be initiated and the cartridge has to be inserted in the device (**h**)

1. Take two waterman filters and wet them shortly in laboratory-grade water. Place them both on a microscope slide (Fig. 2a).
2. Select an appropriate tumor block for the analysis of *EGFR* mutation status according the previously mentioned guidelines.
3. Cut at least one slide of 5 μm thickness using a microtome. Place the slides on one of the wetted waterman filters. Lay the other waterman filter on the cut slide (Fig. 2b–d) (*see* **Notes 4–7**).
4. Take the stack of the FFPE slide between two waterman filters and add it directly in the Idylla™ cartridge. Close the cartridge (Fig. 2e, f).

#### *3.2.2 Preparation of the Idylla™ Cartridge*

To place the device in the Idylla™ system, follow the on-screen steps described by the system. Briefly, log in with your user account and password, scan the QR code on the Idylla™ cartridge (Fig. 2g, h), add sample information, and place the cartridge in the Idylla™ system. All the following steps will be performed automatically by the device and do not need further hands-on time.

### *3.3 Data Analysis*

The Idylla™ system provides the user automatically with a report providing all the important information including the mutation result. Information is organized around the following EGFR mutations: L858R, exon 19 deletions, T790M, G719A/C/S, exon 20 insertions, S768I, and L861Q, respectively. Furthermore, the CQ of an internal control is given for quality control of the test

Idylla™ EGFR Mutation Assay

| EGFR GENOTYPE | |
|---|---|
| G719A/C/S | **MUTATION DETECTED** |
| Protein change | p.Gly719Ala,p.Gly719Ser,p.Gly719Cys,p.Gly719Cys (2) |
| Nucleotide change | c.2156G>C,c.2155G>A,c.2155G>T,c.2154_2155delinsTT |
| L858R | **NO MUTATION DETECTED** |
| EXON 19 DELETION | **NO MUTATION DETECTED** |
| T790M | **NO MUTATION DETECTED** |
| EXON 20 INSERTION | **NO MUTATION DETECTED** |
| S768I | **NO MUTATION DETECTED** |
| L861Q | **NO MUTATION DETECTED** |
| → CQ OF EGFR CONTROL | **18.0** |

**Fig. 3** Sample report of an *EGFR* analysis using Idylla™ of a patient harboring a mutation at position G719. The information on various *EGFR* mutations and the quantitative results of the positive control (black arrow) are given

result (Fig. 3). Consequently, the report gives all the necessary information to the treating oncologist to guide the treatment decision. For detailed information on which treatment is advisable under which circumstances, please refer to the country-specific guidelines.

## 4 Notes

1. Document the time between obtaining the sample and the time it was processed to allow troubleshooting in case any errors occur during the preparation of the specimen.
2. It has been published that it is also possible to directly add isolated DNA from cytological specimen in the cartridge of the Idylla™ device [30]. However, we have not tested this with our hands.
3. If the tumor cell content is lower than 10%, macro or laser microdissection can be performed. Therefore, mark the tumor content areas and remove all surrounding non-cancer tissue to increase the ratio of tumor cells.
4. Usually, one slide of 5 μm thickness is sufficient for the analysis. If enough tumor material is present (e.g., when using surgical specimen), even two slides of 10 μm thickness can be used. Especially cutting of thicker slides (10 μm) might be easier and requires less training.
5. The tumor slide will be usually a rolled up FFPE slide. It can be easily transferred using fine tweezers or using a fine paintbrush (Fig. 2c).

6. Always prepare fresh slices for the analyses. While the genomic content is well prepared in tissue blocks, it might be degraded over long-term storage in fine tissue slices.
7. If preparing different samples, carefully remove all remaining tissue from the microtome to avoid cross-contamination. Mandatorily change the razor blade between each sample.

## Acknowledgments

This work was supported by the French Government (National Research Agency, ANR) through the "Investments for the Future" LABEX SIGNALIFE: program reference # ANR-11-LABX-0028-01.

*Disclosure*: Paul Hofman is an expert member of the Key Expert Forum for Biocartis NV, Mechelen, Belgium. Simon Heeke has declared no conflicts of interest.

### References

1. Didkowska J, Wojciechowska U, Mańczuk M et al (2016) Lung cancer epidemiology: contemporary and future challenges worldwide. Ann Transl Med 4:150
2. Siegel RL, Miller KD, Jemal A (2017) Cancer statistics, 2017. CA Cancer J Clin 67:7–30
3. Ferlay J, Soerjomataram I, Dikshit R et al (2015) Cancer incidence and mortality worldwide: sources, methods and major patterns in GLOBOCAN 2012. Int J Cancer 136: E359–E386
4. Dela Cruz CS, Tanoue LT, Matthay RA (2011) Lung cancer: epidemiology, etiology and prevention. Clin Chest Med 32:1–61
5. Torre LA, Islami F, Siegel RL et al (2017) Global cancer in women: burden and trends. Cancer Epidemiol Biomark Prev 26:444–457
6. Jemal A, Miller KD, Ma J et al (2018) Higher lung cancer incidence in young women than young men in the United States. N Engl J Med 378:1999–2009
7. Rahal Z, El Nemr S, Sinjab A et al (2017) Smoking and lung cancer: a geo-regional perspective. Front Oncol 7:1–7
8. De Matteis S, Heederik D, Burdorf A et al (2017) Current and new challenges in occupational lung diseases. Eur Respir Rev 26. https://doi.org/10.1183/16000617.0080-2017
9. Wong MCS, Lao XQ, Ho K-F et al (2017) Incidence and mortality of lung cancer: global trends and association with socioeconomic status. Sci Rep 7:14300
10. Travis WD, Brambilla E, Nicholson AG et al (2015) The 2015 World Health Organization classification of lung tumors: impact of genetic, clinical and radiologic advances since the 2004 classification. J Thorac Oncol 10:1243–1260
11. Zappa C, Mousa SA (2016) Non-small cell lung cancer: current treatment and future advances. Transl Lung Cancer Res 5:288–300
12. Hellmann MD, Ciuleanu T-E, Pluzanski A et al (2018) Nivolumab plus ipilimumab in lung cancer with a high tumor mutational burden. N Engl J Med 378:2093–2104
13. Gandhi L, Rodríguez-Abreu D, Gadgeel S et al (2018) Pembrolizumab plus chemotherapy in metastatic non-small-cell lung cancer. N Engl J Med 378:2078–2092
14. Zhang Y-L, Yuan J-Q, Wang K-F et al (2016) The prevalence of EGFR mutation in patients with non-small cell lung cancer: a systematic review and meta-analysis. Oncotarget 7:9–13
15. Drug Approval Package: Iressa (gefitinib) NDA #021399. https://www.accessdata.fda.gov/drugsatfda_docs/nda/2003/021399_iressa.cfm
16. Drug Approval Package: Tarceva (Erlotinib) NDA #021743. https://www.accessdata.fda.gov/drugsatfda_docs/nda/2004/21-743_Tarceva.cfm
17. Drug Approval Package: Gilotrif (afatinib). https://www.accessdata.fda.gov/drugsatfda_docs/nda/2013/201292Orig1s000TOC.cfm

18. TAGRISSO (osimertinib) tablets. https://www.accessdata.fda.gov/drugsatfda_docs/nda/2015/208065Orig1s000TOC.cfm
19. Morgillo F, Della Corte CM, Fasano M et al (2016) Mechanisms of resistance to EGFR-targeted drugs: lung cancer. ESMO Open 1: e000060
20. Stewart EL, Tan SZ, Liu G et al (2015) Known and putative mechanisms of resistance to EGFR targeted therapies in NSCLC patients with EGFR mutations — a review. Transl Lung Cancer Res 4:67–81
21. Ettinger DS, Wood DE, Aisner DL et al (2017) Non-small cell lung cancer, version 5.2017, NCCN clinical practice guidelines in oncology. J Natl Compr Cancer Netw 15:504–535
22. Normanno N, Denis MG, Thress KS et al (2017) Guide to detecting epidermal growth factor receptor (EGFR) mutations in ctDNA of patients with advanced non-small-cell lung cancer. Oncotarget 8:12501–12516
23. Ellison G, Zhu G, Moulis A et al (2013) EGFR mutation testing in lung cancer: a review of available methods and their use for analysis of tumour tissue and cytology samples. J Clin Pathol 66:79–89
24. Lin C-C, Huang W-L, Wei F et al (2015) Emerging platforms using liquid biopsy to detect *EGFR* mutations in lung cancer. Expert Rev Mol Diagn 15:1427–1440
25. Xu T, Kang X, You X et al (2017) Cross-platform comparison of four leading technologies for detecting EGFR mutations in circulating tumor DNA from non-small cell lung carcinoma patient plasma. Theranostics 7:1437–1446
26. Uguen A, Troncone G (2018) A review on the Idylla platform: towards the assessment of actionable genomic alterations in one day. J Clin Pathol 71(9):757–762
27. Ilie M, Butori C, Lassalle S et al (2017) Optimization ofEGFRmutation detection by the fully-automated qPCR-based Idylla system on tumor tissue from patients with non-small cell lung cancer. Oncotarget 8:103055–103062
28. Technical sheet Idylla™ EGFR Mutation Test. https://media.biocartis.com/biocartis/documents/A0060-6_EGFR_CE_IVD-ENG_R1.012018.pdf
29. Minca EC, Lanigan CP, Reynolds JP et al (2014) ALK status testing in non-small-cell lung carcinoma by FISH on ThinPrep slides with cytology material. J Thorac Oncol 9:464–468
30. De Luca C, Gragnano G, Pisapia P et al (2017) EGFR mutation detection on lung cancer cytological specimens by the novel fully automated PCR-based Idylla EGFR Mutation Assay. J Clin Pathol 70:295–300

# Part III

## Molecular Imaging

# Chapter 11

# Subcellular Localization of MicroRNAs by MicroRNA In Situ Hybridization (miR-ISH)

**Harley Rose Robinson, Michelle Mei Chih Hill, and Alexandre Santos Cristino**

## Abstract

MicroRNAs (miRNAs) are 22-nucleotide RNA sequences that regulate up to 60% of the mammalian transcriptome. Although canonical miRNA-induced silencing complex-mediated messenger RNA degradation occurs in the cytoplasm, miRNAs have been described in other subcellular compartments with potentially novel functions. Currently, there are limited methodologies for visualizing RNA locations within cells to elucidate mechanisms and pathways of miRNA biogenesis, transport, and function. Here, we describe a simple and rapid miRNA in situ hybridization method that can be combined with standard immunofluorescence procedures for subcellular localization of mature and precursor miRNAs.

**Key words** RNA, MicroRNA, Hybridization, Subcellular localization, Microscopy

## 1 Introduction

The subcellular localization of microRNAs (miRNAs) can provide useful information about their functional roles in the control of gene expression via translational repression and/or degradation of targeted RNA transcripts. Canonically, miRNA genes are transcribed in the nucleus where their primary miRNA transcripts are processed by Drosha to form one or more precursor miRNA hairpin structures, which is recognized by Exportin 5 and transported to the cytoplasm where the hairpin loop region is removed by Dicer to become a short duplex (≈22 nt). Each miRNA duplex is then loaded onto Argonaute proteins which retain only one of the strands known as mature miRNAs. Even though AGO1–4 can form AGO:miRNA complexes, only AGO2 has endonuclease activity thus degrading bound messenger RNAs (mRNAs) via miRNA-induced silencing complex (miRISC) [1]. MiRNAs are master regulators of gene expression by inhibiting of protein synthesis, directly contributing to the regulation of more than 60% of the

Jyotsna Batra and Srilakshmi Srinivasan (eds.), *Theranostics: Methods and Protocols*, Methods in Molecular Biology, vol. 2054, https://doi.org/10.1007/978-1-4939-9769-5_11, © Springer Science+Business Media, LLC, part of Springer Nature 2019

mammalian genome [2]. While this process was originally thought to occur exclusively in the cytoplasm [3], recent studies have identified unusual localization of mature and precursor miRNAs to various organelle compartments including endoplasmic reticulum [4], mitochondria [5], nucleolus [6], endosomes [7, 8], and exosomes [9]. It is unclear whether miRISC is involved at these secondary sites, and therefore the function of microRNAs in these subcellular compartments is still unknown. Additionally, the activities of proteins cooperating with miRNAs in these noncanonical functions could be further understood with co-localization studies [8]. Thus highlighting the necessity for a high-resolution subcellular detection of miRNAs in situ.

Various microscopy techniques allow for the visualization of biological molecules to depict temporal and spatial observations of proteins, lipids, and long nucleic acids. Classic use of fluorescent in situ hybridization (FISH [10]) or the more recent development of RNA hybridization techniques (such as RNAscope [11] or noncommercial methods [12]) rely on Watson-Crick complementarity of fluorescent probes hundreds of bases long to establish specificity and selectivity of the target nucleic acid. However, smaller RNAs such as microRNAs face challenges with microscopic detection due to their short sequence length [13]; miRNA precursors range a mere 60–90 nt in length and mature miRNAs are even shorter still at 16–24 nt [14]. Additionally, as the mature miRNAs are derived and identical in sequence to sections of their precursor miRNA hairpin, the ability to differentiate between the two processing stages remain difficult [13]. Although commercial miRNA in situ hybridization kits are available, use of co-localization with indirect or direct protein detection requires further optimization and may not be compatible [15].

Here, we established a simple and rapid method for in situ detection of mature miRNA and its precursor hairpin in cultured cells with optional simultaneous detection of proteins and other molecules. We demonstrate the application of our method for a mature miRNA and it is corresponding precursor hairpin. This method can be generalized to other miRNA sequences (mature and hairpin) to help reveal distinct distribution patterns within subcellular compartments.

## 2 Materials

### 2.1 Coverslip Preparation

1. Growth medium: Specific for cells of choice. For HEK293 cell lines, add 50 mL of Fetal bovine serum (FBS) to 500 mL Dulbecco's Modified Eagle Medium (DMEM). Store at 4 °C and warm to 37 °C prior to use (*see* **Note 1**).

2. Sterile washing buffer: Make 500 mL 1× Phosphate buffered saline (PBS) with distilled water. Autoclave at 121 °C for 20 min. Store at room temperature and only open in sterile conditions.
3. Collagen treatment solution: Using a fume hood, add 10 μL of glacial acetic acid to 5 mL autoclaved water. Dilute collagen in acetic acid solution to a final concentration of 50 μg/mL. Leave one aliquot at 4 °C for current use and store the remaining aliquots at −20 °C (*see* **Note 2**).
4. Coverslips: Autoclave 10 mm glass coverslips at 121 °C for 20 min (*see* **Note 3**).

### 2.2 miRNA In Situ Hybridization

1. Fixing solution: 100% methanol stored at −20 °C.
2. Washing buffer: 1× PBS.
3. Humidity chamber: either manufactured chamber (Cat: HIC-2, Lab Scientific) or prepare as described (Fig. 1). Line the base of a 15 cm petri dish with tissue paper and saturate with water. Place plastic platform (96-well plate lid or similar) above tissue paper and cover with parafilm. Cover lid with aluminum foil to completely darken chamber once assembled (*see* **Note 4**).
4. Anti-target RNA oligonucleotide probe modified with a small fluorophore moiety at the 5′ or 3′ ends. Probe design should be straightforward as the probe sequence will likely expand the entire miRNA sequence (~20 nt), but there will be cases where

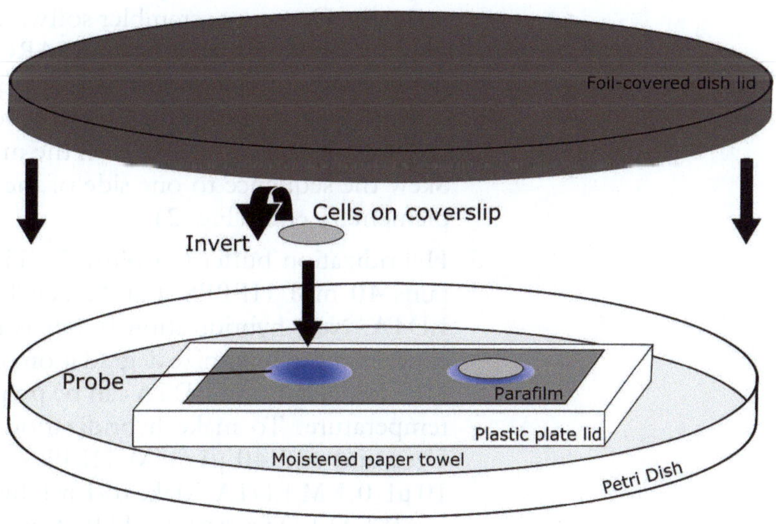


**Fig. 1** Humidity chamber apparatus setup. Opaque container or aluminum foil covered petri dish contains moistened paper towels and a parafilm-covered plastic stand to provide a clean, moist, and light-protected vessel for hybridization

small adjustments are recommended. miR-ISH probe is designed to be complementary to the target miRNA and should cover its entirety. Ideally, G or C at the 3′ end will assist with binding affinity therefore if necessary; design a shorter probe (no shorter than 18 nucleotides) so that G or C is present at the 3′ base (either last or second last base). While miRNAs and their probes are small, secondary structures like hairpins may form and affect efficiency of interaction and cross-linking between miR-ISH probe and target miRNA. However, the hybridization buffer (*see* **item 5**) is designed to minimize these occurrences. To assess the likelihood of hairpin formation, visit the OligoAnalyzer 3.1 (or newer versions) website to predict if the proposed probe sequence may form secondary structures. Information on hairpin and self-dimer formation will be predicted. Generally, strong $\Delta G$ of −9 kcal/mole or more negative indicates probable dimer or hairpin formation, where ideal values should be approximately 0 kcal/mole [16]. If need be, shorten or move the probe sequence a few bases to improve $\Delta G$. Additionally, some miRNAs share very similar sequences; therefore, use miRBase's SSEARCH function to detect specificity to miRNA of choice. Any small fluorophore can be used to tag the oligo probe, e.g., Cy3, Cy5, Cy5.5, and Digoxigenin; however, larger fluorophores may interfere with hybridization dynamics. Additionally, probe fluorophores should not overlap with DAPI or, in the case of protein co-detection, secondary antibody spectra. In the case of low target expression, double tagging both ends of the probe may assist detection. Negative control probes were designed by using the Genscript scrambler software tool [17] for one of the target probes. The miRBase SSEARCH function was used to determine successful scrambling of sequence. To detect precursor miRNAs, design the probe to cover the loop region; often this will overlap somewhat with the mature miRNA sequences. Skew the sequence to one side of the hairpin to prevent complementary ends (Fig. 2).

5. Hybridization buffer (*see* **Note 5**): Final buffer (pH 6.5) contains 40 mM HEPES, 0.4 M NaCl, 2 M Urea, and 1 mM EDTA. New hybridization buffer is made for each miR-ISH experiment to minimize degradation of Urea. Stock solutions of HEPES, NaCl, and EDTA can be prepared and stored at room temperature. To make hybridization buffer, measure 0.12 g Urea and add 40 μL 1 M HEPES, 200 μL 2 M NaCl, and 10 μL 0.1 M EDTA. Make to 1 mL by adding UltraPure water.

   0.1 M EDTA stock: Add 0.29 g of Ethylenediaminetetraacetic acid (EDTA) to 9 mL of UltraPure (RNase/DNase-free) water. Adjust to pH 8.0 with NaOH pellets and stir until EDTA is dissolved. Top up to 10 mL to make a 0.1 M EDTA solution.

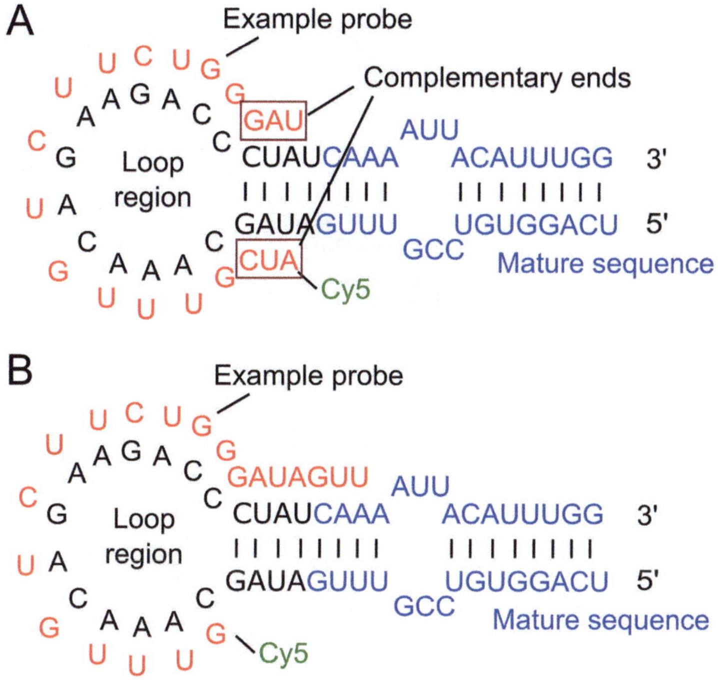


**Fig. 2** Optimal miRNA hairpin probe design. To meet probe length requirements, miRNA loop probe (red) will generally extend longer than loop region. To avoid designing a probe with complementary ends (**a**), skew the probe to one side of the hairpin arms (**b**). The resulting probe may overlap somewhat with the mature miRNA sequences (blue)

1 M HEPES stock: Add 2.38 g of powdered HEPES to 10 mL UltraPure water.

2 M NaCl stock: Dissolve 0.58 g NaCl in 9 mL UltraPure water, top up to 10 mL.

### 2.3 Immunofluorescence

1. Blocking buffer: 1% BSA in PBS. Store at 4 °C.
2. Prolong Diamond (Invitrogen) mounting media (*see* **Note 6**).
3. Glass slides for mounting cover slips.
4. 1 μg/mL DAPI (Invitrogen) in PBS. Store in −20 °C away from light.
5. Antibodies at desired concentrations.
6. Confocal microscope and image analysis.

## 3 Methods

All procedures are to be done at room temperature unless otherwise stated. Handle with care to avoid ribonuclease contamination (*see* **Note 7**). For a workflow of the following steps, *see* Fig. 3.

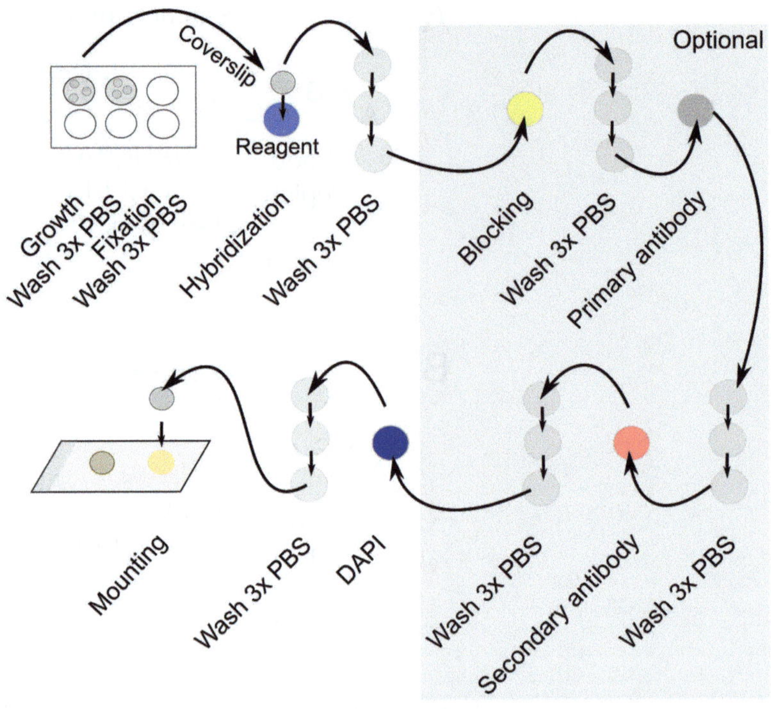


**Fig. 3** Progression of coverslips through miR-ISH process. The steps in Subheading 3.1 are performed in a 6-well plate. All the remaining steps are completed by inverting coverslip and placing onto a drop of reagent on parafilm. Between washing and reagent steps, gently tap edge of coverslip onto Kimwipe to remove excess fluid

### *3.1 Cell Culture to Prepared Coverslips*

1. Place coverslips on the base of a 6-well plate and add 50 μg/mL collagen until just covered (*see* **Note 8**).
2. Incubate at 37 °C for 1 h, remove and wash thrice in sterile PBS. Add prewarmed (37 °C) growth media and press down on coverslips lightly with pipette tip to ensure coverslips are firmly resting on the base of the wells.
3. Seed cells into wells at a density of $0.3 \times 10^6$ cells/mL and grow in a 37 °C incubator with 5% $CO_2$ until 70–80% confluent (*see* **Note 9**).
4. Wash cells three times with sterile PBS. Add 2 mL of cold 100% methanol to washed coverslips and incubate for 2 min (*see* **Note 10**). Wash thrice again with PBS and add 3 mL PBS to each well for storage. Fixed cells can be stored at 4 °C for up to 2 weeks.

### *3.2 miRNA In Situ Hybridization*

1. Dilute 100 pmol fluorophore-tagged probe (i.e., Cy5) in 100 μL hybridization buffer. Place gently onto parafilmed apparatus within humidity chamber (Fig. 1). Ideally, this

should form a bead of solution where the coverslip will lightly sit on top.

2. Invert coverslips and remove excess liquid from coverslip by touching the edge on a Kimwipe. Place coverslip upon probe mixture. Extra care is to be taken from here to avoid light exposure to the fluorophore.
3. Incubate overnight (16–20 h) at room temperature (*see* **Note 11**).
4. Washing. Place three drops of PBS (100 μL) on parafilm and wash the coverslip by sequentially transferring coverslips on each of the drops.
5. Optional (*see* **Note 12**): continue onto Subheading 3.3 for protein detection or skip to Subheading 3.3, **step 5**, for DAPI staining and mounting.

### 3.3 Indirect Immunofluorescence

1. Block coverslips with blocking solution. Incubate for 30 min. Wash with PBS.
2. Dilute antibody in blocking solution to optimized concentration or manufacturer's recommendation. Add coverslips to 50 μL of antibody solution. Incubate for 1 h in the dark.
3. Wash thrice with PBS as per **step 4**, Subheading 3.2.
4. Prepare secondary fluorophore conjugated antibody in blocking buffer. Remove excess liquid from coverslips by tapping on Kimwipe and add to secondary antibody solution. Incubate for 1 h in the dark. Wash with PBS (**step 4**, Subheading 3.2).
5. DAPI staining is recommended. Incubate 1:1000 DAPI solution with coverslips for 10 min.
6. Wash thrice with PBS (**step 4**, Subheading 3.2). Remove excess liquid with Kimwipe. Mount in 8 μL Prolong Diamond Mounting medium and set overnight in a dry dark place. Mounted slides can be stored at room temperature in a dry dark slide box for several days or at 4 °C for several weeks.
7. Visualize on a confocal microscope (*see* **Note 13**).

## 4 Results and Discussion

Mature miRNAs are now known to localize in multiple subcellular compartments in the cytoplasm including endoplasmic reticulum, RNA granules, mitochondria, and multivesicular bodies [7]. However, the functional role and dynamical distribution of mature miRNAs and their precursor miRNA hairpin in distinct subcellular compartments are still unknown. We developed a simple and versatile method to co-localize mature miRNAs and their respective precursor hairpins, which is compatible with standard

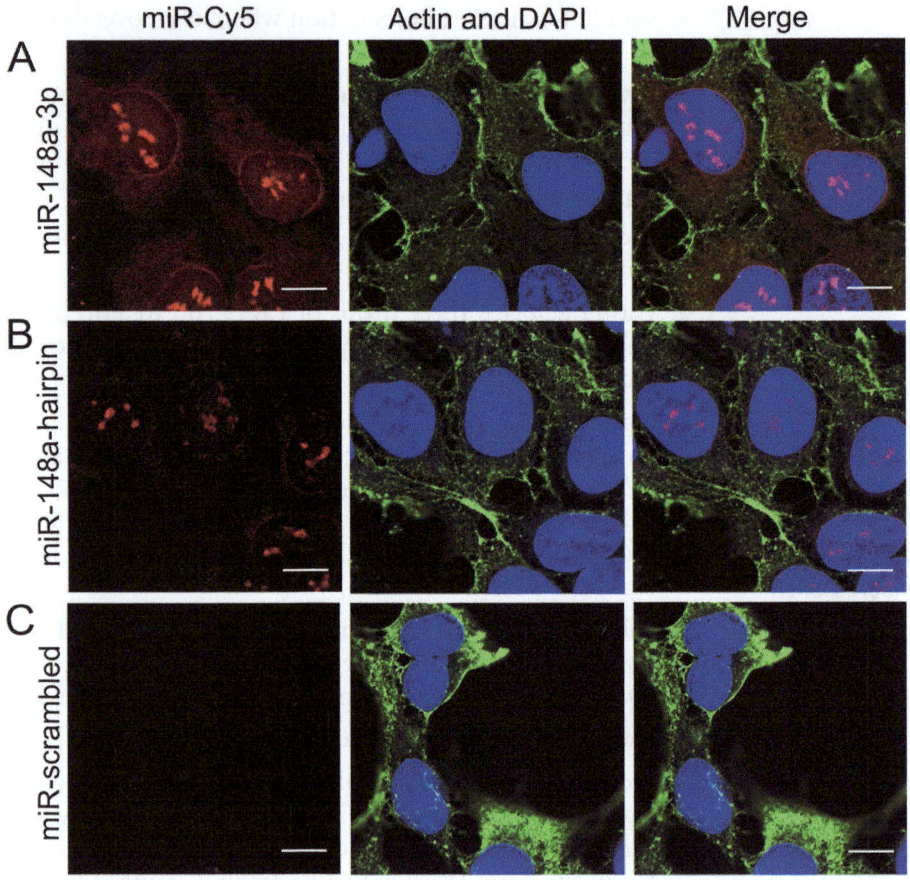


**Fig. 4** Example microRNA in situ hybridization results in HEK293 cells. Cy5-tagged probes (red) used were complementary to miR-148a-3p (**a**), miR-148a hairpin (**b**) and a scrambled miRNA probe (miR-148a-3p randomized sequence, **c**). DAPI nuclear staining (blue) and beta-actin (green; Sigma-Aldrich antibody, number: A1978 [1:200 dilution] with AlexaFluor568 secondary antibody) used to position miRNAs relative to nucleus and cytoplasm (merge)

immunofluorescence and DNA markers. Example of an application of the miR-ISH immunofluorescence assay is shown in Fig. 4 in HEK293 cells, with DAPI staining for DNA to visualize the nucleus and anti-beta actin antibody to visualize the cytoplasm. The mature miRNA probe, miR-148a-3p, localized to both cytoplasm and nucleus with high density areas inside the nucleus and along the nuclear membrane (Fig. 4a), while the precursor hairpin probe showed a different distribution more concentrated in globular structures inside the nucleus and nuclear membrane (Fig. 4b). As a control, a scrambled miRNA probe based on scrambled miR148a-3p sequence showed no specific signal (Fig. 4c). Our combined miR-ISH/immunofluorescence method can be applied to a range of cell types and experimental conditions to determine the subcellular distribution patterns of miRNA mature and hairpin sequences.

## 5 Notes

1. Growth media and cell treatment will differ depending on cells and cell lines used. This method is optimized for cultured HEK293 cells but it should be applicable to other adherent cell lines. Primary cells and frozen sections have also been used with success (data not shown) although further optimization may be required.
2. Other treatment options can be used, such as poly-L-Lysine or fibronectin, to ensure adequate numbers of cells adhered to the coverslips. Follow standard procedure if alternative treatment is used.
3. The volume of buffers, probe, and antibodies described is for a 10 mm diameter coverslip, and should be scaled according to coverslip size.
4. Humidity chamber is used as a precaution to prevent evaporation of hybridization buffer during long incubations.
5. Buffer composition was carefully chosen based on reagent purpose; however, substitutions and additions can be made as required. While urea acts as a destabilizing agent necessary to prevent secondary structure of the probes, formamide may also be used [18]. Formamide is a hazardous solvent and was initially discounted due to its toxicity. HEPES was also used as a denaturant and its use in PCR tends to lower nonspecific interactions of PCR primers [19]. Higher NaCl buffers tend to produce reproducibly low background images due to lowering nonspecific hybridization [20]. EDTA assists in protecting nucleic acids against enzymatic degradation by chelating the cofactor, magnesium [21]. RNAse inhibitors could also be added or to replace EDTA if need be. Additionally, sheared salmon sperm DNA or similar reagents may also reduce nonspecific binding in a pre-hybridization step.
6. An equivalent antifade mounting media can be used.
7. Take care to avoid ribonuclease contamination; this may include using special RNase/DNase-free pipette tips, precleaning work area with RNase-Zap or similar and using UltraPure RNase-free water.
8. Collagen, or similar, treatment of coverslips may not be necessary for all cell preparations.
9. Seeding density, passaging method, and incubation length prior to fixation will depend on cells used. Ideally, cells should not overlap with one another as it interferes with the quality of image.

10. Successful miR-ISH has been achieved with 4% paraformaldehyde (PFA) fixation but it may inhibit probe interaction with stem loop. A separate permeabilization step will be required for PFA fixation and optimization to prevent over-crosslinking, which may inhibit hybridization.
11. While optimal RNA:RNA hybridization occurs at higher temperatures, we achieved sufficient hybridization at room temperature (approximately 22 °C). The urea is unstable at higher temperatures, and can release ammonium gas if heated. Additionally, longer incubation time and urea content of the hybridization buffer may compensate for lower hybridization temperatures [22, 23].
12. Extra fixation steps may be added following hybridization to stabilize the probe–target duplex using 4% PFA in PBS. This could be especially useful if the probe–target interaction is weak and followed by immunofluorescence.
13. Image J (FIJI) contains a variety of tools for analyzing co-localization and general image analysis. Various practical guides and resources are available for co-localization analysis [24, 25].

## References


1. Ha M, Kim VN (2014) Regulation of microRNA biogenesis. Nat Rev Mol Cell Biol 15:509. https://doi.org/10.1038/nrm3838
2. Friedman RC, Farh KK, Burge CB, Bartel DP (2009) Most mammalian mRNAs are conserved targets of microRNAs. Genome Res 19 (1):92–105. https://doi.org/10.1101/gr.082701.108
3. Cougot N, Babajko S, Séraphin B (2004) Cytoplasmic foci are sites of mRNA decay in human cells. J Cell Biol 165(1):31
4. Montgomery Taiowa A, Ruvkun G (2013) MicroRNAs visit the ER. Cell 153 (3):511–512. https://doi.org/10.1016/j.cell.2013.04.014
5. Borralho PM, Rodrigues CM, Steer CJ (2015) microRNAs in mitochondria: an unexplored niche. Adv Exp Med Biol 887:31–51. https://doi.org/10.1007/978-3-319-22380-3_3
6. Li ZF, Liang YM, Lau PN, Shen W, Wang DK, Cheung WT et al (2013) Dynamic localisation of mature MicroRNAs in human nucleoli is influenced by exogenous genetic materials. PLoS One 8(8):e70869. https://doi.org/10.1371/journal.pone.0070869
7. Leung AKL (2015) The whereabouts of miRNA actions: cytoplasm and beyond. Trends Cell Biol 25(10):601–610. https://doi.org/10.1016/j.tcb.2015.07.005
8. Cipolla GA (2014) A non-canonical landscape of the microRNA system. Front Genet 5:337. https://doi.org/10.3389/fgene.2014.00337
9. Zhang J, Li S, Li L, Li M, Guo C, Yao J et al (2015) Exosome and exosomal microRNA: trafficking, sorting, and function. Genomics Proteomics Bioinformatics 13(1):17–24. https://doi.org/10.1016/j.gpb.2015.02.001
10. Van Stedum S, King W (2003) Basic FISH techniques and troubleshooting. In: Fan Y-S (ed) Molecular cytogenetics: protocols and applications. Humana Press, Totowa, NJ, pp 51–63. https://doi.org/10.1385/1-59259-300-3:51
11. Wang F, Flanagan J, Su N, Wang L-C, Bui S, Nielson A et al (2012) RNAscope: a novel in situ RNA analysis platform for formalin-fixed, paraffin-embedded tissues. J Mol Diagn 14 (1):22–29. https://doi.org/10.1016/j.jmoldx.2011.08.002
12. Jensen E (2014) Technical review: in situ hybridization. Anat Rec 297(8):1349–1353. https://doi.org/10.1002/ar.22944
13. Urbanek MO, Nawrocka AU, Krzyzosiak WJ (2015) Small RNA detection by in situ hybridization methods. Int J Mol Sci 16 (6):13259–13286. https://doi.org/10.3390/ijms160613259
14. Starega-Roslan J, Krol J, Koscianska E, Kozlowski P, Szlachcic WJ, Sobczak K et al

(2011) Structural basis of microRNA length variety. Nucleic Acids Res 39(1):257–268. https://doi.org/10.1093/nar/gkq727

15. miRCURY® LNA® miRNA detection probes handbook (2017). Qiagen
16. IDT (2018) How do I use the OligoAnalyzer tool to analyze possible hairpins and dimers formed by my oligo? Integrated DNA Technologies. https://sg.idtdna.com/pages/support/faqs/how-do-i-use-the-oligoanalyzer-tool-to-analyze-possible-hairpins-and-dimers-formed-by-my-oligo. Accessed 14 June 2018
17. Leitzmann MF (2005) Is there a link between macronutrient intake and prostate cancer? Nat Clin Pract Oncol 2(4):184–185. https://doi.org/10.1038/ncponc0127
18. Simard C, Lemieux R, Cote S (2001) Urea substitutes toxic formamide as destabilizing agent in nucleic acid hybridizations with RNA probes. Electrophoresis 22(13):2679–2683. https://doi.org/10.1002/1522-2683(200108)22:13<2679::aid-elps2679>3.0.co;2-l
19. Ahmad A, Ghasemi J (2007) New buffers to improve the quantitative real-time polymerase chain reaction. Biosci Biotechnol Biochem 71(8):1970–1978. https://doi.org/10.1271/bbb.70164
20. Rose K, Mason JO, Lathe R (2002) Hybridization parameters revisited: solutions containing SDS. BioTechniques 33(1):54–56, 58
21. Yagi N, Satonaka K, Horio M, Shimogaki H, Tokuda Y, Maeda S (1996) The role of DNase and EDTA on DNA degradation in formaldehyde fixed tissues. Biotech Histochem 71(3):123–129
22. Durm M, Haar FM, Hausmann M, Ludwig H, Cremer C (1997) Non-enzymatic, low temperature fluorescence in situ hybridization of human chromosomes with a repetitive alpha-satellite probe. Z Naturforsch C 52(1–2):82–88
23. Hutton JR (1977) Renaturation kinetics and thermal stability of DNA in aqueous solutions of formamide and urea. Nucleic Acids Res 4(10):3537–3555
24. Dunn KW, Kamocka MM, McDonald JH (2011) A practical guide to evaluating colocalization in biological microscopy. Am J Physiol Cell Physiol 300(4):C723–C742. https://doi.org/10.1152/ajpcell.00462.2010
25. Schindelin J, Arganda-Carreras I, Frise E, Kaynig V, Longair M, Pietzsch T et al (2012) Fiji: an open-source platform for biological-image analysis. Nat Methods 9:676. https://doi.org/10.1038/nmeth.2019

# Chapter 12

# Digital Holographic Imaging as a Method for Quantitative, Live Cell Imaging of Drug Response to Novel Targeted Cancer Therapies

**Laura V. Croft, Jaimie A. Mulders, Derek J. Richard, and Kenneth O'Byrne**

## Abstract

Digital holographic imaging (DHI) is a noninvasive, live cell imaging technique that enables long-term quantitative visualization of cells in culture. DHI uses phase-shift imaging to monitor and quantify cellular events such as cell division, cell death, cell migration, and drug responses. In recent years, the application of DHI has expanded from its use in the laboratory to the clinical setting, and currently it is being developed for use in theranostics. Here, we describe the use of the DHI platform HoloMonitorM4 to evaluate the effects of novel, targeted cancer therapies on cell viability and proliferation using the HeLa cancer cell line as a model. We present single cell tracking and population-wide analysis of multiple cell morphology parameters.

**Key words** Digital holographic imaging, Noninvasive live cell imaging, Label-free live cell imaging, Apoptosis imaging, Drug effect analysis, HolomonitorM4, Cytotoxicity, Targeted cancer therapies

## 1 Introduction

Over the past decade, digital holographic imaging (DHI) has been developed as a label-free, 3D, live cell imaging technique with the capacity to analyze cells growing in culture for long periods of time [1–4]. DHI uses phase-shift imaging combined with computer algorithms to construct holographic images and enable the simultaneous study of multiple cellular parameters such as cell count, confluence, optical thickness, cell area, irregularity, and volume, at both single cell level and in entire cell populations [5]. Fluorescence-based imaging of live cells requires the use of dyes and high intensity lasers to enable cell detection, which causes phototoxicity and leads to limitations in experimental analysis [6–8]. DHI is label-free and thus a noninvasive method that allows for unbiased, detailed analysis of live cell behavior in multiple settings. Indeed, this technique has been demonstrated to accurately measure cell growth [9, 10], cell motility and migration [11–13], cell death

Jyotsna Batra and Srilakshmi Srinivasan (eds.), *Theranostics: Methods and Protocols*, Methods in Molecular Biology, vol. 2054, https://doi.org/10.1007/978-1-4939-9769-5_12, 

[14–19], cell cycle [20, 21], and drug responses [22–26] in a wide variety of cell lines in the laboratory and in the clinical setting [5]. DHI was demonstrated as a useful screening tool for detection of neoplasia in cervical cancer samples [27], for blood cell analysis to detect malaria infected red blood cells [28], for the characterization of sperm morphology, motility, and concentration [29], and for quantification of inflammation in ulcerative colitis [30]. More recently, DHI has extended its use in GlycoImaging, where a new type of cancer-specific nanoprobes are combined with a patient's blood sample and the cell interaction with the nanoprobe can be visualized in real time using the HoloMonitor platform to confirm the presence of cancer cells in a diagnostic setting [31]. This study aims to develop more sensitive methods to detect and diagnose cancer at an earlier stage than what is possible today. This application can be further developed for use in theranostics, where cancer therapeutics could be conjugated to, or encapsulated within, these nanoprobes, and then using DHI, both the specific targeting to cancer cells and response to various therapies can be visualized and quantified in patient-derived samples. A recent review providing insight into how DHI is an ideal method for studying live cell dynamics of drug effects on cell migration, proliferation, and apoptosis also points out that the use of DHI applications in cancer theranostics is rapidly emerging [32]. DHI would be uniquely well suited for assessing the effects of different therapies on patient-derived cancer cells as the technology is noninvasive, allowing the cells to remain in their original state. This would render the predictive power of such experiments greater than that of the more invasive, fluorescent label-requiring methods. Such methods are often endpoint methods, while in the case of DHI the cells can be retrieved at the end of the imaging so they can be used for additional experiments.

In this chapter we describe the use of the HoloMonitorM4 DHI platform as a fast, automated, and cost-effective evaluation tool for studying the drug response of two lead compounds, DKLS02 and DKLS03, of a new class of targeted cancer drugs using the HeLa cervical cancer cell line as a model. In order to delineate the cellular response to our novel drugs, we compare DKLS02 and DKLS03 with some well characterized cancer drugs: Etoposide, a topoisomerase II inhibitor that induces double-stranded DNA breaks [33, 34]; Staurosporine, a protein kinase inhibitor that causes apoptosis [35] and Cisplatin, a chemotherapeutic drug that induces DNA adduct formation and interferes with DNA-dependent cellular functions [36]. We present representative data of population-based analysis of cell proliferation and cell morphology parameters, as well as single cell tracking analysis of morphological changes in response to the various treatments.

## 2 Materials

HoloMonitor M4 microscope (Phase Holographic Imaging, PHI, Lund Sweden), HoloLid 71120 (Phase Holographic Imaging, PHI, Lund Sweden), 6-well plates, and RPMI media 1640 supplemented with 10% fetal calf serum. $CO_2$ incubator with humidified chamber for cell culture. HeLa cell line (American Tissue Culture Collection). Etoposide (Selleckchem), Staurosporine (Selleckchem), and Cisplatin sterile concentrate 1 mg/mL (Hospira).

## 3 Method

All procedures involving cell culture are carried out in a biosafety cabinet.

1. Plate HeLa cells into a 6-well plate at 90,000 cells per well in 4 mL RPMI media and incubate in the humidified chamber incubator at 37 °C for a minimum of 8 h to allow the cells to adhere. Longer incubation time can be performed as required for individual cell lines. *See* **Note 1**.
2. Clean and sterilize the HoloLids as per the manufacturer's instructions: wash in detergent, rinse with deionized water, and then soak in 70% ethanol for 10–15 min. Inside the biosafety cabinet, remove the HoloLids from the ethanol with sterile forceps, place in large sterile petri dishes, and allow to air-dry for 30 min. Then, cover the petri dishes containing the HoloLids, place inside the incubator, and allow to adjust to 37 °C before use in imaging. *See* **Note 2**.
3. Launch HStudio software and follow the prompts for automatic background image calibration. The stage must be empty for this step. Make sure the three diagnostic values are in the green (Exposure time, Pattern contrast, and Hologram noise).
4. Add the drug treatments to the cells in each well recording the position in the plate for each treatment. In this experiment, the apoptosis-inducing drugs Etoposide (85 μM) and Staurosporine (1 μM) are used as positive controls. The chemotherapeutic drug Cisplatin (20 μg/mL) is used as an additional cytotoxic drug for comparison. The novel investigatory drugs are added according to preestablished conditions. In this case, DKLS02 and DKLS03 single-stranded RNA oligonucleotides are added at 100 nM final concentration by transfection using RNAimax as per manufacturer's instructions.
5. Replace the standard plate lid with the HoloLids, making sure no bubbles are formed between the lids and the media. If bubbles are formed, tilt the plate carefully until the bubbles are removed.

6. Immediately place the plate on the HoloMonitorM4 stage making sure the cut corner of the plate is positioned in the top left corner.
7. In HStudio software, choose the culture vessel that is being used. In this case, the "Sarstedt 6-well plate with PHI lid" plate map is loaded. Click anywhere within the rectangle in well A1 and the stage will move to that position. Check the live image for the appearance of cells, while monitoring the Autofocus on the left. If the bar is in the yellow or red, and the image appears to be out of focus, use the Stage (coarse adjustment) and red circles (intermediate and fine adjustment), to focus the image, until the bar is in the middle of the green (the value should be about 1 or 1.1). Press R to adjust the contrast each time the field of view (FOV) is changed. *See* **Note 3**.
8. When the cells are in focus, click the "Remember" button after each capture position. Continue selecting the image capture positions in each well. *See* **Note 4**.
9. An alternative to the manual selection of image capture positions is the random generation of positions (this saves the *X*–*Y* co-ordinates). *See* **Note 5**.
10. Once focusing has been performed, choose a new project and assign it a name and add a new group name. Tick the boxes for Time-lapse and Capture Pattern. Define the desired length of the time-lapse and generate image capture positions by manually selecting them or generating them randomly. In this experiment images were set to capture every 2 min for 72 h. *See* **Note 6**.
11. Lastly, click on Advanced Storage and ensure data is saved to Multiple Destination Groups. This enables each set of images from one imaging position to be saved as one file for easier analysis process.
12. Start the experiment by clicking Capture, which will automatically focus and acquire the images at the set image capture positions, at the selected interval, for the selected period of time.
13. Following completion of image capture, analysis can be done on a population of cells or for single cells. The View images tab is used to visualize the images, and also to add a color palette for easier identification. The mouse scroll can be used to zoom in/out of the field of view. The measure tool can be used to profile one or a group of cells, which allows determination of height and roughness on the cell surface (or blebbing).
14. Select the Identify Cells tab next, to assign a segmentation algorithm to the cells, so each cell has one mask. Choose an algorithm that seems to suit the cell type best then adjust the

numeric settings for "Background threshold" so that cell area is separated well from background noise and the entire cell is included/highlighted within the mask. This is especially important for accurate area and volume measurements. Adjust the setting number for "Object size" to make sure that each cell has one nucleus and debris is excluded. Users can manually modify cell segments by using "Manual changes," if needed (add, delete, grow, shrink or hand draw cell areas). In this experiment we used the Otsu segmentation algorithm.

15. Population-wide analysis. On the Analyze data tab, click New and add all frames or selected frames of interest. To analyze overall cell growth over the selected period of time, data can be exported to an excel spreadsheet and growth curves can be generated. Here, we selected cell confluence and cell number over time to compare the cell growth rates following each treatment. The data is shown in Fig. 1. *See* **Note 7**.

16. Population-wide analysis. On the Analyze data tab, once the selected frames of interest are added, average values against the various morphological parameters available in HStudio can be visualized as scatter plots. The axes can be easily changed as user-defined morphological parameters of interest. In this case, we used Optical Thickness (avg) vs. Cell Area (Fig. 2). Optical thickness vs. cell area are suitable parameters for analysis of drug responses, to quantify and visualize how cells round up and undergo a cell division, or apoptosis. Unperturbed cells are flat and have a high surface area while cells undergoing cell division or apoptosis have a high optical thickness and low surface area as shown in Figs. 2 and 4. Multiple analyses can be performed by opening new tabs, saving each one under a new name. For drug response studies, including cell-death events, useful parameters to analyze include thickness, area, volume, and roughness. Use Save Plot button to save these plots. *See* **Note 8**.

17. Single cell analysis. For single cell tracking, users must first confirm that the cell segmentation frame-to-frame is correct, to ensure the same cell is tracked over the duration of the time-lapse. Single-cell tracking analysis can be used for migration, motility, apoptosis, and cell cycle information such as time in mitosis.

18. After visualizing the time-lapse or movie, focus on a few cells of interest that remain in the field of view and undergo apoptosis. Go to the Tracking Cells tab, click New, and add all frames (from one imaging spot in one well) or just the frames of interest. Track those cells, it is recommended to track 1–5 cells per group. Ensure the software is following the correct cell throughout (there is a manual changes tab which allows you to correct incorrect calling). Switch to Plot Movement tab to see migration and motility data. Switch to Plot Features tab

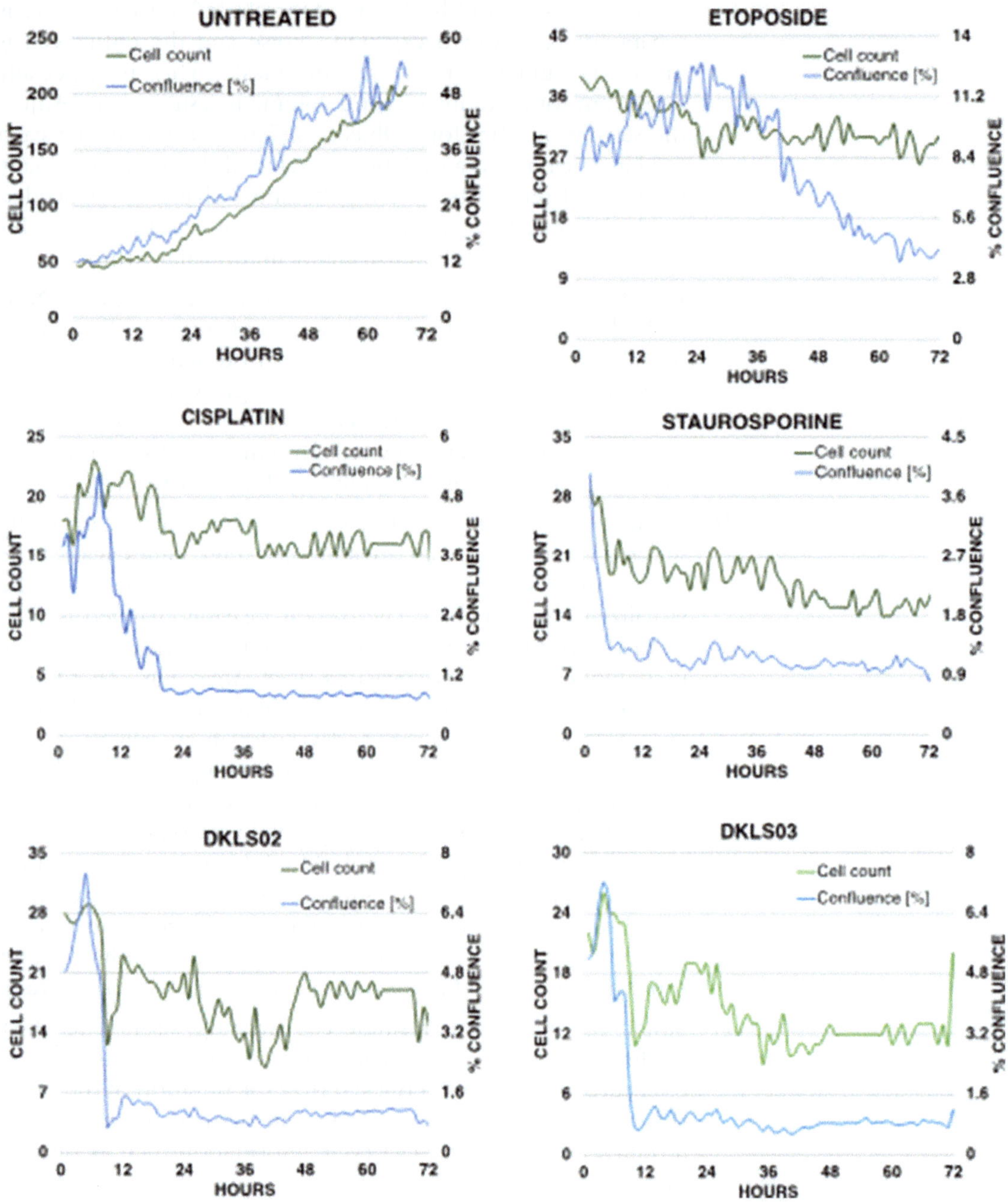


**Fig. 1** Population-wide analysis of HeLa cell count and confluence over time following drug treatments acquired using the HoloMonitorM4. Untreated cells show a continuous increase in cell number and confluence over time. Etoposide-treated cells show an initial increase in confluence due to increase in cell size, followed by a steep drop due to apoptosis onset at around 24–36 h post treatment. Cisplatin-treated cells show an increase in confluence and cell number for about 12 h, also followed by a steep drop in confluence at later timepoints. Staurosporine-treated cells show an immediate decrease in cell number and confluence, suggesting a rapid drug-induced effect. DKLS02 and DKLS03 treated cells show an initial increase in confluence and cell number following drug administration by transfection, followed by a sudden decrease in both cell confluence and number. This pattern is consistent with the cisplatin and staurosporine treatments, suggesting that DKLS02 and DKLS03 have similar drug-induced effects on HeLa cells

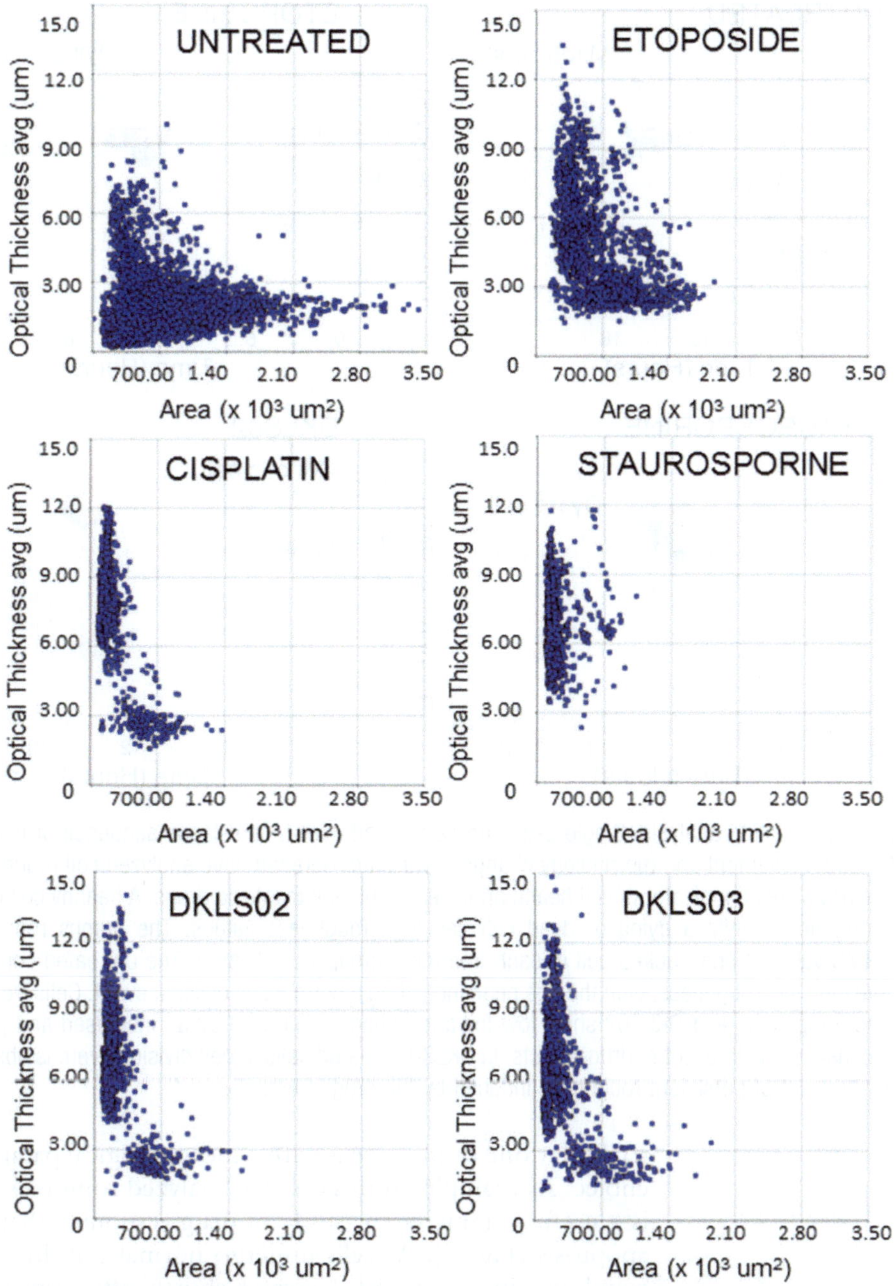


**Fig. 2** Population-wide analysis of HeLa cell optical thickness over cell area (μm$^2$) in response to drug treatments. Scatter plots show the drug-specific effects on HeLa cell optical thickness (average) over cell area (μm$^2$) as acquired using the HoloMonitorM4. The analysis shows the cells in the field of view over selected time-points during the time-lapse imaging. Scatterplot of untreated cells shows a representation of cells during a normal cell cycle. Unperturbed cells have a low thickness and large area in the G1 phase of the cell cycle as defined by a low thickness and large area-left bottom corner. As cells "round up" to divide during mitosis, they have a high thickness and small surface area. As shown in the scatterplots, etoposide, staurosporine, cisplatin, DKLS02, and DKLS03-treated HeLa cells all show a similar pattern of high thickness and low area, which is consistent with the increase in cell size during apoptosis as well as mitosis. Further morphological data is necessary to distinguish between the two processes, e.g. roughness

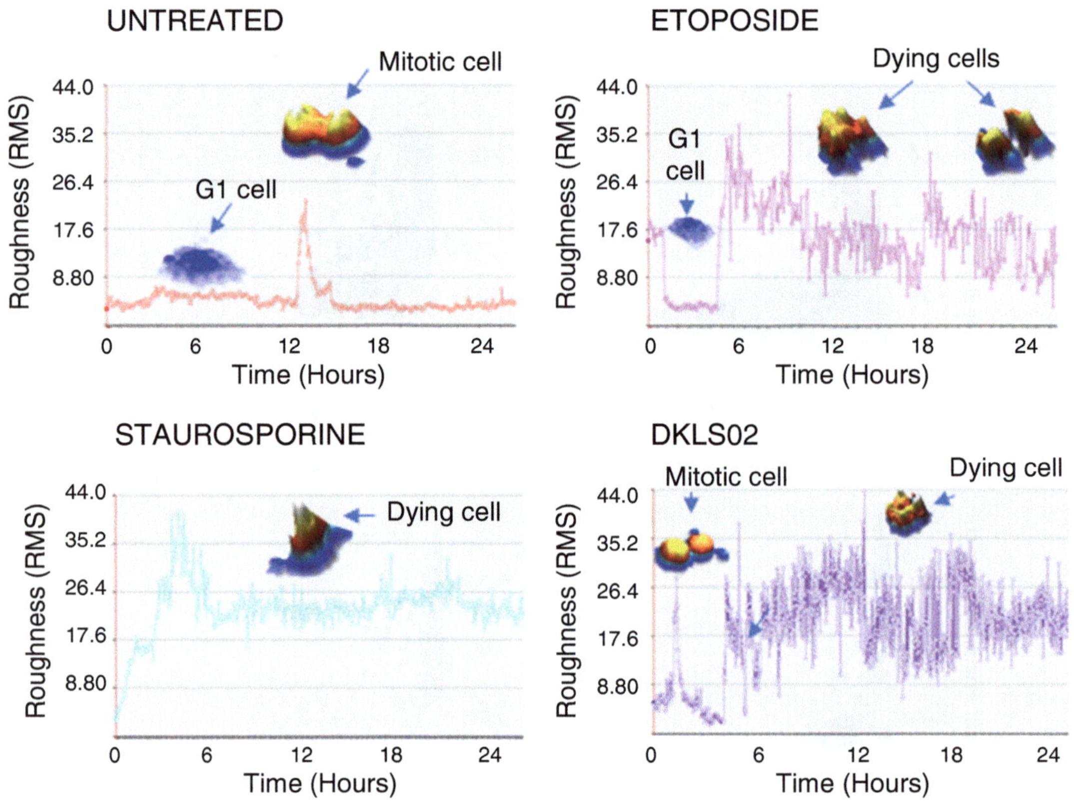


**Fig. 3** Single cell tracking analysis. Single cells can be tracked over a time-lapse sequence of images and analyzed for both movement and morphology changes over time. Here we have analyzed cell roughness root mean square (RMS) as an indicator of cell health after treatment with cytotoxic drugs. A healthy cell has a low degree of roughness, while a dying or dead cell has high roughness values. The graphs represent the roughness RMS value of one tracked cell in each treatment group as indicated. The untreated cell shows a smooth curve with low roughness, with the one peak indicating a normal cell division event. Cells treated with staurosporine, etoposide, and DKLS02 show low initial roughness, followed by an increased and persistent roughness, consistent with onset of drug-effects. In DKLS02-treated cells, a cell division event is observed as a peak prior to onset of persistent roughness induced by the drug effects

to display one or more cells with a morphological parameter of choice. As example, in this case we analyzed roughness, which is a useful parameter for study of drug responses that induce apoptosis (Fig. 3). As cells undergo normal cell division, the roughness increases and decreases shortly after, once mitosis has been completed (Fig. 3, untreated cells). During apoptosis, roughness also increases; however, this persists for prolonged periods of time and can be visualized as shown in Fig. 3, following Etoposide, Staurosporine, and DKLS02 treatments.

19. The Export Images tab can be used to export all or some of the images from any time point during the time-lapse. Images can be exported in many file formats (.tiff, .png, .bmp, etc.). Examples of 3D images captured at the beginning of the

time-lapse and once treatment-related effects are visible are shown in Fig. 4. *See* **Note 9**.

20. The Export Images Tab is also used to generate movies. In View Images tab, ensure that the appearance and perspective are optimal (adjust color, choose 2D or 3D, rotate the FOV, or zoom in on fewer cells), then select "Use for all" for coloring and perspective, export the movie and save as a .avi file. *See* **Note 10**.

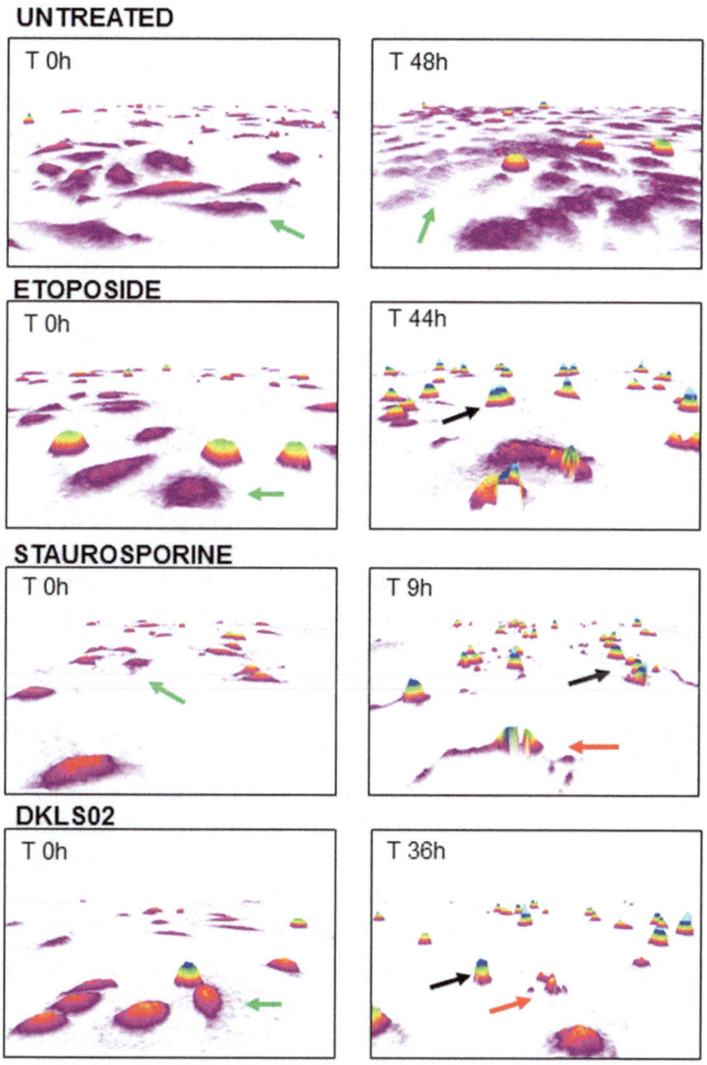


**Fig. 4** Holographic 3D cell images with pseudo coloring indicative of cell thickness. Images are overview of cells in one selected image capture location at the start of the time-lapse at T 0 h and at drug effect onset, at T as indicated. Healthy cells are thin and have a large surface area (green arrows) while dying and dead cells are thick, with low surface area and can also appear fragmented (black and red arrows respectively), as observed at late timepoints following etoposide, staurosporine, and DKL02 treatments

## 4 Notes

1. Here, 4 mL of media/well is used despite the 3 mL/well recommendation by the manufacturer. The higher volume ensures that the observation window remains immersed for the entire imaging time of 72 h. The seeding cell number is another critical factor to consider. Depending on the length of the time-lapse, cells must be seeded at 3–15% confluency to ease the cell segmentation/identification and tracking process.
2. It is important to only soak the HoloLids in ethanol for the recommended time of 10–15 min. Longer incubation may damage the plastic and render the lids unsuitable for imaging. Place the HoloLids in the incubator for at least 30 min prior to imaging to warm them up and prevent condensation when placed on the warm plate. Ensure the incubator water tray is filled to maximum in order to maintaining humidity and reduce evaporation of media over longer time-lapse imaging sessions.
3. If it is difficult to find the cells, it is possible to toggle between the Phase and Amplitude buttons under the Viewer options. It is also possible to toggle between 2D and 3D, and to add a color palette for better contrast in this initial focusing step.
4. The "Remember" button saves the *X*, *Y*, and *Z* coordinates for the image capture position in the well. If the focus is not ideal in another image capture position, it is possible to make fine adjustments to the *Z*/stage position using the red circles, and monitoring the Digital Autofocus on the left side. The correct focus for that image capture position can be saved in Microscope Settings on the right, by selecting a new name.
5. The random capture pattern generator can be a time-saver when setting up the run; however, there is a risk that the selected image capture position may be on the edge of the HoloLid. To ensure sufficient good quality image capture positions, an extra position can be added. This is particularly of value in the set-up of 96 or 24-well plates, which can be a time-consuming process if manual selection is performed.
6. Depending on type of assay, the time-lapse imaging design may vary. In general, three image capture positions per well, with 3–5 min imaging intervals, ensures good data and tracking possibilities. A shorter interval, of 2 min is ideal to capture apoptotic events or cell cycle events. Longer intervals (12 h) can be chosen for proliferation data.
7. The graphs and plots generated within HStudio can be used as is. However, if users wish to perform more sophisticated analysis, or for example, plot graphs with primary and secondary

vertical axes, the data can be exported to excel where it can be more easily manipulated and displayed in the desired format.

8. It is possible to display the data in the scatterplot in different colors. For example, data collected from 0–12 h can be presented in blue, 13–24 h in pink and 25–48 h in yellow. This will show in the scatter plot how the selected morphological parameters for the same group of cells have changed over time. This function can be selected by adding the groups of images and clicking on the blue square in the lower left and choosing a different color.
9. Throughout the software, the camera button can save cell images, cell segmentation images, graphs, or plots.
10. Before exporting movies, the frame rate should be adjusted to slow; this is particularly handy for visualization of cell death.

## References


1. Alm K, Cirenawis H, Gisselsson L, Wingren AG, Janicke B, Molder A, Oredsson S, Persson J (2011) Digital holography and cell studies. IntechOpen, London
2. Alm K, El-Schich Z, Falck M, Gjrloff Wingren A, Janicke B, Oredsso S (2013) Cells and holograms – holograms and digital holographic microscopy as a tool to study the morphology of living cells. In: Holography – basic principles and contemporary applications. https://doi.org/10.5772/54505
3. Kemper B, von Bally G (2008) Digital holographic microscopy for live cell applications and technical inspection. Appl Opt 47(4): A52–A61
4. <1055–1062.pdf>
5. El-Schich Z, Kamlund S, Janicke B, Alm K, Wingren AG (2017) Holography: the usefulness of digital holographic microscopy for clinical diagnostics. In: Holographic materials and optical systems. https://doi.org/10.5772/66042
6. Cox S (2015) Super-resolution imaging in live cells. Dev Biol 401(1):175–181. https://doi.org/10.1016/j.ydbio.2014.11.025
7. Purschke M, Rubio N, Held KD, Redmond RW (2010) Phototoxicity of Hoechst 33342 in time-lapse fluorescence microscopy. Photochem Photobiol Sci 9(12):1634–1639. https://doi.org/10.1039/c0pp00234h
8. Tinevez JY, Dragavon J, Baba-Aissa L, Roux P, Perret E, Canivet A et al (2012) A quantitative method for measuring phototoxicity of a live cell imaging microscope. Methods Enzymol 506:291–309. https://doi.org/10.1016/b978-0-12-391856-7.00039-1
9. Janicke B, Karsnas A, Egelberg P, Alm K (2017) Label-free high temporal resolution assessment of cell proliferation using digital holographic microscopy. Cytometry A 91 (5):460–469. https://doi.org/10.1002/cyto.a.23108
10. Molder A, Sebesta M, Gustafsson M, Gisselson L, Wingren AG, Alm K (2008) Non-invasive, label-free cell counting and quantitative analysis of adherent cells using digital holography. J Microsc 232(2):240–247. https://doi.org/10.1111/j.1365-2818.2008.02095.x
11. Tuerk C, Gold L (1990) Systematic evolution of ligands by exponential enrichment: RNA ligands to bacteriophage T4 DNA polymerase. Science 249(4968):505–510
12. Kamlund S, Strand D, Janicke B, Alm K, Oredsson S (2017) Influence of salinomycin treatment on division and movement of individual cancer cells cultured in normoxia or hypoxia evaluated with time-lapse digital holographic microscopy. Cell Cycle 16 (21):2128–2138. https://doi.org/10.1080/15384101.2017.1380131
13. Langehanenberg P, Ivanova L, Bernhardt I, Ketelhut S, Vollmer A, Dirksen D et al (2009) Automated three-dimensional tracking of living cells by digital holographic microscopy. J Biomed Opt 14(1):014018. https://doi.org/10.1117/1.3080133
14. El-Schich Z, Molder A, Tassidis H, Harkonen P, Falck Miniotis M, Gjorloff Wingren A (2015) Induction of morphological changes in death-induced cancer cells monitored by holographic microscopy. J Struct

Biol 189(3):207–212. https://doi.org/10.1016/j.jsb.2015.01.010

15. Kavitha N, Chen Y, Kanwar JR, Sasidharan S (2017) In situ morphological assessment of apoptosis induced by Phaleria macrocarpa (Boerl.) fruit ethyl acetate fraction (PMEAF) in MDA-MB-231 cells by microscopy observation. Biomed Pharmacother 87:609–620. https://doi.org/10.1016/j.biopha.2016.12.127
16. Kunzelmann K (2016) Ion channels in regulated cell death. Cell Mol Life Sci 73 (11–12):2387–2403. https://doi.org/10.1007/s00018-016-2208-z
17. Ousingsawat J, Cabrita I, Wanitchakool P, Sirianant L, Krautwald S, Linkermann A et al (2017) Ca(2+) signals, cell membrane disintegration, and activation of TMEM16F during necroptosis. Cell Mol Life Sci 74(1):173–181. https://doi.org/10.1007/s00018-016-2338-3
18. Ousingsawat J, Wanitchakool P, Schreiber R, Kunzelmann K (2018) Contribution of TMEM16F to pyroptotic cell death. Cell Death Dis 9(3):300. https://doi.org/10.1038/s41419-018-0373-8
19. Vijayarathna S, Chen Y, Kanwar JR, Sasidharan S (2017) Standardized Polyalthia longifolia leaf extract (PLME) inhibits cell proliferation and promotes apoptosis: the anti-cancer study with various microscopy methods. Biomed Pharmacother 91:366–377. https://doi.org/10.1016/j.biopha.2017.04.112
20. Falck Miniotis M, Mukwaya A, Gjorloff Wingren A (2014) Digital holographic microscopy for non-invasive monitoring of cell cycle arrest in L929 cells. PLoS One 9(9):e106546. https://doi.org/10.1371/journal.pone.0106546
21. Guo P, Huang J, Moses MA (2017) Characterization of dormant and active human cancer cells by quantitative phase imaging. Cytometry A 91(5):424–432. https://doi.org/10.1002/cyto.a.23083
22. Farkas E, Szekacs A, Kovacs B, Olah M, Horvath R, Szekacs I (2018) Label-free optical biosensor for real-time monitoring the cytotoxicity of xenobiotics: a proof of principle study on glyphosate. J Hazard Mater 351:80–89. https://doi.org/10.1016/j.jhazmat.2018.02.045
23. Hackler L Jr, Ozsvari B, Gyuris M, Sipos P, Fabian G, Molnar E et al (2016) The curcumin analog C-150, influencing NF-kappaB, UPR and Akt/notch pathways has potent anticancer activity in vitro and in vivo. PLoS One 11(3): e0149832. https://doi.org/10.1371/journal.pone.0149832
24. Ozdemir A, Yildiz M, Senol FS, Simay YD, Ibisoglu B, Gokbulut A et al (2017) Promising anticancer activity of Cyclotrichium niveum L. extracts through induction of both apoptosis and necrosis. Food Chem Toxicol 109 (Pt 2):898–909. https://doi.org/10.1016/j.fct.2017.03.062
25. Semenas J, Hedblom A, Miftakhova RR, Sarwar M, Larsson R, Shcherbina L et al (2014) The role of PI3K/AKT-related PIP5-K1alpha and the discovery of its selective inhibitor for treatment of advanced prostate cancer. Proc Natl Acad Sci U S A 111(35): E3689–E3698. https://doi.org/10.1073/pnas.1405801111
26. Zhang Y, Sriraman SK, Kenny HA, Luther E, Torchilin V, Lengyel E (2016) Reversal of chemoresistance in ovarian cancer by co-delivery of a P-glycoprotein inhibitor and paclitaxel in a liposomal platform. Mol Cancer Ther 15 (10):2282–2293. https://doi.org/10.1158/1535-7163.Mct-15-0986
27. Benzerdjeb N, Garbar C, Camparo P, Sevestre H (2016) Digital holographic microscopy as screening tool for cervical cancer preliminary study. Cancer Cytopathol 124(8):573–580. https://doi.org/10.1002/cncy.21727
28. Anand A, Chhaniwal VK, Patel NR, Javidi B (2012) Automatic identification of malaria-infected RBC with digital holographic microscopy using correlation algorithms. IEEE Photonics J 4(5):1456–1464. https://doi.org/10.1109/jphot.2012.2210199
29. Di Caprio G, Ferrara MA, Miccio L, Merola F, Memmolo P, Ferraro P et al (2015) Holographic imaging of unlabelled sperm cells for semen analysis: a review. J Biophotonics 8 (10):779–789. https://doi.org/10.1002/jbio.201400093
30. Lenz P, Bettenworth D, Krausewitz P, Bruckner M, Ketelhut S, von Bally G et al (2013) Digital holographic microscopy quantifies the degree of inflammation in experimental colitis. Integr Biol (Camb) 5(3):624–630. https://doi.org/10.1039/c2ib20227a
31. <645042[1].pdf>
32. Kasprowicz R, Suman R, O'Toole P (2017) Characterising live cell behaviour: traditional label-free and quantitative phase imaging approaches. Int J Biochem Cell Biol 84:89–95. https://doi.org/10.1016/j.biocel.2017.01.004
33. Chen AY, Liu LF (1994) DNA topoisomerases: essential enzymes and lethal targets. Annu Rev Pharmacol Toxicol 34:191–218. https://doi.org/10.1146/annurev.pa.34.040194.001203

34. Chen GL, Yang L, Rowe TC, Halligan BD, Tewey KM, Liu LF (1984) Nonintercalative antitumor drugs interfere with the breakage-reunion reaction of mammalian DNA topoisomerase II. J Biol Chem 259 (21):13560–13566
35. <1204436.pdf>
36. Dasari S, Tchounwou PB (2014) Cisplatin in cancer therapy: molecular mechanisms of action. Eur J Pharmacol 740:364–378. https://doi.org/10.1016/j.ejphar.2014.07.025

# Chapter 13

# Luminescent Porous Silicon Nanoparticles for Continuous Wave and Time-Gated Photoluminescence Imaging

Tushar Kumeria, Zhi Qu, Amirali Popat, Tariq Altalhi, and Abel Santos

## Abstract

Luminescent porous silicon nanoparticles (LpSiNPs) display red-orange photoluminescence (PL) that provides large penetration depth for precise deep-tissue imaging and diagnostics. Herein, we describe in detail the fabrication process of porous silicon nanoparticles (pSiNPs), activation of photoluminescence, quantum yield measurement, and PL imaging. LpSiNPs perform as imaging probe for both the continuous wave and time-gated PL imaging.

**Key words** Porous silicon nanoparticles, Time-gated imaging, Photoluminescence, Quantum yield, Quantum dots

## 1 Introduction

Photoluminescence (PL) and fluorescence (FL) imaging have become very popular recently due to easy use of the technique and availability of a wide variety of small molecule PL probes. The issues with the small molecule PL probes has led to more research being directed towards development of nanoparticles-based PL probes with most early efforts concentrated on quantum dots (QDs) [1–3]. However, QDs are typically based on heavy metals like cadmium (Cd) and arsenic (As) and were proven to be highly cytotoxic in the year 2000 by Prof. S. N. Bhatia's group [4]. Once the toxicity of QDs was established, efforts were diverted towards alternative more biocompatible and noncytotoxic PL nanoparticles probes.

Porous silicon (pSi) was first reported by the Uhlir couple in 1956 at the Bell Laboratories, but the interest in pSi exploded in 1990s after the discovery of the quantum confinement effect and red-orange photoluminescence by Gösele and Canham, respectively [5–8]. Soon after, Canham reported the biocompatible and noncytotoxic nature of pSi in 1995 [9]. Therefore, pSi has gathered huge attention in biomedical applications field as therapeutics cargo

Jyotsna Batra and Srilakshmi Srinivasan (eds.), *Theranostics: Methods and Protocols*, Methods in Molecular Biology, vol. 2054, https://doi.org/10.1007/978-1-4939-9769-5_13, 

delivery and imaging agent in last two decades because of its biocompatibility and intrinsic PL coupled with large surface area, tuneable pore architecture, surface chemistry, and particle size [10–12]. Furthermore, tuneable surface chemistry allows for attachment of different targeting moieties (e.g. peptides and antibodies) to enable targeted delivery and noninvasive PL imaging using functionalized LpSiNPs.

In this chapter, we describe in details the process of electrochemical fabrication of porous silicon nanoparticles of 200 nm [13]. We will focus on chemical PL activation process that involves growth of passivating oxide sheath layer leading to quantum confinement of the silicon core in presence of an oxidizing agent like sodium borate [14]. LpSiNPs are physically and chemically characterized using standard techniques like scanning and transmission electron microscopy, Raman, Fourier transform IR spectroscopy, while their PL efficiency is measured as quantum yield (QY) that are described in this chapter. Lastly, we have focused on PL imaging using LpSiNPs as nanoparticle probes [10]. The major challenge for PL imaging is the strong background tissue autofluorescence. However, the long PL decay lifetime (>10 s of μs) of LpSiNPs provides a means to overcome this limitation, combined with an appropriate time-gated PL imaging system [15].

## 2 Materials

### 2.1 Fabrication of LpSiNPs

Single-side polished highly boron-doped (p-type) monocrystalline silicon wafers (100) of resistivity less than 1 mΩ-cm and thickness ~525 μm (Virginia semiconductors or Sil'tronix Silicon Technologies) are used for fabricating LpSiNPs [11]. ACS-grade hydrofluoric acid (48%), potassium hydroxide, and absolute ethanol are required to prepare the etching electrolyte solutions. *Important note: Hydrofluoric acid (HF) is a highly corrosive and toxic acid, thus any exposure to HF should be avoided (see* ***Note 1****).* Thick aluminum foil (1 mm) is used as back contact for Si in the custom-designed Teflon etch cell. A platinum wire (0.5 mm diameter) shaped as a coil is used as the counter electrode. All the aqueous solutions are prepared using Milli-Q water with resistivity of 18 MΩ-cm. A computer-controlled high power source measure unit (power supply; Keithly 2651A or 2461) is required to supply current at controlled rate. A schematic illustration of the electrochemical etching setup for pSi fabrication is demonstrated in Fig. 1.

### 2.2 Quantum Yield Measurements

For relative quantum yield (QY) measurements Rhodamine 6G (R6G) solution (1 mM) is required. A UV-Vis spectrometer and a fluorescence spectrometer are required for measuring the absorbance and photoluminescence of the dye and LpSiNPs at different concentrations [16]. For absolute QY measurement, an extensive

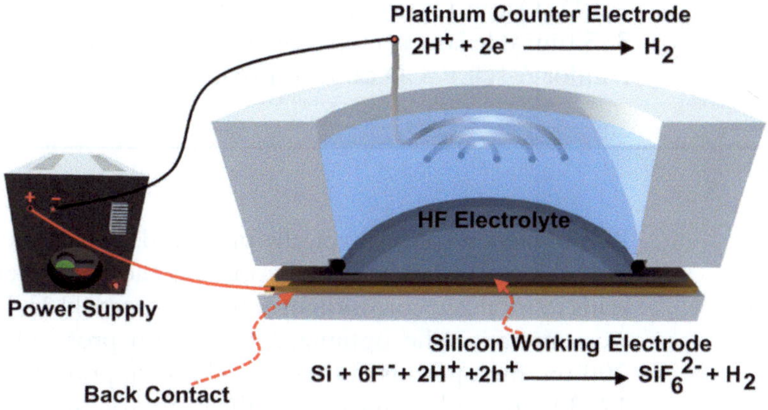


**Fig. 1** A schematic of the electrochemical etching setup. A cut-out of the HF electrolyte-filled Teflon etch cell shows the arrangement of back contact with silicon wafer and the Pt counter electrode. The two half-reactions taking place at Si anode and Pt counter electrode (cathode) are also described

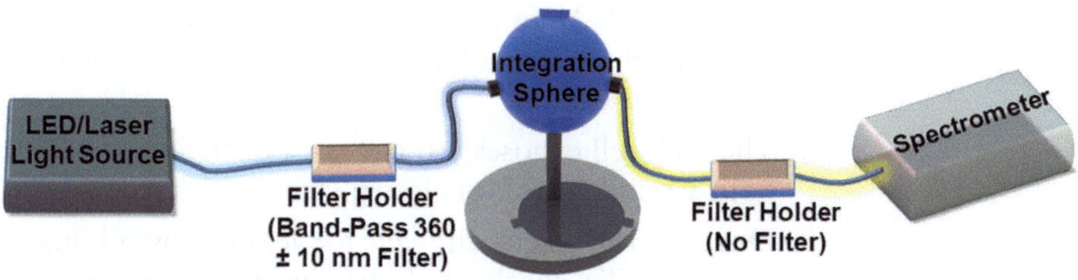


**Fig. 2** Schematic showing the absolute QY measurement setup comprising of a light source (generally LED at 365 nm or diode laser at 370 nm), coupled to an integrating sphere through an in-line optical filter holder, and a high-resolution spectrometer (like Ocean Optics QE pro) connected to the other port of IS

spectrometer setup capable of coupling with integrating sphere as shown in Fig. 2 is needed. The setup illustrated in Fig. 2 consists of a light source, which can be either a Light Emitting Diode (Ocean Optics LLS-365) or a laser that is coupled to an integrating sphere (Labsphere 3P-GPS-030-SF) through an inline optical filter holder (Ocean Optics FHS-UV). The other port of the integrating sphere is connected to a high-resolution and highly sensitive portable spectrometer (Ocean Optics QE PRO) [17].

### 2.3 *In Vivo* Photoluminescence Imaging of LpSiNP

LpSiNPs display strong red-orange PL upon excitation with UV light that enables them to act as biodegradable and highly biocompatible imaging probes. In vivo-injected LpSiNPs have been previously demonstrated for both the continuous wave imaging and time-gated imaging. For continuous wave imaging a commercially available fluorescence imaging system like IVIS® can be used. Whereas, for time-gated imaging a custom-assembled imaging setup is generally used, which consists of a pulsed light source

(LED; Ocean Optics LLS 365 or pulsed laser; Opotek Opolette 355 tuneable laser), a function generator, and a time-gated camera (Andor A-star iCCD camera) [16].

## 3 Methods

### *3.1 Fabrication and Properties of Porous Silicon*

Porous silicon (pSi) is typically produced by electrochemical etching of silicon wafers in hydrofluoric (HF) acid-based electrolyte that leads to localized corrosion of Si to form pores [13]. In this section, we summarize the optimized fabrication protocol for preparing 200 nm mesoporous Si nanoparticles with pores ranging between 12–30 nm and activation of their photoluminescence through chemical route.

1. Break the 4 in. silicon wafer in four quarters and pack one into the etch cell as shown in Fig. 1. Then, cut a 5 cm × 2 cm piece of thick Al foil and place on the base of the etch cell. Next, place the Si wafer on top of Al foil making sure the screw-holes are not covered. Align the screw-holes on the top half of the etch cell with the bottom half with Al foil and Si wafer and clamp with nylon screws.
2. The etch cell exposes 8.6 $cm^2$ area of the packed Si wafer and hold around 22 mL of electrolyte.
3. Once packed check the cell for leakage by adding 5 mL of DI-water. To check for leaks run a piece of paper between visible gaps between the top and bottom halves of the cell. If the paper is wet, open the cell and carefully repack to check the leakage again. Once no leaks are detected, remove the DI-water, dry the cell under nitrogen and place the etch cell in HF compatible fume hood.
4. Add 15 mL of HF: ethanol electrolyte (3:1 v/v) and carry out a sacrificial etch at 50 $mA/cm^2$ current density for 60 s. After 60 s, carefully transfer the used electrolyte to the appropriately labeled HF waste container. Wash the etch cell with ethanol three times followed by three rinse with DI-water.
5. Remove the sacrificial etch layer using 2 M potassium hydroxide (KOH) solution (1–2 mL). Immediately after adding KOH, bubbles should appear in the etch cell and the entire porous layer is chemically dissolved. Once the porous layer is dissolved the bubbles disappear and the Si wafer appears visibly shiny.
6. Once the wafer is clean and shiny, remove the KOH and discard in the properly labeled KOH waster container followed by three rinses with DI-water (*see* **Note 2**). Then, rinse the etch cell with ethanol three times followed by one rinse with HF electrolyte (3:1 HF:ethanol).

7. Add 20 mL HF electrolyte (3:1 v/v HF:ethanol) to the etch cell and carry out the etching using previously demonstrated perforation etch with alternating low and high current densities. In this etch profile, low current density of 50 mA/cm$^2$ is applied for 1.82 s followed by a high current density of 385 mA/cm$^2$ for 0.36 s, and this process is repeated 150 times.
8. Once the etch is complete, transfer the used electrolyte to the HF waste container and rinse three times with ethanol.
9. For removal of the porous layer from the underlying Si substrate, add 10 mL of highly dilute HF electrolyte (1:29 v/v HF: ethanol) and etch at 3 mA/cm$^2$ current density for 240 s. After 240 s, remove the used electrolyte from the cell and rinse carefully with ethanol three times.
10. Carefully transfer the lifted off porous silicon layer into a plastic weighing boat by washing the cell with ethanol.
11. Repeat the **steps 7–10** five times using the same Si wafer to obtain significant amount of porous Si layers. Transfer six of the lifted off porous Si etch layers to a 20 mL glass vial with 6 mL water and sonicate for 18 h (*see* **Note 3**).
12. Rest the sonicated particles for 1 h and collect 1 mL of the supernatant (containing pSi nanoparticles) and centrifuge at 14,400 × *g* for 10 min. Discard the supernatant after centrifugation and suspend the pellet in 1 mL water. To measure the amount (concentration) of particles, pipette out 100 μL of the resuspended pellet into a pre-weighed empty Eppendorf tube and dry in vacuum oven. Reweigh the Eppendorf tube and subtract the mass of the empty tube to obtain the mass of particles in 100 μL.
13. Based on the calculated pSi particle concentration in **step 12**, pipette out 1 mg of particles and centrifuge at 14,400 × *g* for 10 min and discard the supernatant. Suspend the pellet in 1 mL of 0.5 mM aqueous solution of sodium tetraborate and incubate at room temperature on an elliptical shaker for 90 min.
14. After 90 min, wash the pSi particles with water three times and measure their size with dynamic light scattering technique and transmission electron microscopy, and check the PL using a fluorimeter at excitation wavelength 365 nm and emission between 500–900 nm wavelength range.
15. The above 14 steps provide photoluminescent 200 nm mesoporous silicon nanoparticles (LpSiNPs) that can be used for applications like drug delivery, imaging, and diagnostics (*see* **Note 4**). This process is summarized in the schematic presented in Fig. 3.

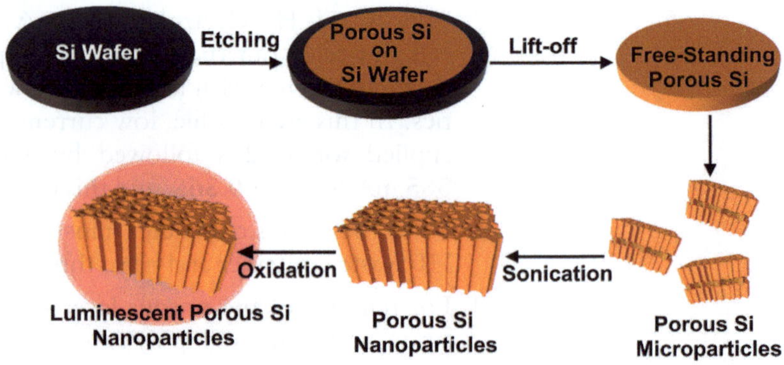


**Fig. 3** Schema of LpSiNP fabrication using electrochemical etching process detailed above followed by activation of photoluminescence through borate oxidation

## 4 Quantum Yield Measurement

Quantum yield (QY), as the name suggests, is the ratio of number of photons emitted to the number of photons absorbed. QY is either measured as relative QY with respect to a fluorophore with known QY (e.g. Rhodamine 6G) or absolute QY by collecting total PL of the material in an integrating sphere.

### 4.1 Relative Quantum Yield (QY) Measurements

1. Prepare 1 mM solution of Rhodamine 6G (R6G) in ethanol and measure its absorbance (Abs) at 365 nm. Dilute the R6G solutions to obtain six R6G concentrations with Abs values of ranging between 0.01–0.1 at 365 nm.
2. Measure the PL value of the six R6G solutions at 365 nm excitation ($\lambda_{ex}$) wavelength and emission ($\lambda_{em}$) between 420–900 nm spectral range. Calculate and note down the integrated PL intensity (i.e. area under curve of the PL spectrum) of all six concentrations of R6G solutions. Note: make sure the PL spectra are corrected.
3. Plot the measured PL values ($y$-axis) against Abs ($x$-axis). The PL vs. Abs plot should result in a curve that appears similar to Fig. 4. Fit the PL vs. Abs plot with linear regression and note down slope ($S_{R6G}$) of the linear fit (*see* **Notes 5** and **6**).
4. Suspend 1 mg of LpSiNPs in 1 mL ethanol using mild sonication for 1 min and measure the Abs at 365 nm. Similar to R6G, dilute the LpSiNP suspension to obtain six different concentrations with Abs values of ranging between 0.01–0.1 at 365 nm.
5. Measure the PL value of the six LpSiNP suspensions with excitation ($\lambda_{ex}$) wavelength at 365 nm and emission ($\lambda_{em}$) between 420–900 nm spectral range. Calculate and note

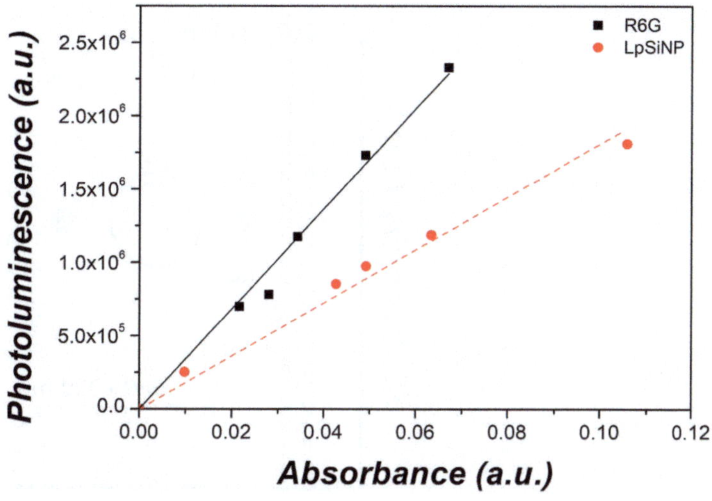


**Fig. 4** A representative PL vs. Abs plot for R6G and LpSiNPs showing a linear relationship between the two for both the fluorophores. The linear fit of R6G data provides a slope of $3.41 \times 10^7$, whereas the slope of the linear fit to LpSiNP data is $1.81 \times 10^7$. This results in a calculated relative QY of LpSiNPs of around 0.51 (only an indicative value from representative curve)

down the integrated PL intensity (i.e. area under curve of the PL spectrum) from fully corrected PL spectra for all six concentrations of LpSiNP suspension.

6. Plot the PL values of the LpSiNP suspension (*y*-axis) against the Abs (*x*-axis) for respective concentration. The PL vs. Abs plot should result in a curve that appears similar to Fig. 4. Fit the PL vs. Abs plot with linear regression and note down slope ($S_{\text{LpSiNP}}$) of the linear fit.
7. The slopes $S_{\text{R6G}}$ and $S_{\text{LpSiNP}}$ are proportional to the QY of R6G and LpSiNPs, respectively. The absolute QY of R6G in ethanol is known (approximately 0.95; i.e. 95%). The QY of LpSiNP is then calculated by comparing the $S_{\text{R6G}}$ to $S_{\text{LpSiNP}}$ using the following equation:

$$\varphi_{\text{LpSiNP}} = \varphi_{\text{R6G}} \left( \frac{S_{\text{LpSiNP}}}{S_{\text{R6G}}} \right) \left( \frac{\eta^2_{\text{LpSiNP}}}{\eta^2_{\text{R6G}}} \right) \tag{1}$$

8. Where, $\varphi_{\text{LpSiNP}}$ and $\varphi_{\text{R6G}}$ are QY of LpSiNPs and R6G, respectively. $\eta_{\text{LpSiNP}}$ and $\eta_{\text{R6G}}$ are the refractive index of the solvent used for PL and Abs measurements of LpSiNPs and R6G, respectively. Since the same solvent is used for our measurements the Eq. 1 can be reduced to Eq. 2.

$$\varphi_{\text{LpSiNP}} = \varphi_{\text{R6G}} \left( \frac{S_{\text{LpSiNP}}}{S_{\text{R6G}}} \right) \tag{2}$$

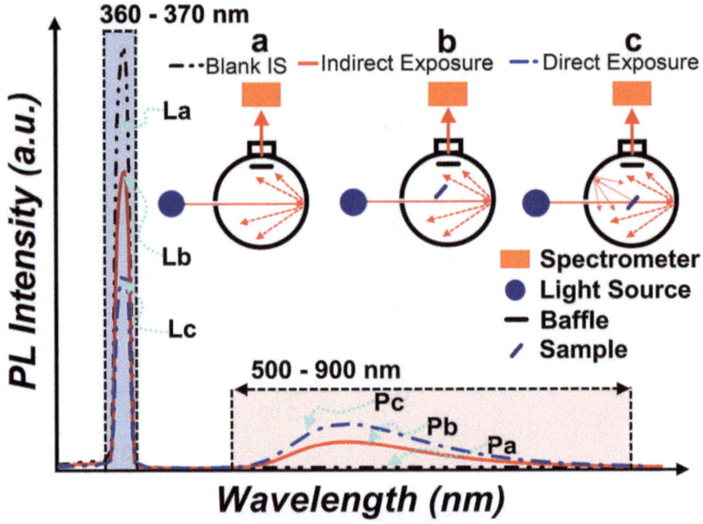


**Fig. 5** A representative PL intensity/wavelength plot obtained using the IS setup described above. The three spectrum curves are obtained under different sample conditions presented in the schemes. The Black curve (dot-dot-dash) corresponds to conditions "a" where no sample is placed inside the IS. The red curve (solid) corresponds to condition "b" where sample is placed inside the IS but not exposed to excitation light directly. The blue curve (dot-dash) is the spectrum obtained when the sample is directly exposed to excitation wavelength inside the IS as shown in condition "c." The blue-shaded region corresponds to "L" region and the pink-shaded region is the "P" component of the spectrum

### 4.2 Absolute QY Measurement Using Integrating Sphere

Absolute QY of LpSiNPs can be measured either in suspension form or as solid flakes depending upon the integrating sphere and available attachments. For typical absolute QY measurements of LpSiNPs a three measurements approach is used as outlined below:

1. After setting up the IS as shown in Fig. 5, record a UV-Vis spectrum of the empty IS (i.e. no sample) in 300–950 nm spectral range (labeled as "Spectrum a").
2. LpSiNP suspension in ethanol with Abs below 0.15 is transferred to a glass vial (e.g. HPLC vial). Cap the vial with a Teflon lid and if a Teflon lid is not available Teflon sealing tape can be used (*see* **Note 7**). Attach the vial to the sample holder of the integrating sphere (IS) and place inside the IS (Fig. 5).
3. The IS sample holder can be positioned in two settings, where the first setting exposes the LpSiNPs directly to the excitation light beam and the other excites the sample indirectly with the secondary reflection of excitation light beam inside the IS (Fig. 4).
4. Record a UV-Vis spectrum in 300–950 nm spectral range with indirect exposure settings and another spectrum in the same range with direct sample exposure setting. This enables

collection of total spectral signal from IS like in Fig. 5. The spectrum with indirect excitation is labeled as "Spectrum b," while the spectrum with direct excitation is labeled as "Spectrum c."

5. The spectra are divided into two components; "L" is the spectral signal from the light source, whereas "P" is the PL emission from the LpSiNPs in the respective wavelength range.
6. La, Lb, and Lc correspond to number of photons supplied by the light source and obtained from integrated signal intensity of the light source within 360–370 nm wavelength range under the three measurement conditions (i.e. empty IS, indirect exposure, and direct exposure), respectively.
7. Similarly Pa, Pb, and Pc, correspond to number of photons emitted by the sample and obtained from integrated PL intensity within 500–900 nm for the three measurement conditions mentioned above (*see* **Notes 8** and **9**).
8. The absolute QY of LpSiNPs can be calculated by fitting the values of La, Lb, Lc, Pa, Pb, and Pc in the following equation:

$$\varphi_{\mathrm{LpSiNP}} = \left(\frac{\mathrm{Pc} - (1-\alpha)\cdot\mathrm{Pb}}{\mathrm{La}\cdot\alpha}\right) \quad (3)$$

9. Where, $\alpha$ is direct absorbance given by the following equation:

$$\alpha = 1 - \frac{\mathrm{Lc}}{\mathrm{Lb}} \quad (4)$$

## 5 PL Imaging Using LpSiNP

LPSiNPs are proven to be biocompatible and biodegradable both in vitro and in vivo by various groups. Furthermore, the intrinsic red-NIR photoluminescence enables visualization of LpSiNPs without labeling with a fluorescent dye. PL imaging using LpSiNPs as probes is typically carried out under continuous wave imaging mode, where the emitted PL signal is collected under continuous excitation with the light source. Despite the red-NIR spectral range of PL from LpSiNPs background tissue autofluorescence remains a significant problem during in vivo imaging. To overcome this issue long PL decay lifetime (>10 s of μs) of LpSiNPs can be utilized for ex vivo/in vivo imaging using the time-gated imaging setup. Time-gated imaging setup collects the PL signal after a set time interval (gate-delay) of switching-off of the excitation light source. Due to short PL decay lifetime (~10 s ns) of the organ tissue the background autofluorescence is suppressed under time-gated imaging mode. This enables clear visualization of the LpSiNPs that luminesce for several microseconds compared to 10 s of ns for tissue background autofluorescence.

### 5.1 Continuous Wave Imaging of LpSiNPs

Continuous wave (CW) PL imaging is the most commonly used method for in vivo imaging using small molecule and QD-based fluorophores including LpSiNPs due to the simplicity, widespread availability of the CW imaging instruments, and ability to generate 3D volumetric fluorophore distribution. Typically, LpSiNPs probes are used for CW imaging of whole animal (i.e. in vivo) or collected tissue (i.e. ex vivo) using commercially available setups like IVIS® Spectrum (Perkin Elmer, USA). For CW imaging, LpSiNPs are generally injected through systemic (intravenous, IV or intraperitoneal, IP) or local (subcutaneous, SC) routes. This section of the chapter describes the steps involved in acquiring CW PL images using LpSiNPs injected systemically or locally using an IVIS$^{(R)}$ imaging system.

1. Carefully place the LpSiNP-injected animal or the organs harvested from the injected animal on a black surface and transfer to the imaging chamber of the IVIS® (*see* **Notes 10** and **11**).
2. Start the Living Image® software to initialize the IVIS® imaging system and allow the charge coupled device (CCD) camera to reach the prescribed temperature (−30 °C), which is indicated by the green system status light on the control panel.
3. For PL imaging select the "Fluorescent and Photograph mode" from the "IVIS Acquisition Control Panel" followed by setting the excitation wavelength ($\lambda_{ex}$) at 440 nm and emission wavelength ($\lambda_{em}$) between 600–700 nm through appropriate excitation and emission filters.
4. From the "Field of View" and "Focus" drop-down lists select the sample region and subject height (adjusted by the arrows on the keyboard), respectively.
5. Acquire the gray scale and CW image by selecting "Acquire" on the front panel.
6. After acquiring preliminary image, adjust the exposure time to a suitable number to enable capturing of sufficient number of emitted photons.
7. Total Radiant Efficiency (TRE) (i.e. intensity) quantification region of interest on the acquired images can be performed by adjusting the color scale in the "Tool Palette" to make sure removal of background autofluorescence from the blank.

## 6 Time-Gated PL Imaging of LpSiNPs

Time-gated PL imaging, as the name suggests, enables gating of the imaging setup to allow for acquisition of PL images few nano- to microseconds after the excitation light source has been switched off. The advantage of LpSiNPs as the imaging probe include ultralong PL lifetime (>10 s of μs) compared to conventional small molecule fluorophores with lifetime of only a few nanoseconds. Thus, using LpSiNP as an imaging agent enables complete suppression of background autofluorescence from the target tissue. However, most reports on time-gated imaging using LpSiNP only show ex vivo imaging due to low penetration depth of the 365 nm excitation light. This section of the chapter describes the steps involved in setting up and acquiring ex vivo time-gated PL images of organs collected from the animals systemically or locally injected with LpSiNPs.

1. Fix the organ tissue to be imaged in 4% formalin for 24–48 h followed by setting up the organs in desired order in a plastic petri dish.
2. Once the Andor iCCD camera sensor has reached the prescribed temperature (−30 °C) start the Andor Solis program and setup the acquisition conditions.
3. Under the "Acquisition" tab select the "Setup data type" and change to "Counts (Bg corrected)" to enable background subtraction, which is generally not selected.
4. Connect the function generator to the LED light source and Andor camera and select the pulse mode with a frequency of 10 Hz and pulse width of 500 μs. Set the high voltage to 5 V low voltage to 0 V and enable hi-Z function. Switch on the function generator and LED to make sure the LED is pulsing.
5. In the Andor Solis front panel select the "Setup Acquisition" and set the number of accumulations to 150 and delay to 5 μs, while the gate width to 400 μs. Make sure the acquisition mode is set to accumulation, the triggering is external and readout is in FVB mode.
6. Once these conditions are established switch the LED source off (not the function generator) and acquire the background signature from the acquisition tab. Make sure the photocathode is on in the Andor Solis software during the background acquisition (*see* **Notes 12–14**).
7. After the background is successfully collected switch the LED light source on and collect the signal (Fig. 6) by clicking the "collect signal" under the acquisition tab.

## 7 Notes

1. While working with HF, make sure to use correct protective gear including a neoprene apron, double set of gloves (inner nitrile and outer neoprene), closed lab shoes, full pants, eye glass, and full-face shield.

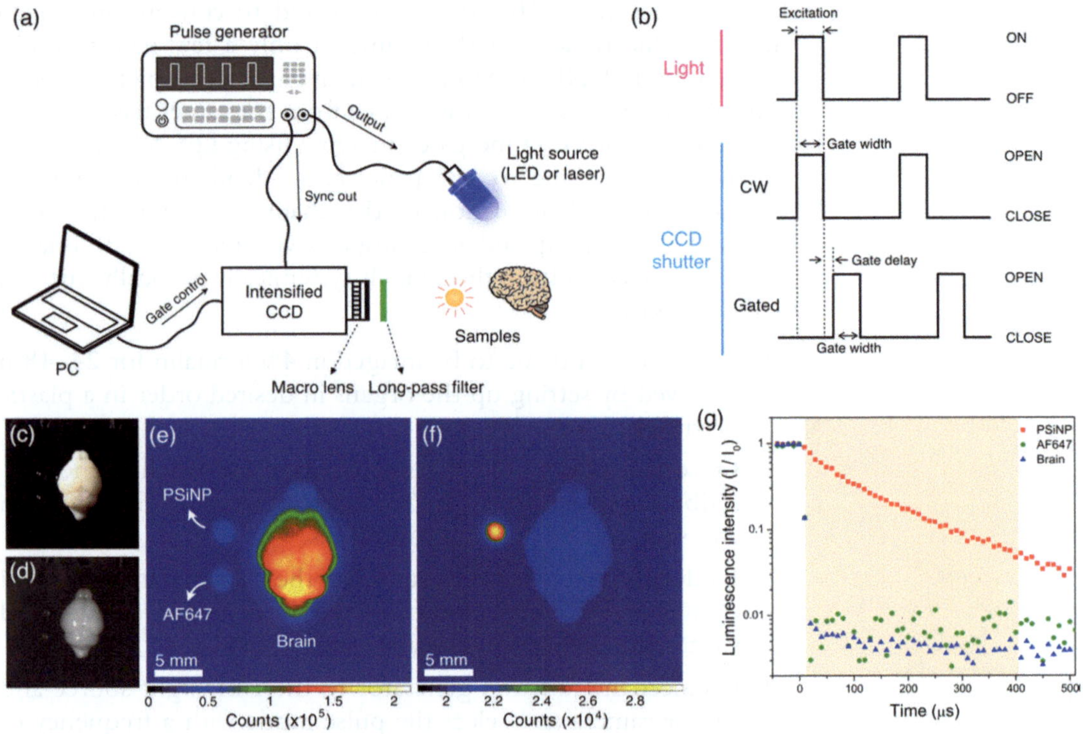


**Fig. 6** Methodology and example of time-gated PL imaging using LpSiNPs. (**a**) Schematic showing the imaging setup. The iCCD camera and the light source are controlled by an external function generator. (**b**) Notional waveforms for illumination and camera gating used to acquire continuous wave (CW) imaging and time-gated PL images. (**c**) Digital color photograph and (**d**) gray scale image (from Andor iCCD) of mouse brain obtained under ambient light. (**e**) CW and (**f**) GLISiN images of the same brain under UV LED excitation ($\lambda_{ex} = 365$ nm, $\lambda_{em} = 460$ nm long-pass filter; gate width, 400 μs, 40 accumulations, gate delay for CW = 0 μs, gate delay for GLISiN = 5 μs). Phantom samples corresponding to 150 ng of porous Si nanoparticles ("PSiNP") and 2.5 ng of the molecular dye Alexa Fluor 647 ("AF647") dropped next to the brain for comparison, as indicated. (**g**) Normalized intensity decay of the photoluminescence/fluorescence signals from the samples in (**e, f**) as a function of time after excitation pulse (gate width, 10 μs; gate step increase, 10 μs; accumulation, 20 times) [16]

2. After removal of the first sacrificial layer using 2 M potassium hydroxide solution, remember to wash thrice with water followed by ethanol.
3. During the 18 h sonication of porous silicon layers cover the glass vial with parafilm and create air escape holes in the parafilm to prevent the vial from breaking.
4. Note, for long-term storage transfer the LpSiNPs to ethanol medium by centrifugation and resuspending.
5. For relative QY measurements, it is recommended to make the PL and Abs measurements by pipetting the different concentrations of R6G and LpSiNPs in a 96-well culture plate. This

enables rapid measurement and reduces the chances of variation due to evaporation of ethanol that is used as the solvent.

6. The intercept of the linear fit should be at the origin ($x = 0$, $y = 0$).
7. For absolute QY measurements, the light source must be point focused. In case the available light source is diffused, collimating or focusing lenses can be attached to the port of the IS.
8. It is recommended that the wavelength range for calculating the integrated signal should be fixed, e.g., for relative QY calculate the integrated PL signal of LpSiNPs in 500–900 nm range, which should be the same for all the concentrations make sure to use the same wavelength range. Similarly, for the absolute QY fix the wavelength range for both "L" and "P" component of the PL for calculating integrated PL signal.
9. Note, when no sample is present in the IS, Pa should be zero. The PL from the LpSiNPs in the indirect exposure mode (i.e. Pb) in the IS is usually relatively low.
10. For CW imaging, avoid using plastic base for animals/tissues due to inherent background fluorescence and make sure to fit samples into selected field of view.
11. Live animals need to be under anesthesia for both the CW and time-gated PL imaging.
12. To achieve the best signal-to-noise ratio during time-gated imaging it is necessary to collect background. In addition, the gain can be adjusted to obtain best PL signal.
13. Make sure appropriate long-pass filter in mounted in front of the iCCD camera for suppression of the excitation source signal.
14. The CW PL images can also be acquired using the described time-gated imaging by setting the gate-delay to 0 s.

## References


1. Gao X, Cui Y, Levenson RM, Ching LWK, Nie S (2004) In vivo cancer targeting and imaging with semiconductor quantum dots. Nat Biotechnol 22:969–976
2. Lee JH, Huh YM, Jun YW, Seo JW, Jang JT, Song HT, Kim S, Cho EJ, Yoon HG, Suh JS (2007) Artificially engineered magnetic nanoparticles for ultra-sensitive molecular imaging. Nat Med 13:95–99
3. Ballou B, Lagerholm BC, Ernst LA, Bruchez MP, Waggoner AS (2004) Noninvasive imaging of quantum dots in mice. Bioconjug Chem 15:79–86
4. Derfus AM, Chan WC, Bhatia SN (2004) Probing the cytotoxicity of semiconductor quantum dots. Nano Lett 4:11–18
5. McInnes SJ, Santos A, Kumeria T (2018) Porous silicon particles for cancer therapy and bioimaging. In: Nanooncology. Springer, Germany, p 305
6. Kumeria T, McInnes SJ, Maher S, Santos A (2017) Porous silicon for drug delivery applications and theranostics: recent advances, critical review and perspectives. Expert Opin Drug Deliv 14:1407–1422
7. Canham LT (1990) Silicon quantum wire array fabrication by electrochemical and chemical dissolution of wafers. Appl Phys Lett 57:1046–1048
8. Lehmann V, Gösele U (1991) Porous silicon formation: a quantum wire effect. Appl Phys Lett 58:856–858

9. Canham LT (1995) Bioactive silicon structure fabrication through nanoetching techniques. Adv Mater 7:1033–1037
10. Park JH, Gu L, Von Maltzahn G, Ruoslahti E, Bhatia SN, Sailor MJ (2009) Biodegradable luminescent porous silicon nanoparticles for in vivo applications. Nat Mater 8:331–336
11. Sailor MJ (2012) Porous silicon in practice: preparation, characterization and applications. John Wiley & Sons, Germany
12. Wang J, Kumeria T, Bezem MT, Wang J, Sailor MJ (2018) Self-reporting photoluminescent porous silicon microparticles for drug delivery. ACS Appl Mater Interfaces 10:3200–3209
13. Qin Z, Joo J, Gu L, Sailor MJ (2014) Size control of porous silicon nanoparticles by electrochemical perforation etching. Part Part Syst Charact 31:252–256
14. Joo J, Cruz JF, Vijayakumar S, Grondek J, Sailor MJ (2014) Photoluminescent porous $Si/SiO_2$ core/shell nanoparticles prepared by borate oxidation. Adv Funct Mater 24:5688–5694
15. Gu L, Hall DJ, Qin Z, Anglin E, Joo J, Mooney DJ, Howell SB, Sailor MJ (2013) In vivo time-gated fluorescence imaging with biodegradable luminescent porous silicon nanoparticles. Nat Commun 4:2326–2332
16. Joo J, Liu X, Kotamraju VR, Ruoslahti E, Nam Y, Sailor MJ (2015) Gated luminescence imaging of silicon nanoparticles. ACS Nano 9:6233–6241
17. Joo J, Defforge T, Loni A, Kim D, Li ZY, Sailor MJ, Gautier G, Canham LT (2016) Enhanced quantum yield of photoluminescent porous silicon prepared by supercritical drying. Appl Phys Lett 108:153111–153115

# Part IV

## Image-Guided Therapy

# Chapter 14

# Nucleic Acid Aptamers as Emerging Tools for Diagnostics and Theranostics

**Ruchi Mutreja, Shahnawaz Ahmad Baba, and Naveen Kumar Navani**

## Abstract

Aptamers are ssDNA or RNA sequences (20–80 nucleotides) generated in vitro by SELEX (Systematic Evolution of Ligands using EXponential enrichment) against diverse range of targets from small molecules to bacteria, viruses, and even eukaryotic cells. Aptamers, also known as chemical bodies, bind to their respective targets with tunable affinity and specificity, making aptamers as potent probes for diagnostics and excellent ligands for drug delivery in therapeutics. In this chapter, we have described the methods for generating DNA aptamers against proteins and their use in theranostics.

**Key words** Aptamers, SELEX, Theranostics, Electrochemical, Probes

## 1 Introduction

Clinical detection of biomarkers, expressed under abnormal health conditions is of immense importance for the early and accurate diagnosis and its timely therapeutic intervention. Theranostics, the coupling of diagnostic tools with therapeutic agents, is gaining popularity among researchers and clinicians. Advanced sophistication in theranostics has led to the development of bio-recognition molecular probes such as antibodies, aptamers, and enzymes. Nucleic acid aptamers, single-stranded DNA (ssDNA) or RNA sequences are emerging as attractive molecular recognition elements (MRE) for designing diagnostics and theranostic platforms [1]. Aptamers as molecular probes is gaining wise due to their diverse properties like small size, flexible folding, high affinity, specificity, stability, ease of modification, and applicability in wide range of environmental conditions [2]. The competent use of aptamers for clinical/field diagnostics and therapeutics has been wisely reviewed and reported in literature by many researchers [3–8]. Aptamers are generated in vitro by iterative process of SELEX (Systematic Evolution of Ligands using EXponential enrichment) and was first reported in 1990 [9, 10]. The word

Jyotsna Batra and Srilakshmi Srinivasan (eds.), *Theranostics: Methods and Protocols*, Methods in Molecular Biology, vol. 2054, https://doi.org/10.1007/978-1-4939-9769-5_14, 

Aptamer is of Latin origin, "aptus" meaning to fit, and "meros," meaning part. Aptamers are screened from DNA/RNA libraries of random nucleotides by SELEX which involves incubation of target analyte with DNA/RNA library, partitioning of binding from non-binding population, recovering the bound DNA population followed by amplification of higher affinity sequences for the target, their purification and conversion into single-stranded DNA (ssDNA) sequences [3, 4, 11, 12]. The process is repeated for a number of rounds, which leads to Darwinian evolution of the random library resulting in the successive enrichment of high affinity sequences at the expense of the low affinity sequences. In vitro SELEX can be carried out by several methods depending upon target such as immobilization, membrane filtration, affinity immobilization, and Flumag. [13–16]. Several groups have reported aptamer selection and their applications, against a wide variety of targets from small molecules [17–23] to macromolecules [24–28]. Different SELEX strategies have been developed in the past three decades for screening aptamers against targets which may undergo mutations or overexpression due to structure switching, interaction with the intra- and extracellular matrix [3, 4], thereby opening new perspectives for theranostics (Fig. 1). Theranostics,

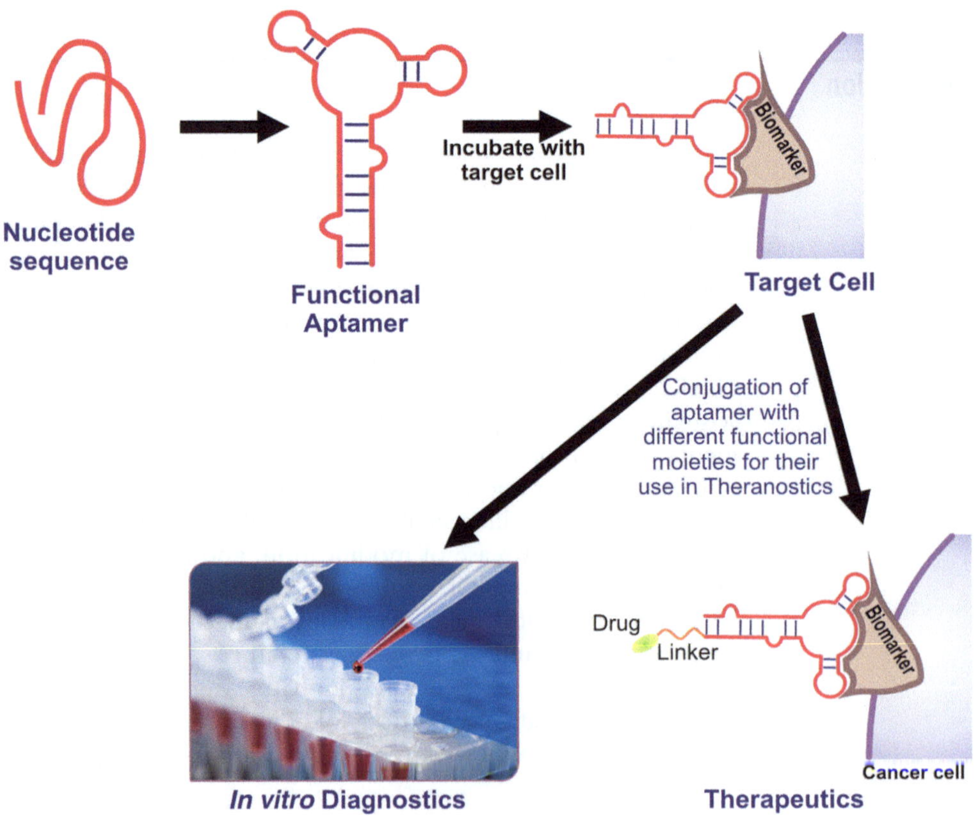


**Fig. 1** Aptamers in theranostics

the coupling of diagnostics and therapeutics, at the individual level for personalized medicine has the potential to provide better health. In this chapter, we have described the detailed method for the in vitro selection of DNA aptamers against a protein target by SELEX with methods for characterization for the binding of aptamer with the target. In addition, we will also describe the generalized methods for the use of aptamers in diagnostics using optical and electrochemical platforms of detection and in delivering drug to the cancer cells.

## 2 Materials

### *2.1 Design, Synthesis, and Purification of DNA Libraries*

1. *DNA libraries and primers*: Random DNA library (RDL) should be of 60–80 nucleotides, synthesized at 1 μmol scale from any commercial source. Both the ends of RDL are fixed as primer binding domains of 16–20 nucleotides (*see* **Note 1**). Random DNA Library: 5′CTTGACTAGTTACGCCGCTA (**$N_{33–40}$**) GCATTCATTCAGTTGGCGCC-3′, is the sequence of a typical DNA library used during aptamer selection.

   For this typical RDL, following fixed end primers can be used: forward primer (FP), 5′-CTTGACTAGTTACGCCGCT A-3′ or FAM labeled FP 5′FAM-CTTGACTAGTTACGCCG CTA-3′ (*see* **Note 2**) and Reverse primer (RP), ribose-modified RP 5′-GGCGCCAACTGAATGAATGrC-3′ or 5′ phosphate-modified RP 5′-phos-GGCGCCAACTGAATGAATGC-3′ or 5′ biotin labeled RP5′-Biotin-GGCGCCAACTGAATGAATG C-3′ (*see* **Note 3**).
2. 10% Denaturing Urea PAGE: Mix 210 g urea (molecular grade) in 50 mL of 10× TBE and 125 mL of 40% acrylamide/bis-acrylamide (add 47.25 g acrylamide and 2.5 g bisacrylamide), add 200 mL deoxygenated double distilled water ($ddH_2O$), dissolve with stirring on magnetic stirrer. Make the volume up to 500 mL with $ddH_2O$ and refrigerate in dark (*see* **Note 4**).
3. APS (Ammonium persulfate): Prepare 10% solution in $ddH_2O$ and refrigerate in dark.
4. TEMED (*N*,*N*,*N*,*N*-tetrametheyl-ethylenediamine,): TEMED is used as a catalyst for polymerization of Acrylamide and Bisacrylamide in the presence of APS (*see* **Note 5**).
5. 10× TBE (Tris-Borate buffer): Add 108 g Tris base, 55 g boric acid in 650 mL of $ddH_2O$. Stir the solution on a magnetic stirrer. As the solution gets transparent, add 40 mL of 0.5 M EDTA (pH 8.0), finally adjust the pH to 8.3 and make up the volume to 1 L. Store at RT.

6. 0.5 M EDTA: Add 18.61 g EDTA to 80 mL $ddH_2O$. Add 5 N NaOH drop by drop, till it dissolves (pH 8.0) and finally make up the volume to 100 mL $ddH_2O$ and refrigerate.
7. DNA Elution buffer for Urea PAGE: Mix 4 mL of 5 M NaCl, 1 mL of 1 M Tris-Cl (pH 8.0), 0.2 mL of 0.5 M EDTA (pH 8.0) and finally make up the volume to 100 mL with $ddH_2O$ and refrigerate.
8. 2× Urea PAGE loading dye: For 40 mL, add 44 g urea to 20 mL $ddH_2O$, 2 mL of 50% glycerol, 400 μL 10% SDS, 4 mL of 10× TBE, 10 mg Bromophenol blue and 10 mg xylene cyanol. Dissolve with mild heat and stirring, make up the volume to 40 mL and refrigerate.
9. Urea PAGE apparatus and accessories: Includes glass plates (17.5 cm × 19 cm), spacers (0.7 mm), combs (0.7 mm), and adjustable clamps, apparatus (CBS Scientific—Model ASG-250, CBS Scientific, USA) with power supply unit (BioRad, USA, PowerPac™ HV).
10. Gel Doc: Gel Doc system from BioRad, USA (Universal Hood II).
11. Nanodrop (Thermo Scientific, USA).

### 2.2 SELEX

1. Immobilization matrix: Choice of immobilization matrix depends upon the nature of ligand against which aptamers are to be selected. For His-tagged recombinant protein, Ni-NTA beads (Qiagen, Germany) are generally used as immobilization/partitioning matrix (*see* **Note 6**).
2. 1× Selection buffer (SB): Mix 1× phosphate buffer saline (PBS) diluted from Dulbecco's 10× PBS (without $MgCl_2$ and $CaCl_2$) with 1 mM $MgCl_2$, 2.7 mM KCl, 150 mM NaCl, 2.5% glycerol and 0.02% Tween-20 in $ddH_2O$. Filter through 0.22 μm syringe-filters (Millipore, USA) and refrigerate.
3. Binding buffer (BB): SB supplemented with yeast tRNA (2 μg/mL, Invitrogen, USA) and Bovine serum albumin (125 μg/mL, Sigma-Aldrich, USA).
4. Elution buffer (EB): SB supplemented with 250–300 mM imidazole (Sigma Aldrich, USA). Filter through 0.22 micro filter and refrigerate.
5. Microcentrifuge tubes (MCTs): low binding MCTs, from Sarstedt, Germany.
6. Rocker shaker (Bestlab Instrumentation, Inc.): refrigerated bench-top centrifuge (Thermo Scientific, USA), dry heating block and Kim wipes.

7. Invitrogen TA Cloning kit, LB agar, Ampicillin, Electroporator (Eppendorf 2510, Germany), electroporation cuvettes—0.2 cm gap (BioRad, USA), SOC broth media.
8. Ni-NTA Agarose Beads (Thermo Fisher Scientific, USA): Binding efficiency 50 mg/mL of His-tagged recombinant protein.

### 2.3 Amplification Using PCR

1. PCR machine: BioRad, USA (T100, Thermal cycler, any other modern PCR machine (thermocycler)) can also be used for amplification.
2. PCR tubes: 200 μL capacity.
3. Filter tips: Low retention filter-barrier micropipette tips of different volumes (0.1–10, 10–20, 20–200, 100–1000 μL).
4. DNA polymerase and 10× Biotools buffer: 750 mM Tris-Cl (pH 9.0), 200 mM $MgCl_2$, 500 mM KCl and 20 mM $(NH_4)_2SO_4$.
5. 100 mM dNTP mix: 2 mM each of dATPs, dTTPs, dGTPs, dCTPs.
6. 3% Agarose gel: Add 3 g of low EEO agarose to 100 mL of 1× TBE. Heat in microwave till it becomes transparent, cool the solution and pour in casting tray. Stain with 0.05% EtBr before observing on Gel Doc system.

### 2.4 Ethanol Precipitation

1. Absolute ethanol.
2. 0.25 N NaOH: Dissolve 1 g NaOH in 90 mL of $ddH_2O$ and make up the volume to 100 mL.
3. 3 M sodium acetate pH 5.2: Add 20.4 g of sodium acetate salt in 40 mL of $ddH_2O$, adjust pH using glacial acetic acid and makeup the volume to 50 mL.

### 2.5 Generation of Single-Stranded DNA (ssDNA) from Double-Stranded DNA (dsDNA)

1. λ exonulease and 10× λ exonulease buffer (670 mM glycine-KOH (pH 9.4), 25 mM $MgCl_2$, 0.1% v/v Triton X 100). 10× buffer should be diluted to 1× for setting up the reaction.
2. For alkaline treatment: 0.25 N NaOH, 3 M sodium acetate.
3. Magnetic beads (Dynabeads™ Streptavidin C1): 10 mg/mL (Thermo Scientific, USA).

### 2.6 Purification of ssDNA

1. QIAEX II Gel Extraction Kit (Qiagen).
2. UREA denaturing PAGE assembly and buffers.
3. Ethanol precipitation reagents (Ethanol, sodium acetate pH 5.2, 0.25 N NaOH).

### 2.7 Evaluation of Binding Affinity of Enriched DNA Population

Native Gel-Electrophoresis Electro Mobility Gel-Shift Assay (EMSA):

1. 8% Native PAGE Mix: To 100 mL of 40% Acrylamide/bis-acrylamide, add 25 mL of 1× TBE, 7.5 mL of 100% glycerol and make volume up to 500 mL with $ddH_2O$ and refrigerate in dark at 4 °C.
2. Native gel running buffer: Dilute 10× TBE stock to 0.5× TBE.
3. EMSA Gel Loading dye: Add 2.5% glycerol, 0.1% bromophenol blue and 0.1% xylene cyanol to 50 mM Tris-Cl.
4. Vertical gel electrophoresis apparatus, phosphorimager screens (GE-Amersham), Phosphosrimager (Typhoon FLA 9000 Imager, GE, USA), 10% APS, TEMED, etc.

### 2.8 Optical Detection

1. 96 black well flat bottom plates.
2. ELISA plate washer.
3. Blocking agent: 10% Skim Milk in PBST (PBS supplemented with 0.1% Tween 20), mix with stirring or 5% Bovine serum albumin (BSA) in PBST.
4. Plate reader with blue laser 490/520 filter.

### 2.9 Electrochemical Detection

Electrochemical workstation: Autolab 204 and Screen printed carbon electrodes (SPE) (Metrohm AG from Metrohm India), Potassium Ferricyanide, Potassium Ferrocyanide, 1-Ethyl-3-(3-dimethylaminopropyl) carbodiimide (EDC), N-Hydroxysuccinimide (NHS), graphene, 70% Nitric acid.

### 2.10 Therapeutics

1. $FeCl_3{\cdot}6H_2O$, $N_2$ gas, $NH_4OH$, NaOH, carboxylated polyethylene glycol (PEG), doxorubicin, dialysis bag.
2. Magnet-Dynal MPC™-S (Thermo Scientific, USA).
3. Bath sonicator.

## 3 Methods

### 3.1 Determination of Purity DNA Libraries

1. Wash the glass plates and comb thoroughly, followed by wiping with 70% Ethanol and air-drying. Fix the spacers between the glass plates and hold them tightly, followed by adjusting the clamps diagonally and lay the assembled plates horizontally on some flat platform.
2. To Prepare Urea PAGE, mix 35 mL of 10% Urea acrylamide mix, add 350 μL of 10% APS and 25 μL of TEMED. Mix thoroughly, quickly and carefully pour it to the assembled gel plates to avoid air bubbles. Gently insert the comb between the gel plates and wait till gel polymerizes.

3. Meanwhile, dissolve the library and primers to make 100 μM stock in nuclease-free water (NFW), incubate at 40 °C for 45 min in dry bath, followed by short spin. Store the stocks at −20 °C.
4. Check the residual Urea acrylamide mix for polymerization (usually gel polymerizes in 30–40 min). Gently remove the comb, set the polymerized gel plates into the apparatus and wash the wells with running buffer using syringe needle. Add 1× TBE to the space provided for buffer, set the electrodes and start the pre-run at constant 350 V till it warms up to ~50–55 °C (around 1 h) (*see* **Note 7**).
5. Take 10 μL of RDL (~50 ng) and add equal volume of 2× Urea gel loading dye. Heat at 95 °C for 5 min, cool down to RT followed by short spin.
6. Pause the gel, clean the wells thoroughly to remove deposited urea. Load the samples (RDL and primers) prepared above and restart the gel at 350 V (constant voltage), run it till the tracking dye migrates to almost three-fourth of the gel length.
7. Turn off the power supply, remove the plates, and disassemble them by gently pressing off the notched plate with a Biorad gel cutter.
8. Visualize the gel under gel doc system. Depending upon the amount, the DNA band will appear as bright band in the gel. FAM labeled DNA can be visualized directly (*see* **Note 8**) and for visualizing unlabeled primers or libraries, gel can be stained with EtBr for 10 min (*see* **Note 9**) or SYBR green under dark (*see* **Note 10**).
9. For DNA extraction from Urea PAGE, cut the desired band with a sterile scalpel. Transfer the band in low binding microcentrifuge tube, crush it with the sterile 1 mL tip to the smallest possible pieces, and add 500 μL of prewarm elution buffer with 0.3 M sodium acetate (pH 5.2), followed by heating at 65 °C and vortexing (*see* **Note 11**).
10. Centrifuge the tube at 2350 × *g* for 5 min to spin down the gel and recover the eluted DNA from supernatant. Add 3 volume of cold 100% ethanol, incubate at −20 °C for 2–3 h followed by spin at 16,000 × *g* for 20 min at 4 °C. Wash the pellet twice with 70% ethanol, dry it and add 50 μL of NFW to dissolve DNA. Determine the concentration of DNA by checking absorbance at 260 nm. Store at −20 °C.

### 3.2 SELEX

1. Take 180 pmol of DNA Library (RDL) in microcentrifuge tube, add 1 mL of 1× SB, denature it by heating at 95 °C for 10 min, immediately transfer to ice for 10 min and then to room temperature for 20 min.

2. Meanwhile, prepare the immobilization matrix, for His-tagged proteins, Ni-NTA beads should be used.
3. For protein conjugation, take 1 mL of Ni-NTA agarose beads, wash thrice with SB, add 2 mg protein and make final volume to 4 mL, incubate at 4 °C for 30 min, remove supernatant and resuspend protein-bound Ni-NTA beads in 4 mL of SB. Store at 4 °C.
4. Add 50 μL of Ni-NTA bead-bound protein to prepared RDL, incubate for 60 min on rocker shaker at room temperature. Separate the bound and unbound RDL sequences by using 1 mL column (BioRad) (*see* **Note 12**).
5. Wash the column thrice with 1 mL SB (*see* **Note 13**).
6. For elution, add 100 μL elution buffer to the column, incubate for 2–5 min (*see* **Note 14**) with intermittent gentle tapping.
7. Collect the eluted sequences in microcentrifuge tubes, protein-bound enriched DNA sequences can be separated from the protein by using DNA purification kits with high affinity for the smaller DNA sequences such as QIAGEN II kit or by ethanol precipitation and resuspend in NFW. This DNA is used as the template for PCR for next round of selection. Schematic is represented in Fig. 2. After 3–4 rounds of positive

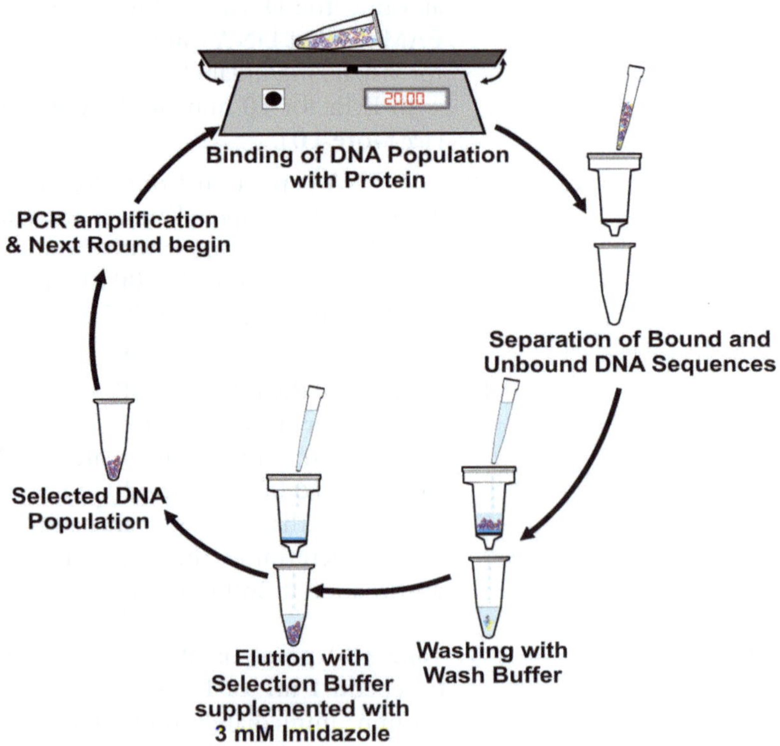


**Fig. 2** Diagrammatic representation of SELEX process for aptamer selection against His-tagged protein

**Table 1**
**Composition of PCR reaction**

| Reagents | Stock concentration | Final concentration | Volume (μL) in 100 μL reaction |
|---|---|---|---|
| Primer Forward (FP) | 10 μM | 0.2–0.3 μM | 2.0 |
| Primer Reverse (RP) | 10 μM | 0.2–0.3 μM | 2.0 |
| 10× Buffer | 10× | 1× | 10.0 |
| dNTPs | 2 mM | 0.2–0.4 mM | 10.0 |
| 3B DNA Polymerase | 5 U/μL | 0.02 U | 0.5 |
| Template DNA | 6 ng/μL | 5–20 ng | 3.0 |
| Nuclease-free water | – | – | 72.5 |
| | | *Total* | *100 μL* |

selection, counter or negative selection should be performed before every round of positive selection (*see* **Note 15**). Set up PCR in 100 μL reaction using primers specific for the library using 3B DNA polymerase with following conditions, Initial Denaturation at 94 °C (3 min), followed by 30 cycles of Denaturation at 93 °C (30 s), Annealing at 55–65 °C (20 s), Elongation at 72 °C (5–10 s). Before adding template to the mixture, remove 25 μL of master mix and label it as control, add template to the remaining master mix and label it as test reaction. Composition of reagents for setting up the PCR reaction is shown in Table 1.

8. Run both test and no template control on 3% agarose gel with 50 bp ladder as shown in Fig. 3.
9. Purify the PCR product using either ethanol precipitation or DNA purification kit (according to manufacturer's protocol).
10. Preparation of ssDNA from dsDNA: ssDNA can be prepared either with lambda exonuclease or alkaline treatment followed by Urea PAGE or biotin labeled reverse primer. To check the purity of RDL, add 10 μL (50 ng) of RDL with 10 μL 2× urea PAGE loading dye and heat at 95 °C, cool and load to the Urea denaturing PAGE (Fig. 4) (*see* **Notes 16** and **17**).
11. For elution of ssDNA from Urea PAGE, crush the gel containing band with 1 mL tip, (*see* **Note 18**), add 500 μL EB, heat at 65 °C for 5 min, vortex, repeat the heat and vortex step to increase the yield.
12. Collect the supernatant by centrifugation at 2350 × *g* for 5 min. Carefully collect the supernatant as gel particles can easily mix with the supernatant. Precipitate the DNA by ethanol precipitation. Determine the concentration of DNA

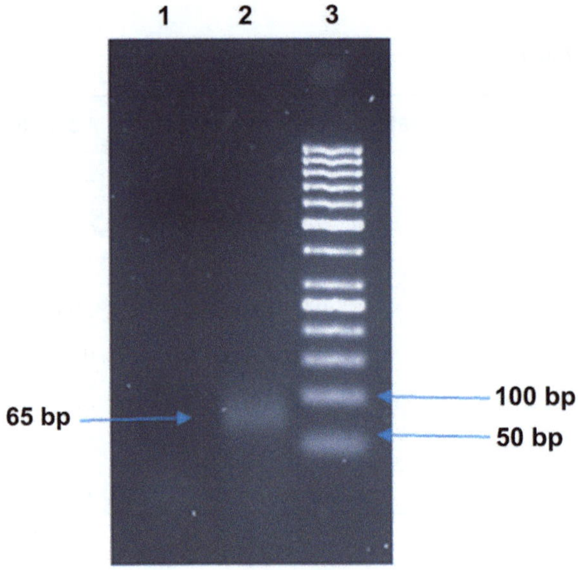


**Fig. 3** PCR product run on 3% agarose gel and was visualized with SYBR green or EtBr staining. Lane 1: no template control. Lane 2: PCR product. Lane 3: 50 bp ladder

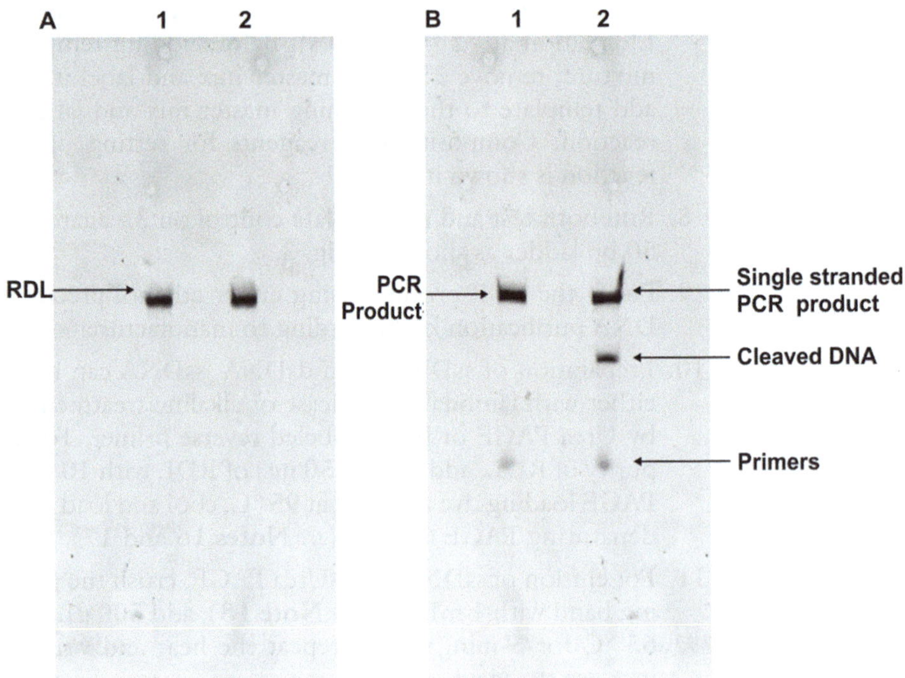


**Fig. 4** 10% denaturing UREA-PAGE: 50 ng RDL run on 10% Urea denaturing PAGE, visualized under typhoon scanner, Lane 1, 2: HPLC purified RDL (**a**); PCR product run on 10% Urea denaturing PAGE before (Lane 1) and after ribose treatment (Lane 2) for preparation of ssDNA (**b**)

by measuring absorbance at 260 nm. Store at −20 °C. Eluted purified ssDNA is used for next round of SELEX.

For protein SELEX, at least ten rounds of selection should be carried out. Following tenth round, assessment of binding can be carried out using EMSA as described below:

### 3.3 Electrophoretic Mobility Gel Shift Assay (EMSA)

1. Wash the gel plates (10 cm × 8 cm), clean with ethanol, use 1.5 mm spacer plate with its glass plate fix, adjust to the assembly. Make sure that plates are tightly fixed to avoid any leakage from the plates.
2. Take 10 mL of the 8% native PAGE mix, add 100 μL of 10% APS and 6 μL of TEMED. Mix well and pour into the set plates. Wait for the gel to polymerize. Add 0.5× TBE buffer and start the run at 100 V. Run the gel till the current stabilizes to 8–10 mA. Flush the wells of pre-run native gel with syringe.
3. Meanwhile, prepare the following binding reaction
   (a) FAM Labeled DNA ~5–10 ng.
   (b) Protein ~20–500 ng.
   Total volume = 20 μL with SB.
4. For sample preparation, heat denature ss FAM (6-Flourescein amidite (6-FAM)) labeled enriched population of DNA at 95 °C for 5 min, snap cool and add different concentration of proteins, mix well, incubate at 37 °C for 15–30 min, add EMSA dye to final concentration of 1×. Prepare control as DNA only (*see* **Note 19**). Load the sample. Run the gel at constant current of 10 mA for another 1.5 h and capture the image on Typhoon scanner.
5. After monitoring the progress of sequences enrichment, a decision can be made to clone the population after no further enrichment is seen. A typical gel showing good progress in selection against a protein biomarker is shown in Fig. 5.
6. After the saturation of the population, final PCR should be performed with the normal primers by Taq polymerase. Purify the PCR product with DNA purification kits as per manufacturer's instructions (*see* **Note 20**).
7. For cloning, add 50 ng of PCR product to 500 ng of pTZ57R/T plasmid, ligate at 22 °C for 2 h, deactivate ligase by heating at 70 °C for 5 min.
8. Transform 1 μL of the ligated product to 50 μL of *E. coli* DH5α electro competent cells through electroporation using an electroporator.
9. Add prewarmed (at 37 °C) SOC recovery media and incubate for 45 min on 37 °C under shaking and spread 50 μL on LB amp (100 μg/mL) agar plate, incubate at 37 °C overnight.

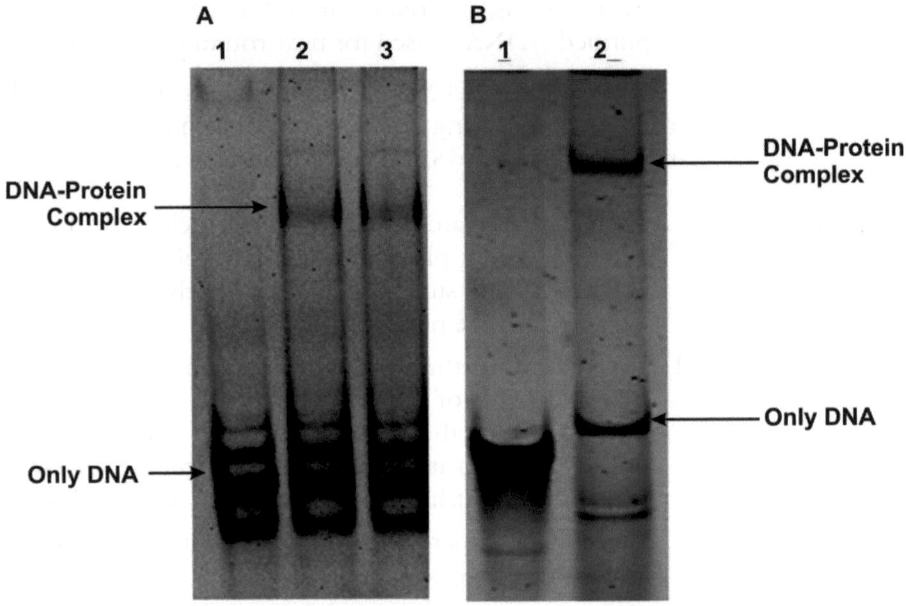


**Fig. 5** EMSA of FAM labeled aptamers on 8% native PAGE. Incubate aptamer with protein in binding buffer for 30 min, load incubated samples on pre run gel, and examine under typhoon scanner: Lane1: 20 ng of sixth round FAM-labeled enriched population (control), sixth round population with 500 ng target protein (Lane 2), and eighth round population with 500 ng target protein (Lane 3) (**a**), 10 ng FAM labeled aptamer (Lane 1), 10 ng FAM labeled aptamer with 500 ng of protein (Lane 2) (**b**)

10. Screen for the recombinants using blue white screening on LB agar plates.
11. Patch approximately 50–70 white colonies on LB agar amp plate. The plasmids of individual colonies should be sequenced; the sequences should be categorized on the basis of their homology. Online software like BioEdit, ClustalW, Clustal omega, and T-COFFEE can be used for alignment. Binding affinity of all the putative aptamers from every group should be tested either by EMSA, MST (Microscale Thermophoresis), ITC (isothermal titration calorimetry), FACS (Fluorescence Activated Cell Sorting), or surface plasmon resonance.
12. One or few aptamers showing good binding affinities can be chosen for their diagnostics or theranostics applications.

### 3.4 Aptamers in Diagnostics

Different assays can be performed for the detection of cancerous cells using aptamers. Researchers have reported various optical [29–31] and electrochemical assays [32–38] for the detection of cancerous cells by selecting aptamers against a particular and specific aptamer. Common assays are discussed below:

#### 3.4.1 Optical Assay

1. Coat the polystyrene black fluorescent plates with 100 μL protein (2.5 μg/mL) in each well, incubate at 4 °C overnight or

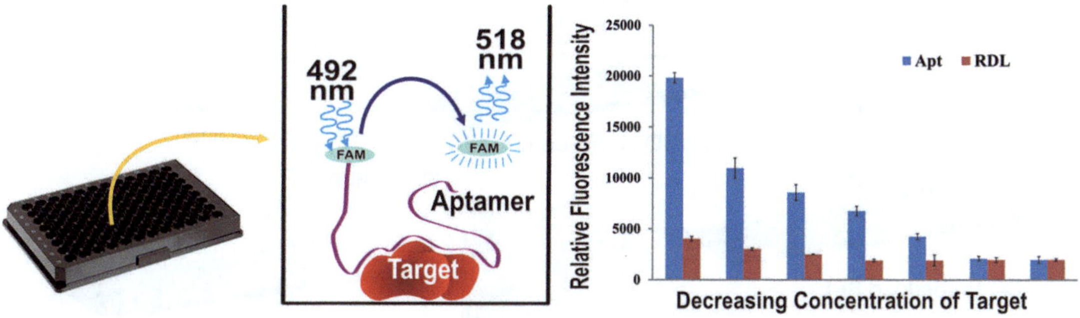


**Fig. 6** Schematic of Optical detection using FAM labeled aptamer

37 °C for 1 h or RT for 1.5 h, wash with PBST either manually or using plate washer.

2. Block the coated plates with 10% skim milk or 5% BSA in 10 mM PBST at RT for 1 h, wash with 10 mM PBST.
3. Add FAM labeled aptamers (screened against protein) to the coated and blocked plates, incubate at RT for 1 h, wash, excite at 492 nm and emission can be measured at 520–600 nm (Fig. 6).

*3.4.2 Electrochemical Assay*

1. Graphene-Graphene oxide (G-GO) nanostructure preparation: Add 5 mg Graphene to 50 mL $HNO_3$ for the functionalization with carboxyl group, reflux for 24 h at 70 °C [39], filtered and vacuum dry, store in vacuum till further use.
2. Development of Biosensing platform: EDC/NHS crosslinking chemistry can be used for the immobilization of amine labeled aptamer on rG-GO (reduced G-GO) SPE [39]. Graphene oxide can be reduced by reductive scans in PBS, resulting in increased conductivity of the platform, followed by addition of 10 μL of 50 mM of EDC and 10 μL of 50 mM of NHS, incubated for 1 h at 37 °C [40], wash to remove excess EDC and NHS (*see* **Note 21**).
3. Drop cast amine-modified DNA aptamer to the activated rG-GO nanostructure on SPE, incubate for 1 h at 37 °C, washed with $ddH_2O$ to remove unbound aptamers sequences. Block the modified SPE with 1% BSA solution for 1 h to prevent nonspecific binding.
4. Add protein in different concentrations to the bound aptamer for 10 min, washed with $ddH_2O$ and Electrochemical impedance spectroscopy (EIS) scans can be measured in 5 mM potassium ferro/ferri-cyanide at each step of aptasensor development (Fig. 7). Square wave voltammetry, differential scan voltammetry, cyclic voltammetry, etc., can also be performed.

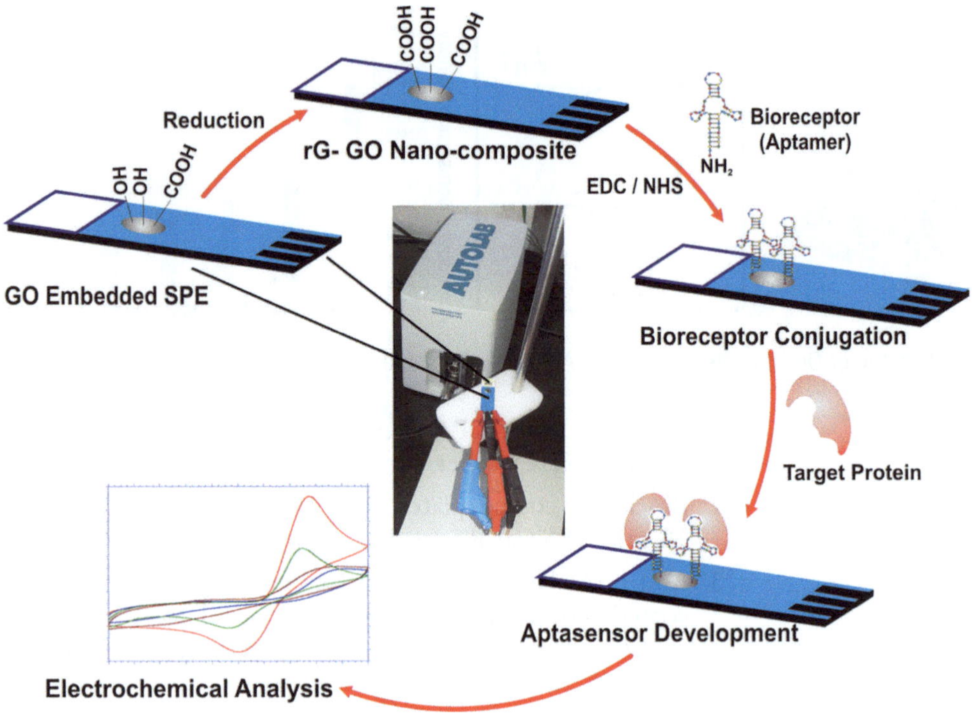


**Fig. 7** Electrochemical detection using amine labeled aptamer

### 3.5 Aptamers in Therapeutics

Aptamers selected against various cancerous targets can be used for the purpose of theranostics just like antibodies. But unlike antibodies, aptamers are advantageous for applications like imaging, drug delivery, diagnostics, and miRNA detection. These days, essentiality of imaging of cancerous cells with higher resolution for therapeutics is of immense importance and aptamers wisely fit into the requirement. One of the commercially used and FDA approved aptamer, Macugen is a great success to the aptamers as therapeutics. Many such aptamers are in clinical trials for their application, and in the near future aptamers are going to play a leading role in theranostics. Aptamers can be conjugated with nanoparticles or quantum dots (QDs) or micelles for increasing the efficiency and specificity of drug delivery to the cancer cells [41–45]. One such strategy for delivering drugs to cancer cells using aptamer-conjugated iron nanoparticles is shown in Fig. 8.

#### 3.5.1 Aptamer Cross-Linked Superoxide Iron Nanoparticle for Drug Delivery and Imaging

Synthesis of carboxyl-modified superparamagnetic iron oxide nanoparticle:

1. Take 500 mg of $FeCl_3{\cdot}6H_2O$ and 184 mg of $FeCl_3{\cdot}4H_2O$, dissolve in 30 mL under degassing by $N_2$ bubbling for 30 min.
2. Add 7.5 mL of 28% $NH_4OH$, stir vigorously under nitrogen atmosphere (pH increases from 2.3 to 10). Change in color to deep black indicates the formation of iron superoxide

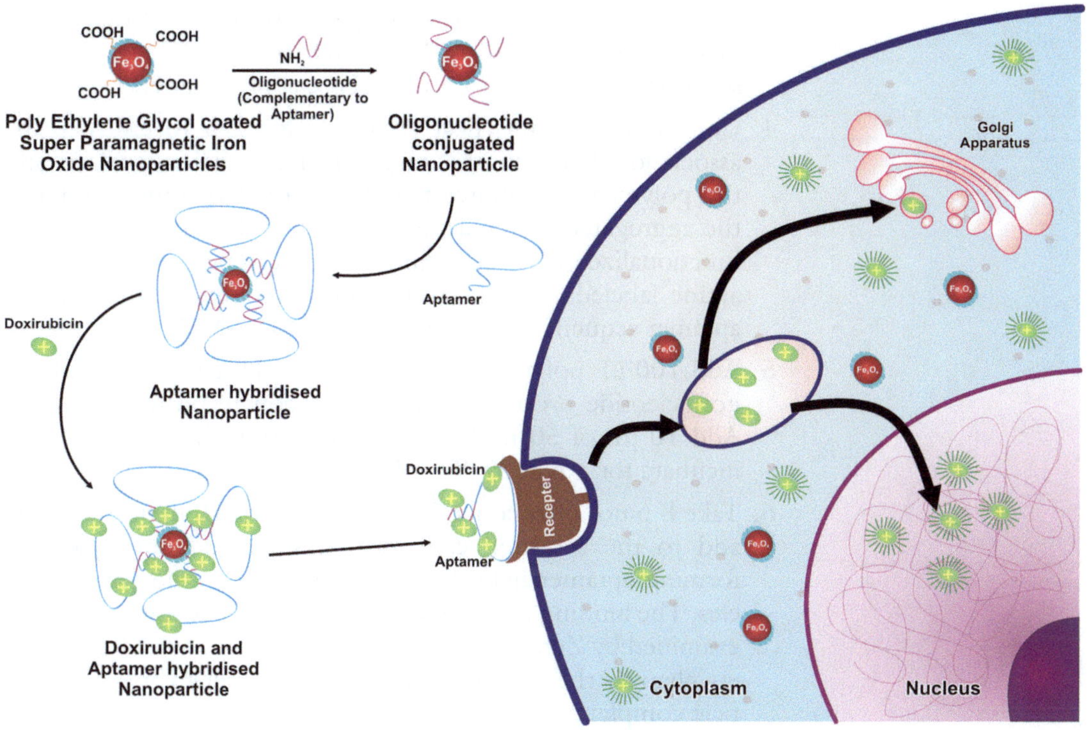


**Fig. 8** Schematic of aptamer use in therapeutics

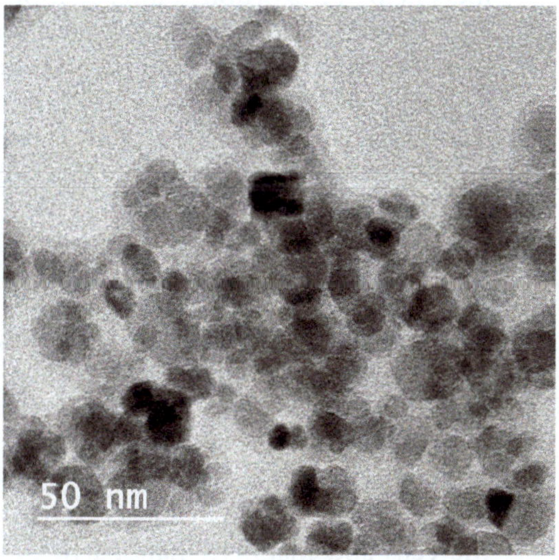


**Fig. 9** TEM image of Superoxide nanoparticles

nanoparticles. Transmission electron microscopy (TEM) images are shown in Fig. 9 [41].

3. Apply external magnet and remove supernatant containing extra salts. Wash with $ddH_2O$ thrice, and resuspend particles

in 30 mL of dd$H_2O$, add 250 mg of carboxylated PEG, stir on magnetic stirrer, wash with dd$H_2O$ to remove extra polymer solution, sonicate in 25 mL dd$H_2O$, apply magnet.

4. Collect supernatant, heat at 80 °C for 2 h to enhance the association between the magnetic nanoparticle surface and the polymer, centrifuge at 9,500 × *g* for 10 min to remove the aggregated materials, if left. Collect the supernatant. The functionalized magnetic nanoparticles can be conjugated with amine labeled oligonucleotide probe (complementary to the aptamer sequence) with spacer using EDC-NHS chemistry.
5. Add 100 μL polymer-coated nanoparticles to 80 pmol of oligonucleotide probe in sodium bicarbonate buffer (pH 8.0). Add 10 μL of 50 mM of EDC and 10 μL of 50 mM NHS, incubate for 1 h at 37 °C, followed by dialysis.
6. Take 8 pmol of aptamer, heat at 95 °C for 10 min, snap cool, add to the activated carboxylated superoxide nanoparticles forming aptamer hybrid functionalized iron oxide nanoparticles. The binding of oligonucleotides probe and aptamer can be examined by Zeta potential since addition of any charged moiety changes the zeta potential of the aptamer hybrid nanoparticle complex.
7. Add doxorubicin to aptamer-conjugated magnetic nanoparticles (electrostatic interaction and GC-rich sequences of aptamers help in doxorubicin intercalation) in different ratios (1:2, 1:4, 1:6, 1:8, 1:10) and check the fluorescence at 550 nm for calculating loading efficiency and better intercalation. For checking the drug release profile, drug intercalated aptamer should be put in dialysis bag and then in media (rat plasma). Solutions should be removed at regular intervals for fluorescence analysis.
8. Allow aptamer hybrid nanocomplex to bind to the cancerous cells expressing the biomarkers against the screened aptamer.
9. For checking in vitro therapeutic efficacy: positive and negative cell lines grown under appropriate media conditions along with required supplements and antibiotics should be treated with drug-Apt-polymer-NP complex, incubate for 3 h at 37 °C in humidified 5% $CO_2$, wash and fix with 4% formaldehyde. Aptamer binds to its cognate target, so that the drug releases inside the cells and magnetic nanoparticles pass through the membrane, which can help in the imaging of the cancerous cells by MRI. Antitumor activity of the drug-Apt-polymer-NP complex can be accessed via standard MTT [3-(4,5-dimethylthiazol-2-yl)-2,5-diphenyltratrazolium bromide] assay by reading optical density at 570 nm using a standard microplate reader like Bio-Tek Inc., SynergyH1 model, USA.

10. For in vivo studies, xenografted mouse model should be prepared, for example male nude athymic BALV/c-nu/nu mice. Mice need to be fed and treated as under standard conditions as per animal ethics and care approvals and norms of the respective institutions. Xenografted mice are prepared by inoculating the respective tumor cells (~8 × $10^6$ cells per animal) along with a commercial basement membrane matrix like "matrigel" (BD Biosciences, USA) or "culturex BME" (Trevigen, USA). Following this, in vivo antitumor activity should be assessed after the tumors grow to 100–150 $mm^3$ volume. Intravenously administer appropriate dosage of drug-Apt-polymer-NP complex along with negative controls (glucose 5%) and free drug or aptamer-NP complex only and check for change in tumor size (width and length) by in vivo MRI.

## 4 Notes

1. Primers should be designed to avoid any complementarity between forward and reverse sequences. Also, strong secondary structures should be avoided and melting temperature of both the primers should be in close range with only 2–5 °C difference.
2. FAM labeled primers/DNA should be stored in dark to avoid fluorescence bleaching.
3. ssDNA can be generated in a couple of ways. This can be achieved either by labeling the reverse primer with phosphate group at the 5′ prime end to yield phosphorylated primer, followed by digestion of the PCR product with lambda exonuclease to yield ssDNA. Alternatively, ribose residue can also be inserted at the 3′ end of the primer to generate an alkali cleavable ribonucleotide linkage. Cleavage at the ribo-linkage upon alkali treatment and further purification by Urea denaturing PAGE results in ssDNA. As a third option, ssDNA can also be generated by labeling the reverse primer with Biotin at 5′ end and separation can be done with the help of streptavidin-coated magnetic beads after denaturation of dsDNA.
4. While handling acrylamide, wear personal protection equipment, as unpolymerized acrylamide is known to be a neurotoxin. Always store acrylamide in amber-colored bottles.
5. TEMED is harmful if inhaled; therefore always open the bottle containing TEMED in fumehood.
6. Different immobilization matrices can be used depending upon the target, for the selection of aptamers against proteins, protein should be immobilized and for small molecules, preferably DNA population should be immobilized.

7. The temperature of the plates should not rise above 65 °C or above as it may lead to the breaking of the glass plates.
8. Depending on the amount of the DNA present, the brightness of the band varies. Both the tracking dyes have a particular migration rates on different percentage of gels. For 10% Urea denaturation PAGE, xylene cyanol dye (upper band) migrates at a size of around 55 nucleotides and bromophenol blue (lower band) at approximately 12 nucleotides. These tracking dyes help to know the distance to which one needs to run the gel. DNA can be labeled with different dyes available commercially. One of the most commonly used fluorochrome for labeling of the DNA is 6-carboxy fluorescein (6-FAM). 6-FAM is economical, reactive, and stable compared to other fluorescent dyes.
9. EtBr is carcinogenic; it should not be disposed of in to the environment directly. EtBr-stained gel should be stored in containers designated for the harmful materials and should be incinerated before discard, and liquid containing EtBr should be passed through activated charcoal followed by incineration.
10. For visualization of unlabeled DNA, staining can be done either with EtBr or SYBR green. SYBR green is safer than EtBr; therefore, SYBR green should always be preferred.
11. Vortexing and heating should be done repeatedly for better yield.
12. Incubation time should gradually decrease as the selection progresses, to increase the specificity and stringency.
13. The number of washes should be increased from 3 to 15 times as the selection progresses. This will lead to selection of aptamers with high affinity and specificity.
14. Elution of the bound DNA sequences to their target molecule can be performed either by low pH elution buffer or by the using 250–300 mM imidazole supplemented selection buffer.
15. Initially, for negative selection, use only Ni-NTA beads. When selection progresses, some other His-tagged protein of similar size should be used. The negative selection cycle should be performed after third or fourth round of selection.
16. Urea PAGE is preferred over other methods of purification, as it does not contribute to background signal and gives much cleaner DNA population.
17. The well of the gel should be washed thoroughly by syringe as urea and other salts deposited in the wells hamper the loading of the samples into the wells.

18. Crush the gel to a fine powder as the elution of the PCR product is obtained by simple diffusion process, so finer the gel is crushed, higher will be the DNA yield.
19. The concentration of protein target and the aptamer must be titrated. The protein concentration can be varied to measure the affinity of the selected DNA pool. Depending upon the size of protein and putative aptamer complex, assessment of the migration can be monitored with the help of bromophenol blue and xylene cyanol present in the EMSA loading dye.
20. High-fidelity Taq polymerase is used for the cloning of the putative aptamers after final selection round. Taq polymerases add template-independent Adenine at the 3′ end of the PCR product, which binds to T of the vector and thus can be used during TA cloning.
21. Depending upon the availability of functional group on the bioreceptor surface, different conjugation chemistries can be used. If the aptamer is amine-labeled then EDC/NHS chemistry is used for conjugating the carboxyl-activated surface with amine aptamer. If the aptamer is thiol-labeled then, either gold electrodes are used or SPE can be electroplated with gold to utilize thiol chemistry. Similarly, if the aptamer is biotin-labeled, the well-known biotin/streptavidin chemistry can be used.

## Acknowledgments

The aptamer work in Navani group is supported by Department of Biotechnology grant, India and DST, India. The authors acknowledge Shubham Jain for his assistance in drawing schematics. R.M. is supported by SERB-NPDF.

### References

1. Parashar A (2016) Aptamers in therapeutics. J Clin Diagn Res 10(6):BE01–BE06. https://doi.org/10.7860/JCDR/2016/18712.7922
2. Song K-M, Lee S, Ban C (2012) Aptamers and their biological applications. Sensors (Basel) 12 (1):612–631. https://doi.org/10.3390/s120100612
3. Cibiel A, Dupont DM, Ducongé F (2011) Methods to identify aptamers against cell surface biomarkers. Pharmaceuticals 4 (9):1216–1235. https://doi.org/10.3390/ph4091216
4. Dua P, Kim S, Lee DK (2011) Nucleic acid aptamers targeting cell-surface proteins. Methods 54(2):215–225. https://doi.org/10.1016/j.ymeth.2011.02.002
5. Gopinath SC (2007) Methods developed for SELEX. Anal Bioanal Chem 387(1):171–182. https://doi.org/10.1007/s00216-006-0826-2
6. Allali-Hassani A, Pereira MP, Navani NK, Brown ED, Li Y (2007) Isolation of DNA aptamers for CDP-ribitol synthase, and characterization of their inhibitory and structural properties. Chembiochem 8(17):2052–2057. https://doi.org/10.1002/cbic.200700257
7. Kumar P, Lambadi PR, Navani NK (2015) Non-enzymatic detection of urea using unmodified gold nanoparticles based

aptasensor. Biosens Bioelectron 72:340–347. https://doi.org/10.1016/j.bios.2015.05.029

8. Navani NK, Li Y (2006) Nucleic acid aptamers and enzymes as sensors. Curr Opin Chem Biol 10(3):272–281. https://doi.org/10.1016/j.cbpa.2006.04.003
9. Tuerk C, Gold L (1990) Systematic evolution of ligands by exponential enrichment: RNA ligands to bacteriophage T4 DNA polymerase. Science 249(4968):505–510
10. Ellington AD, Szostak JW (1990) In vitro selection of RNA molecules that bind specific ligands. Nature 346:818. https://doi.org/10.1038/346818a0
11. Sefah K, Shangguan D, Xiong X, O'Donoghue MB, Tan W (2010) Development of DNA aptamers using Cell-SELEX. Nat Protoc 5:1169. https://doi.org/10.1038/nprot.2010.66
12. Sypabekova M, Bekmurzayeva A, Wang R, Li Y, Nogues C, Kanayeva D (2017) Selection, characterization, and application of DNA aptamers for detection of Mycobacterium tuberculosis secreted protein MPT64. Tuberculosis 104:70–78
13. Liss M, Petersen B, Wolf H, Prohaska E (2002) An aptamer-based quartz crystal protein biosensor. Anal Chem 74(17):4488–4495
14. Misono TS, Kumar PK (2005) Selection of RNA aptamers against human influenza virus hemagglutinin using surface plasmon resonance. Anal Biochem 342(2):312–317
15. Song K-M, Lee S, Ban C (2012) Aptamers and their biological applications. Sensors 12 (1):612–631
16. Stoltenburg R, Nikolaus N, Strehlitz B (2012) Capture-SELEX: selection of DNA aptamers for aminoglycoside antibiotics. J Anal Methods Chem 2012:415697
17. Jiang L, Suri AK, Fiala R, Patel DJ (1997) Saccharide-RNA recognition in an aminoglycoside antibiotic-RNA aptamer complex. Chem Biol 4(1):35–50
18. Patel DJ, Suri AK (2000) Structure, recognition and discrimination in RNA aptamer complexes with cofactors, amino acids, drugs and aminoglycoside antibiotics. Rev Mol Biotechnol 74(1):39–60
19. Rowe AA, Miller EA, Plaxco KW (2010) Reagentless measurement of aminoglycoside antibiotics in blood serum via an electrochemical, ribonucleic acid aptamer-based biosensor. Anal Chem 82(17):7090–7095
20. Kase D, Kulp JL, Yudasaka M, Evans JS, Iijima S, Shiba K (2004) Affinity selection of peptide phage libraries against single-wall carbon nanohorns identifies a peptide aptamer with conformational variability. Langmuir 20 (20):8939–8941
21. Famulok M (1999) Oligonucleotide aptamers that recognize small molecules. Curr Opin Struct Biol 9(3):324–329
22. Zayats M, Huang Y, Gill R, Ma CA, Willner I (2006) Label-free and reagentless aptamer-based sensors for small molecules. J Am Chem Soc 128(42):13666–13667
23. Kumar P, Ramulu Lambadi P, Kumar Navani N (2015) Non-enzymatic detection of urea using unmodified gold nanoparticles based aptasensor. Biosens Bioelectron 72:340–347. https://doi.org/10.1016/j.bios.2015.05.029
24. McCauley TG, Hamaguchi N, Stanton M (2003) Aptamer-based biosensor arrays for detection and quantification of biological macromolecules. Anal Biochem 319 (2):244–250
25. Xiang Y, Zhang Y, Qian X, Chai Y, Wang J, Yuan R (2010) Ultrasensitive aptamer-based protein detection via a dual amplified biocatalytic strategy. Biosens Bioelectron 25 (11):2539–2542
26. Liu G, Yu X, Xue F, Chen W, Ye Y, Yang X et al (2012) Screening and preliminary application of a DNA aptamer for rapid detection of Salmonella O8. Microchim Acta 178 (1–2):237–244
27. Joshi R, Janagama H, Dwivedi HP, Kumar TS, Jaykus L-A, Schefers J et al (2009) Selection, characterization, and application of DNA aptamers for the capture and detection of Salmonella enterica serovars. Mol Cell Probes 23 (1):20–28
28. Queirós RB, de-Los-Santos-Álvarez N, Noronha J, Sales MGF (2013) A label-free DNA aptamer-based impedance biosensor for the detection of E. coli outer membrane proteins. Sens Actuators B Chem 181:766–772
29. Medley CD, Smith JE, Tang Z, Wu Y, Bamrungsap S, Tan W (2008) Gold nanoparticle-based colorimetric assay for the direct detection of cancerous cells. Anal Chem 80(4):1067–1072
30. Wang G, Wang Y, Chen L, Choo J (2010) Nanomaterial-assisted aptamers for optical sensing. Biosens Bioelectron 25 (8):1859–1868
31. Liu X, Wang F, Aizen R, Yehezkeli O, Willner I (2013) Graphene oxide/nucleic-acid-stabilized silver nanoclusters: functional hybrid materials for optical aptamer sensing and multiplexed analysis of pathogenic DNAs. J Am Chem Soc 135(32):11832–11839
32. Feng L, Chen Y, Ren J, Qu X (2011) A graphene functionalized electrochemical

aptasensor for selective label-free detection of cancer cells. Biomaterials 32(11):2930–2937

33. Wang J (2006) Electrochemical biosensors: towards point-of-care cancer diagnostics. Biosens Bioelectron 21(10):1887–1892
34. Min K, Song K-M, Cho M, Chun Y-S, Shim Y-B, Ku JK et al (2010) Simultaneous electrochemical detection of both PSMA (+) and PSMA (−) prostate cancer cells using an RNA/peptide dual-aptamer probe. Chem Commun 46(30):5566–5568
35. Lai RY, Plaxco KW, Heeger AJ (2007) Aptamer-based electrochemical detection of picomolar platelet-derived growth factor directly in blood serum. Anal Chem 79 (1):229–233
36. Taleat Z, Cristea C, Marrazza G, Mazloum-Ardakani M, Săndulescu R (2014) Electrochemical immunoassay based on aptamer–protein interaction and functionalized polymer for cancer biomarker detection. J Electroanal Chem 717:119–124
37. Eivazzadeh-Keihan R, Pashazadeh-Panahi P, Baradaran B, Maleki A, Hejazi M, Mokhtarzadeh A et al (2018) Recent advances on nanomaterial based electrochemical and optical aptasesnors for detection of cancer biomarkers. Trends Anal Chem. https://doi.org/10.1016/j.trac.2017.12.019
38. Rauf S, Mishra GK, Azhar J, Mishra RK, Goud KY, Nawaz MAH et al (2018) Carboxylic group riched graphene oxide based disposable electrochemical immunosensor for cancer biomarker detection. Anal Biochem 545:13–19
39. Sharma P, Tuteja SK, Bhalla V, Shekhawat G, Dravid VP, Suri CR (2013) Bio-functionalized graphene-graphene oxide nanocomposite based electrochemical immunosensing. Biosens Bioelectron 39(1):99–105. https://doi.org/10.1016/j.bios.2012.06.061. S0956-5663(12)00420-4 [pii]
40. Mutreja R, Jariyal M, Pathania P, Sharma A, Sahoo DK, Suri CR (2016) Novel surface antigen based impedimetric immunosensor for detection of Salmonella typhimurium in water and juice samples. Biosens Bioelectron 85:707–713. https://doi.org/10.1016/j.bios.2016.05.079
41. Yu MK, Kim D, Lee IH, So JS, Jeong YY, Jon S (2011) Image-guided prostate cancer therapy using aptamer-functionalized thermally cross-linked superparamagnetic iron oxide nanoparticles. Small 7(15):2241–2249
42. Bagalkot V, Zhang L, Levy-Nissenbaum E, Jon S, Kantoff PW, Langer R et al (2007) Quantum dot-aptamer conjugates for synchronous cancer imaging, therapy, and sensing of drug delivery based on bi-fluorescence resonance energy transfer. Nano Lett 7 (10):3065–3070
43. Savla R, Taratula O, Garbuzenko O, Minko T (2011) Tumor targeted quantum dot-mucin 1 aptamer-doxorubicin conjugate for imaging and treatment of cancer. J Control Release 153 (1):16–22
44. Gray BP, Kelly L, Ahrens DP, Barry AP, Kratschmer C, Levy M et al (2018) Tunable cytotoxic aptamer–drug conjugates for the treatment of prostate cancer. Proc Natl Acad Sci U S A 115(18):4761–4766
45. Deng R, Qu H, Liang L, Zhang J, Zhang B, Huang D et al (2017) Tracing the therapeutic process of targeted aptamer/drug conjugate on cancer cells by surface-enhanced raman scattering spectroscopy. Anal Chem 89 (5):2844–2851

# Chapter 15

# Aptamer Selection for Detecting Molecular Target Using Cell-SELEX (Systematic Evolution of Ligands by Exponential Enrichment) Technology

**Kimberly D. Stewart, Weihong Tan, and Jong Y. Park**

## Abstract

Cell-SELEX is a live cell-based in vitro selection method that generates functional oligonucleotides, or aptamers. Often referenced as the chemist's antibody, aptamers bind to select targets with high affinity and can be utilized in a number of applications, including biomedicine, bioimaging, and biosensing. Here we describe the cell-SELEX technique and discuss this methodology's unique merit(s)—namely the ability to isolate highly selective aptamer panels with no prior knowledge of cellular signatures. This strategy thus presents as a technology that has the potential to enhance the precision of molecular medicine and targeted therapeutics.

**Key words** Cell-SELEX, Aptamer, Chemical antibody, Nucleic acid, Molecular science, Targeted therapy, Cancer, Theranostics, Functional oligonucleotide, Precision medicine

## 1 Introduction

An aptamer—from the Latin aptus, "to fit," and the Greek meros, "part,"—is comprised of single-stranded DNA or RNA molecules that have regions which fit upon or maintain affinity toward select targets [1]. In contrast to traditional Watson-Crick base pairing, these oligonucleotides pair with their respective targets through sophisticated folding mechanisms. These mechanisms have been characterized via nuclear magnetic resonance studies which revealed the unique nature of aptamer folding; each structurally conforms to bind with its cognate target [2, 3]. In this manner, aptamers closely mimic the binding mechanisms of antibodies, which too are reliant upon highly selective molecular recognition events. Given their capacity for target-based interaction and binding, DNA and RNA oligonucleotides are often referenced as the chemist's antibody, with select merits, as seen in Fig. 1 [4]. A primary strength of the aptamer is its in vitro selection process,

Jyotsna Batra and Srilakshmi Srinivasan (eds.), *Theranostics: Methods and Protocols*, Methods in Molecular Biology, vol. 2054, https://doi.org/10.1007/978-1-4939-9769-5_15, 

**ADVANTAGES OF APTAMERS**

- Chemical synthesis technique affords facile modification and reproducibility;
- Production not susceptible to viral or bacterial contamination (as within antibody synthesis);
- Non-immunogenic;
- Small size allows penetration into varied biological compartments (potentially BBB accessible);
- Facilitates selection of targets without prior knowledge of cellular signatures;
- Can be reversibly denatured (tolerate and recover from variable temperatures and pH).

**LIMITATIONS OF APTAMERS**

- Pharmacokinetic interaction(s) not well understood, thus retarding inclusion within biological applications;
- Small size can result in excretion from biological compartments (highly susceptible to renal filtration);
- Unmodified aptamers subject to degradation at physiological pH;
- Biophysical characteristics of nucleic acid(s) dictate potential targets (hydrophobic or negatively charged binding interactions met with challenge);
- Intellectual property restrictions may limit access to / use of aptamer technologies.

**Fig. 1** Advantages and limitations of aptamers [4]

which, unlike antibodies, requires no prior knowledge of the molecular target. Each targeting moiety is instead selected via a method termed SELEX (*S*ystematic *E*volution of *L*igands by *EX*ponential Enrichment) which takes advantage of discrete molecular variances between a target and nontarget pair [5]. These selections can thus proceed in the presence of diverse targets, including ions, small molecules, peptides, purified proteins, and whole cells—the latter of which is discussed in this chapter.

## 2 Materials

Outlined below are requisite reagents, materials, and equipment, as tailored for a live cell-based DNA aptamer selection. Note that this guide reflects the procedural steps previously outlined by our laboratory at University of Florida [6], with modification(s) where necessary.

### 2.1 Cell Culture

1. Target Cell Line.
2. Counter Cell Line.
3. Appropriate Cell Culture Media.
4. Heat-Inactivated Fetal Bovine Serum.

5. Penicillin Streptomycin (Pen Strep).
6. Trypsin-EDTA Solution 1×.
7. Cell Dissociation Solution Nonenzymatic 1×.
8. Trypan Blue Stain, 0.4%.
9. Hemocytometer, 0.1 nm deep.
10. Ethanol, 200 Proof.
11. Ethanol, 70 v/v % (300 mL of nanopure water and 700 mL of absolute ethanol).
12. Tissue Culture Flasks, 25 and 75 $cm^2$.

### 2.2 Binding and Washing Buffer

1. 1× Binding Buffer: To 1 L of Dulbecco's Phosphate Buffered Saline (DPBS) add 4.5 g of D-(+)-Glucose, 100 mg of tRNA, 1 g of Bovine Serum Albumin (BSA) and 5 mL of 1 M $MgCl_2$. Store at 4 °C for up to 1 month.
2. 10× Binding Buffer: To 1 L of 10× PBS add 45 g of D-(+)-Glucose, 1 g of tRNA, 10 g of BSA, and 50 mL of 1 M $MgCl_2$. Store in 1 mL aliquots at −20 °C for up to 1 year.
3. Washing Buffer: To 1 L of DPBS, add 4.5 g of D-(+)-Glucose and 5 mL of 1 M $MgCl_2$. Store at 4 °C for up to 3 months.

### 2.3 Polymerase Chain Reaction (PCR)

1. 5′ FITC-Labeled Forward Primer.
2. 5′ Biotin-Labeled Reverse Primer.
3. 5′ FITC-Labeled Randomized Oligonucleotide Library (*see* **Note 1**).
4. TaKaRa Taq Hot Start Version.
5. Water, DNA Grade.
6. PCR Solution Tubes, 0.2 and 0.5 mL.

### 2.4 Agarose Gel Electrophoresis

1. Agarose Powder.
2. DNA Step Ladder Molecular Weight Markers, Including 6× Blue/Orange Loading Dye.
3. Ethidium Bromide Solution, 10 mg/mL (*see* **Note 2**).

### 2.5 ssDNA Elution

1. Needleless Syringe, 5 mL.
2. Illustra NAP-5 Columns Sephadex G-25 DNA Grade.
3. Empty Synthesis Columns, 1 μm Expedite Style.
4. Sodium Hydroxide.
5. Sodium Hydroxide Solution, 20 mM, 0.7998 g of NaOH to 1 L of nanopure water to prepare 20 mM solution.
6. Streptavidin Sepharose High Performance.
7. Water, DNA Grade.

### 2.6 Flow Cytometry

1. 5 mL Polystyrene Round-Bottom Tube(s), 12 × 75 mm style
2. 1× Binding Buffer (*see* Subheading 2.2, **item 1**).
3. Sheath Fluid.

### 2.7 Miscellaneous

1. Centrifuge Tubes 15, 50 mL.
2. 1.7 mL Microcentrifuge Tubes
3. PureLink Quick Gel Extraction Kit.
4. Cell Scraper, 220 mm Long, 11 mm Blade, Sterile.
5. Thermometer.
6. Bleach.

### 2.8 Equipment

1. Microscope (Olympus, IX70).
2. Centrifuge (Eppendorf, 5810 R).
3. Microcentrifuge (Eppendorf, 5417 R).
4. Dri-Bath Incubator (Barnstead Thermolyne 17600).
5. Microwave (Kenmore).
6. Weight Scale (Denver Instrument, SI-602).
7. Analytical Balance (Denver Instrument, SI-64).
8. Accuri C6 Flow Cytometer (Becton Dickinson).
9. Series II Water-Jacketed $CO_2$ Incubator (Thermo Electron Corporation).
10. Class II Biological Safety Cabinet, SterilGARD III Advance (The Baker Company, SG603A).
11. Labquake Shaker (Barnstead Thermolyne).
12. T100 Thermal Cycler (Bio-Rad).
13. Purifier Filtered PCR Enclosure (Labconco).
14. Electrophoresis Power Supply (Bio-Rad, Power PAC 200).
15. Mini-Sub Cell GT System (Bio-Rad).
16. ImageQuant Variable Mode Imager System (GE Healthcare).
17. Digital Graphic Printer (Sony, UP-D897).
18. Spectrophotometer (Bio-Rad, SmartSpec Plus).
19. Speedvac Concentrator (Thermo, DNA120OP-115).
20. Water Bath (Fisher Scientific, Isotemp 105).
21. Autoclave (Tuttnauer, Cat2007).

## 3 Methods

The cell-SELEX protocol [5], as conceptualized by Sefah and coworkers, provides an exhaustive workflow and should be referenced in parallel with this source. Note, however, that each cell-SELEX must be tailored with regard to experimental parameters—these including the target and nontarget cell pair, library and primer design, and also operational technique(s). As such, we provide herein an overview of the cell-SELEX procedure, alongside fundamental wet-lab directives, so that its strategies may be readily adapted and modified for each selection.

### 3.1 Oligonucleotide Library and Primer Design

*See* **Note 3** for key considerations.

1. Prepare primer stock(s) by bringing each to a concentration of 100 μM, using DNA-grade water as the diluent. In a separate tube, mix 10 μL of the forward primer (FP), 10 μL of the reverse primer (RP) and 80 μL of DNA-grade water to formulate a PCR primer master mix.
2. Prepare oligonucleotide library by bringing it to a concentration of 100 μM, using DNA-grade water as the diluent. In a separate tube, prepare a 10 μL aliquot at a concentration of 10 pM for use in the primer optimization assay.
3. Create a stock solution containing the prepared primer and library aliquots according to the following recipe: DNA-grade water (64.8%), 10× PCR Buffer ($Mg^{2+}$ plus) (10%), dNTP Mixture (10%), Template/Library (10%), Primer(s) (5%), and TaKaRa Taq HS (0.2%). Prepare negative control in separate PCR solution tube prior to the addition of template.
4. Distribute the above mixture into PCR solution tubes and place in thermocycler heating block. Note: primer annealing should be investigated within the temperature range of 50–65 °C.

   Analyze results via agarose gel electrophoresis (*see* Subheading 3.4 for directives).

### 3.2 Cell Line Recovery and Subculture

*See* **Note 4** for key considerations concerning cell line selection(s) and maintenance.

#### 3.2.1 Recovery

1. Spray and wipe biological safety cabinet surfaces with 70 v/v % ethanol prior to performing cellular work. Note: all reagent bottles and materials must also be sprayed with 70 v/v % ethanol before being placed within the cabinet.
2. Incubate cell culture media and established cell line stock (as contained within a cryogenic tube) in a 37 °C water bath for 10–15 min.
3. Transfer cell line stock, requisite media, and associated subculture materials to cabinet.
4. Remove cell solution from original tube and transfer to centrifuge tube. Add 2–3 mL of cell culture media and gently mix by pipetting.

5. Centrifuge sample at 400 × *g*/ 3 min. Thereafter, remove supernatant.
6. Resuspend pelleted cellular bodies in 5–7 mL of media and transfer to 25 $cm^2$ tissue culture flask. Incubate within water-jacketed $CO_2$ incubator for 24–48 h or until confluent.

#### 3.2.2 Subculture

1. Spray and wipe biological safety cabinet surfaces with 70 v/v % ethanol prior to performing cellular work. Note: all reagent bottles and materials must also be sprayed with 70 v/v % ethanol before being placed within the cabinet.
2. Incubate cell culture media, Trypsin-EDTA Solution 1×, and DPBS in a 37 °C water bath for 10–15 min.
3. Transfer cell line stock, requisite media, and associated subculture materials to cabinet.
4. Add 5 mL of DPBS to tissue culture flask(s) (containing cells) and gently agitate to ensure even distribution.

#### 3.2.3 Adherent Cell Line Guide

1. After subculture (Subheading 3.2.2), remove DPBS from cell dish by aspiration and insert 5 mL of Trypsin-EDTA Solution 1×. Allow to incubate at room temperature for 1–3 min. Note: these values may need to be adjusted based on flask size and attachment characteristics of each cell line.
2. Add 1–2 mL of media to neutralize the trypsin solution. 'Rinse' flask with contained media and trypsin to collect cells. Thereafter, transfer flask contents to centrifuge tube and centrifuge at 400 × *g*/ 3 min.
3. Remove supernatant and resuspend in an appropriate volume of cell culture media, based upon the number of subculture flasks. Note: cells are effectively 'split' or subcultured by distributing cell solution to multiple culture flasks.

#### 3.2.4 Suspension Cell Line Guide

1. After subculture (Subheading 3.2.2), transfer flask contents to centrifuge tube and centrifuge at 400 × *g*/ 3 min.
2. Remove supernatant by aspiration and resuspend in an appropriate volume of cell culture media, based upon the number of subculture flasks. Note: cells are effectively 'split' or subcultured by distributing cell solution to multiple culture flasks.

### 3.3 Live Cell-Based In Vitro Selection Method(s)

*See* **Note 5** for key considerations.

#### 3.3.1 Initial Binding Round

1. Prepare the oligonucleotide library for selection by bringing it to a concentration of 20 nmol and a volume of 700 μL. DNA-grade water should be used as the diluent to reach target concentration(s) (*see* **Note 6**).

2. Snap cool the above solution by placing within a dri-bath incubator for 3 min at 95 °C and then storing on ice for 5 min. Add 70 μL of 10× binding buffer to 630 μL of the snap cooled DNA solution prior to incubation. Note: this solution can remain on ice until ready for use (*see* **Note 7**).
3. Quantify cells and, thereafter, wash three times with unsupplemented DPBS solution. Note: adherent cultures can be washed in situ while suspension cultures must be collected by centrifugation following each wash (*see* **Note 8**).
4. Incubate the prepared solution (700 μL, total) with the target cell at 4 °C for 45 min–1 h. Note: larger volumes may be used with appropriate adjustment of 10× binder buffer addition (*see* **Note 9**).
5. Wash cells with a small volume of chilled washing buffer (0.5–1 mL). Centrifuge and remove supernatant. Resuspend in 500 μL of washing buffer, and transfer to microcentrifuge tube. This step varies according to cell type (*see* **Note 10**).
6. Place the microcentrifuge tube within a dri-bath incubator for 15 min at 95 °C. Note: at this temperature the cell-DNA complex will denature, and candidates bound to the surface proteins will be eluted into solution.
7. Centrifuge the solution at 10,000 × *g*/ 5 min and collect the supernatant. This solution may be directly used as the template for PCR amplification or stored at −20 °C until ready for use.

#### *3.3.2 Iterative Binding and Counter Selection Rounds*

*See* **Note 11** for key considerations.

##### Iterative Binding Rounds

1. Prepare 700 μL of the 'Round 1' ssDNA library in a microcentrifuge tube. Note: solution should be brought to a concentration of 250 nM using DNA-grade water as the diluent (*see* **Note 12**).
2. Snap cool the above solution by placing within a dri-bath incubator for 3 min at 95 °C and then storing on ice for 5 min. Add 70 μL of 10× binding buffer to 630 μL of the snap cooled DNA solution prior to incubation (*see* **Note 13**).
3. Prepare the target cell culture for incubation by washing three times with unsupplemented DPBS, bearing in mind operational requirements for adherent vs. suspension cells (*see* **Note 14**).
4. Incubate the prepared oligonucleotide library with the target cell at 4 °C, bearing in mind operative techniques based on the cell-type (*see* **Note 15**).

5. Wash the cell population with washing buffer, using increased volume and repetition as the selection progresses. Note: DNA pool elution and recovery techniques are identical to those previously outlined (Subheading 3.3.1, **steps 5–7**).

Counter Selection Rounds

*See* **Note 16** for key considerations.

1. Immediately following the recovery of the DNA pool, as eluted from the target cell incubation (above), place the microcentrifuge contents on ice.
2. Prepare counter cell dish by washing three times with DPBS (using appropriate operative techniques based on cell type) (*see* **Note 17**).
3. Transfer the DNA pool to the counter cell dish and incubate at 4 °C for 30 min (*see* **Note 18**).
4. After incubating, collect the supernatant and transfer to a microcentrifuge tube. Adherent and suspension cell solutions contained within centrifuge tubes must be pelleted by centrifugation prior to supernatant collection. The supernatant can thereafter be directly used as a template within PCR. Note: Do NOT perform cell sample washing in counter selection rounds.

### 3.4 Enrichment of DNA Pool(s) via PCR

*See* **Note 19** for key considerations.

#### 3.4.1 Pool Optimization

All PCR mixtures must be prepared within a purifier filter PCR enclosure system.

1. Using the previously prepared 100 μM (FP) and 100 μM (RP) stock solutions, formulate a PCR primer master mix as described above (*see* Subheading 3.1).
2. Create a PCR stock solution according to the following recipe: DNA-grade water (64.8%), 10× PCR Buffer ($Mg^{2+}$ plus) (10%), dNTP Mixture (10%), Template/Round 'X' Eluate (10%), Primer(s) (5%), and TaKaRa Taq HS (0.2%). Prepare negative control within a separate PCR solution tube prior to the addition of template. Note: as this solution is to be used for optimization, only, one may wish to prepare a minimal volume to conserve resources (i.e. ~30 μL/sample).
3. Distribute the above mixture into PCR solution tubes, place in thermocycler heating block, and amplify samples according to previously optimized PCR conditions (*see* **Note 20**).
4. Analyze results via agarose gel electrophoresis (*see* below for directives).

*3.4.2 Analysis via Gel Electrophoresis*

1. Weigh 1.25 g of agarose powder and combine with 50 mL of 1× TBE buffer within an Erlenmeyer flask. Microwave the mixture for 60 s or until agarose powder is fully solubilized.
2. Add 2.5 μL of ethidium bromide solution and mix by gently swirling. CAUTION: EtBr is a suspected mutagen and, at high concentrations, may prove toxic. Proper protective equipment, including nitrile gloves, must be worn during handling and disposal.
3. Cast the gel contents within a sub cell system and allow to solidify for 30 min.
4. At the conclusion of the PCR run, combine 10 μL of each sample with 2 μL of 6× loading dye, load into previously prepared gel and run for 30 min at 100 V.
5. When complete, immediately remove gel from the sub cell system, visualize under UV light (using ImageQuant Variable Mode Imager System or similar), and retain printed image of gel for determination of optimum cycle number. Note: the appropriate cycle number is defined as the band having intensity at least equal that of the DNA ladder, without the presence of double band(s) (indicative of nonspecific amplification).

*3.4.3 Pool Amplification*

All PCR mixtures must be prepared within a purifier filter PCR enclosure system.

1. Using the previously prepared 100 μM (FP) and 100 μM (RP) stock solutions, formulate a PCR primer master mix as described above (*see* Subheading 3.1).
2. Create a PCR stock solution according to the following recipe: DNA-grade water (64.8%), 10× PCR Buffer ($Mg^{2+}$ plus) (10%), dNTP Mixture (10%), Template/Round 'X' Eluate (10%), Primer(s) (5%), and TaKaRa Taq HS (0.2%). Prepare negative control within a separate PCR solution tube prior to the addition of template. Note: as this solution is to be used for ssDNA elution, a 1 mL stock must be formulated.
3. Distribute the above mixture into PCR solution tubes, place in thermocycler heating block, and amplify samples according to optimum cycle number (using previously determined PCR conditions—i.e. melting temperature, annealing temperature, etc.).
4. Thereafter, combine PCR aliquots into a single microcentrifuge tube and analyze amplification efficiency via agarose gel electrophoresis (*see* above for directives). Note: amplicon should be stored on ice or at 4 °C while performing gel analysis.

### 3.5 Recovery of ssDNA Binding Candidates

*See* **Note 21** for key considerations.

1. Insert a filter into an empty DNA synthesis column. Note: this may be facilitated by using a pipette tip to drive the filter securely into one end of the column.
2. Attach the column to a needleless 5 mL syringe and secure the entire apparatus on a clamp stand. Before each removal of the syringe plunger the column must be detached in order to prevent displacement of the inserted filter. That said, before insertion of each reagent into the syringe: (1) detach the column, (2) remove the syringe plunger, and (3) reattach the column.
3. Insert 120 μL of streptavidin sepharose high performance (SAV) beads into the syringe. Push contents through the column (slowly!) using the syringe plunger and allow excess solution to drain into a waste beaker. Note: at this point SAV beads are securely compressed within the column walls.
4. Pretreat the column-contained SAV beads by washing thrice with 20 mM NaOH. Insert 500 μL of NaOH (ea. wash) into the syringe, push through the column using syringe plunger, and allow excess solution to drain into a waste beaker (*see* **Note 22**).
5. Wash the column-contained SAV beads three times with DPBS. Insert 1 mL of DPBS (ea. wash) into the syringe, push through the column using syringe plunger, and allow excess solution to drain into a waste beaker.
6. Insert the amplified pool (1 mL) into the syringe. Push through the column (slowly!) using syringe plunger and collect contents as they drain from column. Repeat twice. At this point, the DNA pool is bound to the SAV beads in the column.
7. Rinse the column-contained amplicons with DPBS three times. Insert 500 μL of DPBS (ea. wash) into the syringe, push through the column using syringe plunger, and allow excess solution to drain into a waste beaker.
8. Insert 500 μL of NaOH into the syringe. Push through the column (slowly!) using syringe plunger and collect contents as they drain from column. Repeat twice. The eluted solution should be set aside for desalting to follow (*see* **Note 23**).
9. Immobilize an Illustra NAP-5 Column (Sephadex G-25 DNA Grade) upon a clamp stand. Note: column volume can be increased by inserting a 10 mL pipette tip into opening.
10. Fill the column with nanopure water and allow draining into waste beaker.
11. Fill the column with DNA-grade water and allow draining into waste beaker.

12. Remove pipette insert (if utilized) and add 500 μL sample (as collected in **step 8**, above) directly into opening of desalting column. Allow any excess solution to drain into waste beaker.
13. Insert 1 mL of DNA-grade water directly into desalting column opening and collect filtered contents as they drain from column. Note: absorbance and resultant concentration can be determined by UV-vis analysis of ssDNA product.
14. Detach column from syringe. Remove SAV beads from column by pipetting DPBS into column opening. Discard solution. Repeat as needed. Note: the column must be treated with a *new* volume of SAV beads when performing each ssDNA separation.

### 3.6 Monitoring Binding Progression

*See* **Notes 24-26** for key considerations.

1. Prepare the pool(s) of interest for incubation by formulating 250 nM samples, at a volume of 50 μL. Note: all sample concentrations *must* be equivalent and ≥250 nM to properly monitor fluorescence via flow cytometry.
2. Snap cool the above solution(s) by placing within a dri-bath incubator for 3 min at 95 °C and then storing on ice for 5 min. Add 5 μL of 10× binding buffer to 45 μL of the snap cooled DNA pool prior to incubation.
3. Prepare the cell culture(s) of interest for incubation by washing three times with unsupplemented DPBS, bearing in mind operational requirements for adherent cells. Suspension cells. In either case, cell populations should be washed thrice with washing buffer (1 mL ea. wash) and resuspended in an appropriate amount of washing buffer for the incubation step(s) to follow. Note: in this assay, adherent cells *must* be dissociated from their original dish. This is to allow for cell quantitation, a step which must be done (via hemocytometer) to ensure equivalent cell densities in each sample.
4. Transfer 150 μL of each cell solution to individual polystyrene round-bottom tubes. Centrifuge at 1350 rpm for 3 min and remove supernatant thereafter (cells should be firmly adsorbed to the bottom of the tube(s) at this point).
5. Insert 50 μL of the DNA stock into each round-bottom polystyrene tube. Vortex. Place upon a rotary shaker and incubate at 4 °C. Note: each sample should be incubated in accordance with chosen cell-SELEX conditions. For example, should the target incubations have been carried out for 30 min periods, as should be the binding assay incubation(s).
6. Following incubation, wash cells three times to remove nonbinders and nonspecific binders. Add 1 mL of washing buffer and vortex. Thereafter, centrifuge at 1350 rpm/3 min and remove supernatant. Repeat two additional times.

7. Add 100 μL of 1× binding buffer and investigate fluorescent behavior via flow cytometry. Note: fluorescence can be alternatively monitored in one of two ways; (1) For investigation of target cell affinity: resuspend cells in 100 μL of washing buffer and transfer cell samples to a microcentrifuge tube following the final wash. Precipitate fluorescently tagged binders by heating within a dri-bath incubator at 95 °C for 15 min. Centrifuge said solution (14,000 rpm/5 min), collect the supernatant and screen using a spectrofluorometer. (2) For investigation of nontarget cell affinity: Do NOT wash cell samples following incubation. Rather, collect the supernatant solution and screen using a spectrofluorometer.
8. A cell-SELEX is considered to be at its termination point when a significant fluorescence intensity shift is noted between the background and enriched DNA pools, while minimal change is recorded between 2–3 mature stage pools. The selected DNA pool presenting with the greatest fluorescent intensity/affinity for the target cell and lowest fluorescent intensity/affinity for the counter cell can then be prepared for sequencing and, thereafter, aptamer characterization. Directives for such have been thoroughly outlined [6] and further reported on (with pertinent modifications) by Shangguan et al. [9]

## 4 Notes

1. Oligonucleotide Synthesis and Purification: The randomized DNA library and primers may be synthesized and purified on-site (if laboratory facilities permit) or purchased commercially. Purification, as performed in the Tan Research Lab, is completed by way of reversed-phase HPLC with a C18 column (250 mm × 4.6 mm, 5-μm particle size) using a 0.1 M triethylamine acetate (TEAA) and acetonitrile solution as the eluent.
2. CAUTION: EtBr is a suspected mutagen and, at high concentrations, may prove toxic. Proper protective equipment, including nitrile gloves, must be worn during handling and disposal.
3. Each cell-SELEX begins with the design and synthesis of a randomized oligonucleotide library. Said libraries consist of a random 20–80 nucleotide region flanked by forward and reverse primer sequences. Primer design software, such as the PrimerQuest Tool available via Integrated DNA Technologies (IDT), is recommended for the development of primer sequences that afford specific amplification and high yield. In general, primer length(s) are between 18–30 bp and are designed so as to minimize hairpin, self-dimer, and heterodimer formation. Prior to use, the primer pair must be optimized to determine PCR amplification conditions. These steps are

outlined in Subheading 3.1 and should be performed within a purifier filtered PCR enclosure. Previously optimized primers [6] are shown below: Forward Primer (FP): 5′-(FITC)-ATCCAGAGTGACGCAGCA-3′ Reverse Primer (RP): 5′-(Biotin)-ACTAAGCCACCGTGTCCA-3′. Note: the 5′end of the forward primer must be labeled with a fluorescent dye (as are FAM, TMRA, and Cy5) to allow for target-ligand affinity monitoring throughout the SELEX process. The biotinylated 3′ end is so-labeled in order to afford single-strand DNA displacement (refer to Subheading 3.5).

A randomized oligonucleotide library which incorporates the above primers would thus reflect the following sequence, with the complementary strand of the biotin-labeled primer at the 3′ end. 5′-ATCCAGAGTGACGCAGCA-$N_{(20-80)}$—TGGACACGGTGGCTTAGT-3′. As indicated, the randomized portion of the library is between 20–80 nucleotides in length, with the DNA pool diversity defined by the formula $4N$. It then follows that larger $N$-values allow for a more diverse library and increase the structural complexity of the starting pool. Alternatively, libraries that are shorter in length generally result in higher synthesis yields and are more cost-effective. The length of each library should therefore be long enough to maximize binding interactions with diverse molecular targets, as are present in cell-SELEX, while also being mindful of synthesis limitations.

4. For each cell-SELEX, established (or immortalized) cell lines are used so as to ensure a stable performance and biomarker expression throughout the entire assay. At present, such cannot be verified in primary cell cultures, given their inability to be maintained over extended periods of time, as generally required in each selection. Cell-SELEX investigations typically utilize two cell types that are closely related and differ only in discrete molecular variances (i.e. cancer cell vs. homologous noncancerous cell). In theory, cell-SELEX can be used to develop a panel of aptamers against any target and nontarget cell pair. It has been noted, however, that this technique is the most practical when investigating select targets. These include molecules that are (1) positively charged, (2) hydrogen bond donors or acceptors, and/or (3) aromatic-containing groups [1]. Moreover, selections are most successful when the target of interest, however unknown, is abundantly expressed on the cell surface membrane. That said, discriminant election of a target—nontarget cell pair does not ensure successful aptamer panel generation. To prevent operator-induced discrepancies, proper maintenance of cell cultures throughout the cell-SELEX progression is vital. Should the investigator fail to monitor the condition of target—nontarget cell cultures, one may select for

contaminants as introduced due to poor culturing conditions. As such, it is recommended that cell populations be periodically tested for mycoplasma contamination, and that cell stocks be repurchased/replaced at select intervals (this owing to changes in cellular expression after extended periods of subculture) [7]. In order to properly preserve cell populations, each must be maintained according to its respective subculture method (Note: for cell lines ordered via ATCC, this information is listed on product website).

5. The successful generation of aptamer panels is directly dependent upon the operator's ability to properly monitor and enhance the quantity of specific versus nonspecific binders throughout the selection process. Such is contingent upon a number of wet-lab techniques and considerations, as discussed within Subheading 3.3. Note that additional directives may be found in the detailed protocol published by Sefah and coworkers [5].
6. At this concentration the oligonucleotide pool is equivalent to $10^{13}$–$10^{16}$ randomized sequences and thus affords maximal binding diversity.
7. At 95 °C DNA fully denatures and immediately cooling the solution thereafter prevents reformation of dsDNA sequences. Binding buffer is added in order to prevent random binding events. Such contains tRNA and BSA, both purposed to compete with nonspecific binders.
8. The target cell culture dish must be 80–90% confluent—or monolayered within a 75 $cm^2$ dish—prior to the initial binding round (*see* [6] for additional detail concerning viable cell quantitation). A dense target cell population is critical in the initial round in order to capture a maximum amount of binding candidates. Should select sequences be 'lost' in this round, they cannot be recovered; in theory, there is a single copy of each sequence in the initial library. Note: cell washing indicates the following workflow (1) add washing reagent, (2) centrifuge at 1350 rpm/3 min, and (3) remove supernatant.
9. Operative techniques may differ herein, based on the cell-type. For adherent cells—incubation may proceed directly on the cell dish or within a centrifuge/polystyrene tube after detaching cells with cell dissociation solution nonenzymatic 1× (5 min) and washing thrice with DPBS. Suspension cells must be centrifuged into a pellet and resuspended into the DNA pool solution (after supernatant is removed). In either case, target cell incubation should proceed at 4 °C (higher temperatures may encourage internalization vs. surface binding of the sequences). Samples should be placed upon a rotary shaker to ensure even distribution throughout the incubation. Note:

incubate the target cell with the DNA pool for 45 min–1 h to allow maximal binding events in the initial round.

10. The cell population, if washed, should *not* be washed with stringency in the initial binding round. Again, this is to ensure the recovery of singly present binding candidates. Thereafter, collection of cellular bodies will be dependent upon the cell type and operational selections. (1) In situ adherent cell cultures: remove the supernatant solution from the culture flask immediately following incubation. Wash cells with a small volume of chilled washing buffer (0.5–1 mL). Add an additional 250 μL of washing buffer and, using a cell scraper, scrape cells from dish. Collect cell solution and transfer to a microcentrifuge tube. Repeat once, beginning with 250 μL washing buffer addition. (2) Centrifuge/polystyrene tube incubation (s)—adherent and suspension cell cultures: Pellet cellular bodies by centrifugation and remove supernatant thereafter. Wash cells with a small volume of chilled washing buffer (0.5–1 mL). Centrifuge. Remove supernatant, resuspend in 500 μL of washing buffer and transfer to a microcentrifuge tube.

11. Following amplification and single-strand separation of the 'Round 1' DNA pool (*see* Subheadings 3.3 and 3.5 for directives) sequential binding rounds must take place. These iterative rounds make for the bulk of the cell-SELEX procedure and are purposed to increasingly enrich the number of selectively binding candidates within the oligonucleotide pool(s) (*see* Fig. 2 [6]). The strategies used within each successive round are similar to the initial binding round, excepting counter cell selection considerations.

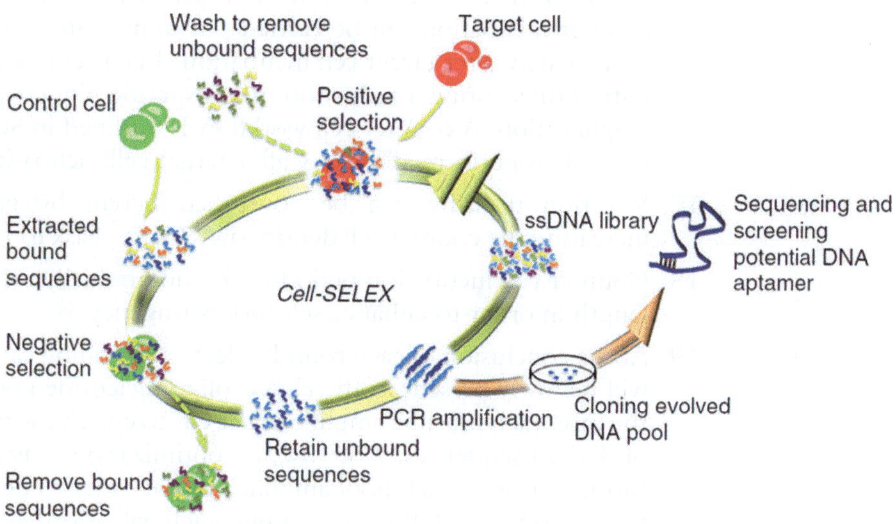


**Fig. 2** Cell-SELEX schematic [6]

12. Owing to differing experimental parameters—such as cell type (s), initial libraries, and operational techniques—the concentration used within successive rounds will vary. That notwithstanding, a starting concentration at or around 250 nM is recommended for each round. If the ssDNA pool is greater than this threshold, incubation(s) can proceed without dilution. However, dilute pools *must* be pre-concentrated in order to achieve maximal binding events.
13. Pre-concentration of said pool(s) to ≥250 nM may proceed by way of (1) reamplifying the template (as within the previous round) and strand-separating into a second ssDNA binding pool (to be combined with the existing sample) or (2) concentrating the existing pool using a speed vacuum concentrator. Note that option (2) may require the solution to be suspended in a smaller volume. If such is the case, 10× binding buffer volume must be adjusted accordingly.
14. In order to generate aptamer panels high in affinity and specificity, the selection stringency must gradually increase in each successive round. As such, it is recommended that a decreased target cell density be used as the cell-SELEX progresses (i.e. 25 $cm^2$ flask vs. 75 $cm^2$ flask).
15. Selection stringency should be successively increased in each round by reducing the target cell incubation period from 60 to 30 min or less.
16. A counter selection should be introduced within the first five rounds in order to reduce nonspecific binders, and at least once every five rounds thereafter. Note that a target cell selection round can proceed without the introduction of a counter cell incubation. However, each counter cell selection round *must* be performed in parallel with a target cell incubation—the counter incubation can be carried out immediately before or immediately after target cell incubation. In either case, such is critical in ensuring a reduction in nonspecific binders prior to amplification. A counter cell workflow is outlined in Subheading 3.3, as performed directly after target cell incubation.
17. Selection pressure can be controlled herein by gradually increasing the counter cell density (i.e. 25 $cm^2$ flask to 75 $cm^2$).
18. Counter cell incubation period(s) should gradually increase in length in order to enhance selection stringency.
19. At the conclusion of each round, selectively binding candidates will be lowly present in the eluted oligonucleotide pool. As is the case, each must be amplified via PCR to enrich the presence of desired sequences. Note that an optimization must be performed prior to each pool amplification in order to determine the appropriate PCR cycle number; such will afford a maximal

recovery of the binding candidates while preventing excess replication of undesired sequences.

20. Each eluted DNA pool *must* be optimized prior to amplification, excepting the 'Round 1' pool. This template is not optimized but instead directly amplified (20 cycles). In subsequent rounds, amplification efficiency is typically investigated at an interval of every three cycles (i.e. cycle no. 6, 9, 12, 15).
21. After confirming the amplification and recovery of the desired oligonucleotide pool, the double-stranded PCR product must be single-strand separated for use within the next round. Note that there are a number of techniques for ssDNA elution, one such is noted in Subheading 3.5.
22. Reports indicate that 20 mM NaOH is sufficient in the selective elution of ssDNA without dsDNA contaminants [8, 9]. This after SAV beads are pretreated with NaOH at the same concentration, as within the Subheading 3.5 workflow.
23. Addition of NaOH into the column induces alkaline denaturation of DNA base pairs, and thus dissociates the forward and reverse strands. The biotinylated (reverse) strand remains bound to the SAV beads within the column, while the FITC-tagged (forward) strand is eluted.
24. The development of an aptamer panel is highly dependent upon the selective enrichment of sequences that bind to the target versus nontarget cell line. As is the case, it is important to monitor binding progress throughout the course of the selection. While enhanced binding or selectivity towards the target may be minimal in primary rounds, such should be evident (in comparison to the nontarget population) in latter rounds. If this is not the case, cell-SELEX operations must be reevaluated and select rounds may have to be repeated and/or adjusted. Note that one may also increase the stringency of cell-SELEX proceedings by increasing selection pressure, as previously described. While the monitoring of binding progression is typically performed using flow cytometry—given that this method maintains high sensitivity and statistical precision, alongside good reproducibility—such can also be investigated via standard fluorescence quantitation (should only a rudimentary analysis be desired). The evaluative binding assay is similar to that of cell-SELEX incubation rounds and is further outlined in Subheading 3.6.
25. The unique and intrinsic merit of cell-SELEX is the ability to carry out selections against whole, live, cells without a predefined target. While this presents as a way in which to explore innate cellular complexities, it likewise introduces a great deal of challenges. Of significant importance is the routine monitoring of binding candidate enrichment(s) over the course of

the selection. It is common to observe a backshift in fluorescence intensity, with regard to the target cell and binding pool, at some point throughout the cell-SELEX procedure. In such cases, the operator should return to the round wherein the greatest fluorescent enhancement has been noted, reamplify the binding candidates from said round, and repeat the succeeding incubation step(s). Such will serve to reduce the quantity of nonspecific binders while enriching the desired binding candidates within the oligonucleotide pool. Identification of novel biomarkers is likewise a key attraction in the generation of aptamer panels. However, only a limited number of targets have been identified to date [10–13]. Such is hindered by the operational inabilities to purify and separate aptamers from their bound targets—rather, a sizable number of proteins are often determined to be potential biomarkers at the conclusion of each selection. Established techniques such as AptaBiD [13] may facilitate these proceedings and are met with an increasing number of newly developed methods, such as the chromatographic strategy published on by Drabik et al. [14].

26. Given the lengthy period(s) of time required for the selection of a single aptamer panel, high operational costs, and impacts of user variability, cell-SELEX studies must be undertaken with careful considerations. That notwithstanding, this technique presents as one that—with meticulous investigation—can generate highly selective targeting moieties and potentially enhance theranostic outcomes.

## References

1. Jin C, Zheng J, Li C, Qiu L, Zhang X, Tan W (2015) Aptamers selected by Cell-SELEX for molecular imaging. J Mol Evol 81 (5–6):162–171. https://doi.org/10.1007/s00239-015-9716-6
2. Patel DJ, Suri AK, Jiang F, Jiang L, Fan P, Kumar RA et al (1997) Structure, recognition and adaptive binding in RNA aptamer complexes. J Mol Biol 272(5):645–664. https://doi.org/10.1006/jmbi.1997.1281
3. Keefe AD, Pai S, Ellington A (2010) Aptamers as therapeutics. Nat Rev Drug Discov 9(7): 537–550. https://doi.org/10.1038/nrd3141
4. Tuerk C, Gold L (1990) Systematic evolution of ligands by exponential enrichment: RNA ligands to bacteriophage T4 DNA polymerase. Science 249(4968):505–510
5. Sefah K, Shangguan D, Xiong X, O'Donoghue MB, Tan W (2010) Development of DNA aptamers using Cell-SELEX. Nat Protoc 5 (6):1169–1185. https://doi.org/10.1038/nprot.2010.66
6. Sun HJ, Bahk YY, Choi YR, Shim JH, Han SH, Lee JW (2006) A proteomic analysis during serial subculture and osteogenic differentiation of human mesenchymal stem cell. J Orthop Res 24(11):2059–2071. https://doi.org/10.1002/jor.20273
7. Wilson R (2011) Preparation of single-stranded DNA from PCR products with streptavidin magnetic beads. Nucleic Acid Ther 21 (6):437–440. https://doi.org/10.1089/nat.2011.0322
8. Ruff P, Pai RB, Storici F (2012) Real-time PCR-coupled CE-SELEX for DNA aptamer selection. ISRN Mol Biol 2012:939083. https://doi.org/10.5402/2012/939083
9. Shangguan D, Cao Z, Meng L, Mallikaratchy P, Sefah K, Wang H et al (2008) Cell-specific aptamer probes for membrane protein elucidation in cancer cells. J Proteome Res 7(5):2133–2139. https://doi.org/10.1021/pr700894d

10. Van Simaeys D, Turek D, Champanhac C, Vaizer J, Sefah K, Zhen J et al (2014) Identification of cell membrane protein stress-induced phosphoprotein 1 as a potential ovarian cancer biomarker using aptamers selected by cell systematic evolution of ligands by exponential enrichment. Anal Chem 86(9):4521–4527. https://doi.org/10.1021/ac500466x
11. Mallikaratchy P, Tang Z, Kwame S, Meng L, Shangguan D, Tan W (2007) Aptamer directly evolved from live cells recognizes membrane bound immunoglobin heavy mu chain in Burkitt's lymphoma cells. Mol Cell Proteomics 6 (12):2230–2238. https://doi.org/10.1074/mcp.M700026-MCP200
12. Yang M, Jiang G, Li W, Qiu K, Zhang M, Carter CM et al (2014) Developing aptamer probes for acute myelogenous leukemia detection and surface protein biomarker discovery. J Hematol Oncol 7:5. https://doi.org/10.1186/1756-8722-7-5
13. Berezovski MV, Lechmann M, Musheev MU, Mak TW, Krylov SN (2008) Aptamer-facilitated biomarker discovery (AptaBiD). J Am Chem Soc 130(28):9137–9143. https://doi.org/10.1021/ja801951p
14. Drabik A, Ner-Kluza J, Mielczarek P, Civit L, Mayer G, Silberring J (2018) Advances in the study of aptamer-protein target identification using the chromatographic approach. J Proteome Res 17(6):2174–2181. https://doi.org/10.1021/acs.jproteome.8b00122

# Chapter 16

# Fluorescence In Situ Hybridization and Rehybridization Using Bacterial Artificial Chromosome Probes

**Elzbieta Stankiewicz, Tianyu Guo, Xueying Mao, and Yong-Jie Lu**

## Abstract

Fluorescence in situ hybridization (FISH) method enables in situ genetic analysis of both metaphase and interphase cells from different types of material, including cell lines, cell smears, and fresh and paraffin-embedded tissue. Despite the growing number of commercially available FISH probes, still for large number of gene loci or chromosomal regions commercial probes are not available. Here we describe a simple method for generating FISH probes using bacterial artificial chromosomes (BAC). Due to genome-wide coverage of BAC clones, there are almost unlimited possibilities for the analysis of any genomic regions using BAC FISH probes.

**Key words** Fluorescence in situ hybridization, Rehybridization, Probes, Bacterial artificial chromosome, Tissue pretreatment

## 1 Introduction

Fluorescence in situ hybridization (FISH) method enables the detection of genomic structure changes not only in metaphase but also in interphase cells, which was impossible to achieve prior to the FISH technique, when chromosome G-banding cytogenetic technique on metaphase slides was mainly used. FISH has become widely used owing to easy interpretation of results, gene mapping, possibility of analyzing the large stock of archival formalin-fixed paraffin-embedded (FFPE) material, and easiness of combining with other cytogenetic and molecular techniques [1]. In cancer research, FISH method is widely used for single gene locus analysis as well as for detection of chromosomal alterations including copy number (aneuploidy) and structure (duplication, amplification, deletion, and translocation) changes [2–5].

FISH is based on the hybridization of complementary, single-stranded fluorescently labeled specific DNA probe to fixed target genetic material (i.e. metaphase chromosomes or interphase nuclei), allowing identification of specific sequences in a structurally

Jyotsna Batra and Srilakshmi Srinivasan (eds.), *Theranostics: Methods and Protocols*, Methods in Molecular Biology, vol. 2054, https://doi.org/10.1007/978-1-4939-9769-5_16, © Springer Science+Business Media, LLC, part of Springer Nature 2019

preserved cell [6]. Prior to FISH, low-resolution chromosome G-banding karyotyping of metaphase nuclei was routinely used in clinic for chromosome structural aberrations studies. FISH not only increased the ability to identify abnormal chromosomes and the resolution of chromosome breakpoints but also made cell karyotyping simpler and much more reliable through the introduction of 24-color karyotyping methods either by spectral karyotyping (SKY) [7] or multiplex FISH (M-FISH) [8, 9], leading to the detection of a large number of novel genomic translocations, gene fusions and characterization of complex chromosomal rearrangements. These methods used whole-chromosome painting probes which are DNA probes derived from a single or multiple chromosomes, amplified and labeled by degenerate oligonucleotide polymerase chain reaction. These probes allowed simultaneous visualization of all chromosomes in different colors [10]. In addition, the development of comparative genomic hybridization (CGH) [11–15] and comparative expressed sequence hybridization (CESH) [1, 16–18] enabled the analysis of the whole genome for genomic copy number changes and gene expression abnormalities. CGH technique relied on simultaneous hybridization of differentially labeled test DNA and normal reference DNA to normal chromosome spreads. Genomic abbreviations were then seen as changes in the ratio of the intensities of the two fluorochromes along the target chromosomes [11]. Similarly, in CESH RNA from the test and the reference cells was simultaneously hybridized to normal metaphase chromosomes, detecting genome-wide relative expression pattern changes according to chromosomal location [1, 16]. The hybridization and analysis principle of CGH and CESH was subsequently used to develop the microarray technology for high-resolution genomic alterations and gene expression changes, where metaphase chromosomes are replaced by an array of thousands or millions of bacterial artificial chromosome (BAC) clones, cDNAs, or oligonucleotides [19, 20].

Although CGH and CESH were replaced by microarray technology and then next generation sequencing, single locus FISH continue to be one of the most common methods to study genomic abbreviations such as gain, loss, translocation and amplification within cells in situ [3, 21–23]. For assessing copy number changes, control probes are used along with probes for the loci of interest. Centromeric probes, which are the satellite sequences flanking the centromeres of human chromosomes are commonly used as control probes due to the simple process to produce this type of probes and their great signal to noise ratio. For the gene of interest assessment, locus-specific probes targeting unique sequence on the chromosome are used. For commonly studied genes commercial FISH probes exist. For novel alterations there is a necessity to generate locus-specific FISH probes from DNA segments that have been incorporated into BACs, yeast artificial chromosome (YAC) or bacterial P1 vectors (PAC) [24]. BAC clones are generally preferred due to large insert capability and stable plasmid distribution among daughter cells. BAC clones come in a form of bacterial culture (*E. coli*) with sequence of interest inserted into plasmid. *E. coli* with specific BAC clone are grown on selective agar plates, single colony is then picked and expanded, and plasmid DNA is extracted, amplified, and labeled using standard protocols.

Here, we describe a detailed FISH protocol for detection of locus-specific genetic alterations in a variety of human cytological and histological material, including cell lines, tissue imprints, and fresh frozen and formalin-fixed paraffin-embedded tissue using BAC probes. The protocol includes generation of high-quality BAC clone-based locus-specific FISH probes, different cell line and tissue pretreatment methods, probe hybridization and posthybridization washes, and FISH signal removal for rehybridization with same or different set of FISH probes when it is required.

## 2 Materials

### 2.1 Bacterial Culture

1. *E. coli* bacteria containing chosen BAC clone (Luria broth (LB) agar stab culture).
2. Autoclave oven.
3. LB agar: 5.0 g tryptone, 2.5 g yeast extract, 5.0 g NaCl, 7.5 g agar-agar, 500 ml distilled water. Autoclave. Cool down at room temperature. Store at 4 °C.
4. Sterile 9 cm petri dishes.
5. Chloramphenicol antibiotic stock solution: dissolve 25 mg of chloramphenicol in 1 ml of pure ethanol. Store at −20 °C.
6. Microwave oven.
7. 37 °C incubator.
8. LB agar plates with 12.5 μg/ml of chloramphenicol: Liquefy LB agar by gentle microwaving. Cool it down to ~50 °C and add 500 μl of chloramphenicol stock solution to final concentration of 12.5 μg/ml. Pour enough of the liquid agar into sterile 9 cm diameter petri dishes to cover their bottoms. Allow the agar to solidify at room temperature. LB agar plates can be stored up to one month in the dark at 4 °C. Before use, prewarm the agar plates in 37 °C incubator for 30 min.
9. Sterile plastic inoculation loops.
10. Sterilized LB broth: 10 g tryptone, 5 g yeast extract, 10 g NaCl, 1 L of distilled water, pH 7.5. Autoclave. Cool down at room temperature. Store at 4 °C.
11. 50 ml falcon tubes.

12. 37 °C shaking incubator.
13. −20 °C Freezer.

### 2.2 DNA Isolation from BAC Clones

1. P1: 50 mM Tris, pH 8.0, 10 mM EDTA, 100 μg/ml RNase A.
2. P2: 0.2 N NaOH, 1% SDS.
3. P3: 3 M KOAc, pH 5.5 solutions.
4. 1.5 ml Eppendorf tubes.
5. Ice cold iso-propanol.
6. 80% ethanol.
7. TE buffer: 10 mM Tris-Cl, 1 mM EDTA, pH 8.0.
8. Centrifuge.
9. NanoDrop UV-Vis Spectrophotometer.
10. Bio-Rad submerged horizontal agarose gel electrophoresis system.
11. GelRed dye.
12. Microwave.
13. 10× TBE buffer stock solution: 0.89 M Tris-base, 0.89 M Boric Acid, 20 mM EDTA, pH 8.3.
14. 1× TBE buffer: 50 ml of 10× TBE buffer stock solution, 450 ml of distilled water.
15. UltraPure Agarose powder.
16. 10× BlueJuice Gel Loading Buffer.
17. 1 kb Plus DNA Ladder.
18. −20 °C freezer.
19. 1% agarose gel: add 0.5 g of agarose powder to 50 ml of 1× TBE buffer. Microwave until agarose powder is completely dissolved. Cool down the solution to 65 °C and add 2 μl of GelRed. Cast the gel using small gel tray. Place the comb in the gel and wait for agarose to solidify. Move the gel tray to the electrophoresis tank filled with 1× TBE buffer. Buffer should completely cover the gel. Remove the comb. Before loading the samples, 3 μl of each sample should be mixed with 1 μl of sample loading dye. 3 μl of DNA ladder should be loaded along the samples.

### 2.3 BAC DNA Amplification and Purification

1. Illustra GenomiPhi V2 DNA Amplification Kit.
2. BAC DNA.
3. Distilled water.
4. Microcentrifuge.
5. PCR Thermal Cycler.

6. 1% agarose gel and gel electrophoresis system (*see* Subheading 2.2, **step 19**).
7. TE buffer: 10 mM Tris-Cl pH 7.5, 1 mM EDTA.
8. Phenol solution equilibrated with 10 mM Tris HCl pH 8.0, 1 mM EDTA.
9. Chloroform.
10. 4 M NaCl.
11. 100% ethanol.
12. 80% ethanol.
13. NanoDrop UV-Vis Spectrophotometer.
14. −80 °C freezer.

### 2.4 BAC DNA Labeling

1. Purified amplified BAC DNA.
2. BioPrime™ DNA Labeling System.
3. Digoxigenin-11-dUTP, alkali-stable.
4. Tetramthylrhodamin-5-dUTP.
5. Fluorescein-12-dUTP.
6. Special 10× dNTP mix (1 mM dATP, dGTP, dCTP, 0.65 mM dTTP): 5 μl of 100 mM stock solutions of dATP, dGTP, dCTP, 3.25 μl of dTTP stock solution, 500 μl of distilled water. Store at −20 °C.
7. 1 μg/μl COT human DNA.
8. 10 μg/μl herring sperm DNA.
9. Illustra microspin G50 columns.
10. TE buffer.
11. 4 M NaCl.
12. 100% ethanol.
13. 80% ethanol.
14. Distilled water.
15. PCR Thermal Cycler.
16. Microcentrifuge.
17. 1% agarose gel and gel electrophoresis system (*see* Subheading 2.2, **step 19**).
18. NanoDrop UV-Vis Spectrophotometer.

### 2.5 Cells and Tissue Pretreatment

1. For pelleting cultured cells: trypsin-EDTA solution, PBS, full growth culture medium.
2. 37 °C incubator.
3. Plastic pipettes.
4. Forceps.

5. Glass microscopic slides.
6. 0.075 M KCl.
7. Methanol:acetic acid 3:1 fixative solution: 150 ml of methanol, 50 ml of acetic acid.
8. Centrifuge.
9. Glass staining dishes/coplin jars.
10. 70% acetic acid.
11. 70% ethanol.
12. 90% ethanol.
13. 100% ethanol.
14. Xyline.
15. Distilled water.
16. Collagenase.
17. SPOT-Light Tissue Pretreatment kit.
18. Hot plate.
19. 28 °C water bath.
20. 100 °C water bath.
21. Coplin jar.
22. Microwave.
23. Rectangular floating tube foam rack.
24. −80 °C freezer.

### 2.6 Probe Hybridization

1. DIG or/and biotin labeled probes.
2. 20× saline-sodium citrate buffer (20× SSC): 175.3 g of NaCl (3 M), 88.2 g of sodium citrate (0.3 M), 1 L of distilled water. Adjust pH to 7.0.
3. Hybridization buffer: 3 ml of DI formamide, 600 mg of dextran sulfate, 600 μl of 20× SSC, 5 ml of distilled water, 1.4 μl of 500 mM EDTA pH 8.0 and 200 μl of 10 mg/ml of salmon sperm. Mix well all the components and allow the buffer solution to completely dissolve overnight at 4 °C. Keep the hybridization buffer aliquots at −20 °C.
4. Block heater.
5. Hot plate.
6. 37 °C incubator.
7. Humidity box: line the plastic box with wet tissue and close the lid. Prewarm it in the 37 °C incubator for 30 min before use.
8. Forceps.

**2.7 Posthybridization Washes**

1. 42 °C water bath.
2. Forceps.
3. Coplin jars.
4. Glass microscopic slides and coverslips.
5. 20× SSC buffer: 175.3 g NaCl, 88.2 g sodium citrate, 1 L distilled water. Adjust pH to 7.0.
6. SSC-Tween 20 (SSCT) buffer (4× SSC, 0.2% Tween): 100 ml of 20× SSC, 1 ml of Tween 20, 400 ml of water.
7. SSCT-milk (SSCTM) buffer: 0.5 g of skimmed milk powder, 10 ml of SSCT buffer. Mix well and filter through 0.45 μm syringe filter. Always prepare the buffer freshly before use.
8. 50% formamide/2× SSC buffer: 50 ml of formamide, 10 ml of 20× SCC, 40 ml of water. Prepare the buffer freshly before use.
9. Anti-Digoxin–FITC antibody.
10. Streptavidin-Cy3.
11. PBS.
12. 70% ethanol.
13. DAPI containing mounting medium.
14. Humidity box: line plastic slide box with wet tissue.
15. 37 °C incubator.

**2.8 Removing the Hybridized Probes for FISH Rehybridization**

1. Fume hood.
2. 68 °C water bath.
3. Coplin jars.
4. 2× SSC solution
5. 70% formamide/2× SSC: 35 ml of formamide, 5 ml of 20× SSC, 10 ml of distilled water.
6. SSCT.
7. Forceps.

## 3 Methods

**3.1 Bacteria Culture and DNA Isolation from BAC Clones**

1. Generation of single bacterial colonies. Take out from the fridge one LB plate containing 12.5 μg/ml chloramphenicol per BAC clone and prewarm it in 37 °C incubator for 30 min. Remove your BAC culture stock from −80 °C and thaw it on ice. Dip sterile inoculation loop into BAC stock and spread the bacteria on the agar. Incubate at 37 °C overnight.
2. Next day examine the plate for single colonies. Using sterile inoculation loop pick one single colony from LB agar plate and place it into 50 ml falcon tube containing 5 ml of Broth liquid

culture medium and 2.5 μl of stock chloramphenicol solution (final concentration 12.5 μg/ml). Incubate the bacterial culture overnight at 37 °C on a shaker-incubator at 220 rpm.

3. Next day aliquot 1 ml of bacterial culture into 1.5 ml eppendorf tube. Spin down your cells at 13,000 rpm (19000 × *g*) for 5 min at room temperature. Carefully discard the supernatant (*see* **Note 1**) and proceed to alkaline lysis (*see* **Note 2**).
4. Resuspend the cells in 150 μl of P1 solution.
5. Add 150 μl of P2 solution to lyse the cells and very gently mix by inversion. Incubate on ice for 5 min. Do not exceed the incubation time as it will lead to irreversible DNA denaturation. During this stage SDS perforates the cell membrane while sodium hydroxide loosens the cell walls and denatures and releases both, the plasmid DNA and sheared cellular DNA.
6. Slowly add 150 μl of P3 solution to each tube and very gently shake during addition to renature the plasmid DNA. A thick white precipitate of *E. coli* cell debris will form. Incubate the samples on ice for 10 min.
7. Centrifuge the tubes at 13,000 rpm (19000 × *g*) for 20 min to remove cell debris and cellular bacterial DNA. The plasmid DNA remains in the supernatant.
8. Transfer the supernatant to a fresh tube and centrifuge at 13,000 rpm (19000 × *g*) for 5 min.
9. Precipitate the plasmid DNA by alcohol precipitation. Transfer the supernatant to a fresh tube and add 0.8 volume of ice cold iso-propanol. Mix well and place the tubes at −20 °C for 30 min.
10. Centrifuge the samples at 13,000 rpm (19000 × *g*) for 15 min at 4 °C.
11. Discard the supernatant. Rinse the pellet (plasmid DNA) with 0.5 ml of fresh 80% ethanol and centrifuge at 13,000 rpm (19000 × *g*) for 5 min at 4 °C.
12. Remove the supernatant. Air-dry the extracted plasmid DNA and resuspend it in 30 μl of TE buffer. Check the DNA concentration using the nanodrop spectrophotometer. Store the samples at −20 °C.
13. To check the quality of extracted BAC DNA run 1 μg of each sample on the agarose gel (*see* Subheading 2.2, **step 19**). Run the gel at 120 V for 30–40 min. Check the DNA bands size with UV box. Good quality sample should mainly contain supercoiled plasmid BAC DNA (strong band) and only faint higher bands of nicked and linear plasmid (Fig. 1).

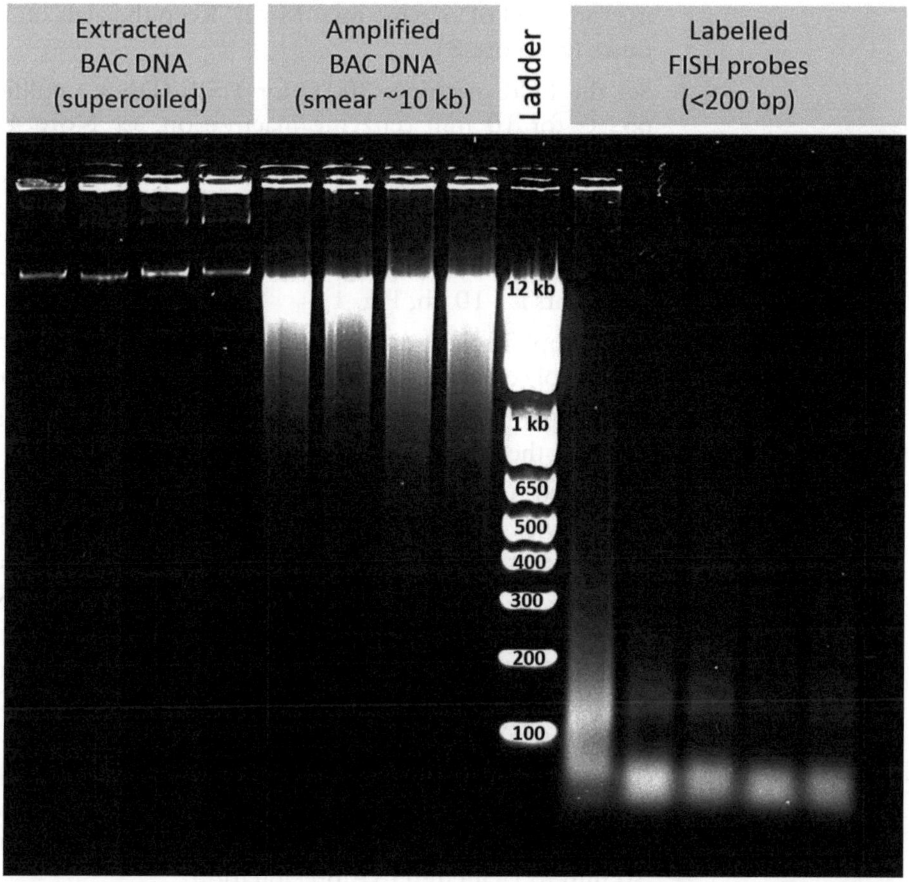


**Fig. 1** BAC DNA quality control steps during FISH probe preparation. Quality BAC plasmid DNA should give strong band of supercoiled DNA with faint bands of nicked and linear plasmid, which are visible above supercoiled band. Amplified BAC DNA should have average size around 10 kb, while after labeling, FISH probe fragments should be small between 100–200 bp. The ladder size is shown in bp, unless specified otherwise

### *3.2 DNA Amplification*

1. Heat the BAC DNA on a hit block to 55 °C for 5 min to completely dissolve the DNA. Centrifuge at 13,000 rpm (19000 × *g*) for 10 min.
2. Place the reagents from Illustra GenomiPhi V2 DNA Amplification Kit: sample buffer, reaction buffer and enzyme mix on ice and let them dissolve.
3. Mix 9 μl of sample buffer with 1 μl of BAC DNA (10 ng–1 μg of DNA) in the PCR tube. Centrifuge the sample mixture briefly in the minicentrifuge.
4. Denature BAC DNA by heating the samples to 95 °C for 3 min using PCR thermocycler. Cool the samples on ice for 5 min.
5. Prepare the master mix for each amplification reaction: add to the above sample mixture 9 μl of reaction buffer (mix before

use) and 1 μl of enzyme mix. Mix it. Keep the vials on ice at all times (*see* **Note 3**).

6. Set the PCR machine: 30 °C for 1.5 h (DNA amplification); 65 °C for 10 min (enzyme inactivation, *see* **Note 4**); 4 °C forever (end of incubation).
7. Load 2 μl of sample on an agarose gel (*see* Subheading 2.2, **step 19**). Run the gel for 30–40 min at 120 mV. Check the DNA smear under UV (average expected size of amplified DNA fragments is ~10 kb, Fig. 1).
8. Store amplification reactions at −20 °C or immediately proceed to phenol/chloroform purification described in steps below.
9. Add 80 μl of TE buffer to 20 μl of each amplified DNA sample. Perform the subsequent steps in the fume hood.
10. Add 100 μl (1 volume) of 1:1 phenol-chloroform. Mix intensely by flipping the samples in your hand for 1 min. Spin for 2 min at 13,000 rpm (19000 × *g*) at room temperature.
11. Transfer the top aqueous layer containing extracted DNA into fresh, labeled tube. Be careful not to disturb the layer below. Add 100 μl of chloroform (1 volume). Mix vigorously for 1 min. Centrifuge the sample for 2 min at 13,000 rpm (19000 × *g*) at room temperature.
12. Transfer the top layer containing the DNA into fresh tube. Add 5 μl (1/20 volume) of 4 M NaCl and 250 μl of 100% ethanol (2.5 volume). Mix well and leave the samples at −20 °C for 30 min to facilitate DNA precipitation.
13. Centrifuge the precipitated samples at 4 °C for 20 min at 13,000 rpm (19000 × *g*).
14. Remove the supernatant.
15. Wash the DNA pellet with 1 volume of 80% ethanol to remove remains of salt from DNA. Spin at 4 °C for 10 min at 13,000 rpm (19000 × *g*).
16. Remove the supernatant. Leave the lids of the tubes open to allow the DNA pellet to air-dry.
17. Resuspend the purified BAC DNA pellet in 20–30 μl of TE buffer.
18. Measure the quantity of DNA with nanodrop spectrophotometer.

### 3.3 Labeling of Purified BAC DNA and Probe Stock Preparation

1. Place the BioPrime DNA Labeling System components: random primers, 10× dNTP, Klenow enzyme, stop buffer, and all other reagents on ice. Dilute 300 ng of purified BAC DNA in water to the final volume of 24 μl. Add 20 μl of random primers.

2. Denature the samples in the PCR thermocycler for 5 min at 95 °C, then cool them on ice for 10 min.
3. Probe labeling:
   (a) Direct labeling: add 5 μl of the special 10× dNTP mix and 1.75 μl of Tetramthylrhodamin-5-dUTP (red probe) or 1.75 μl of Fluorescein-12-dUTP (green probe).
   (b) Indirect labeling: for biotynylated FISH probes add 5 μl of 10× dNTP, and for DIG labeled FISH probes, add 1.75 μl of DIG-11-dUTP and 5 μl of the special 10× dNTP mix.
4. Add 1 μl of Klenow enzyme to each sample.
5. Place the samples in the PCR thermocycler for 3 h at 37 °C to allow labeling.
6. After 3 h of labeling, add 5 μl of stop buffer to each sample. You can run 2 μl of the samples on the agarose gel at 120 V for 30–40 min to check the labeled probe size (*see* Subheading 2.2, **step 19**). The DNA fragment size should be 150–200 bp (Fig. 1).

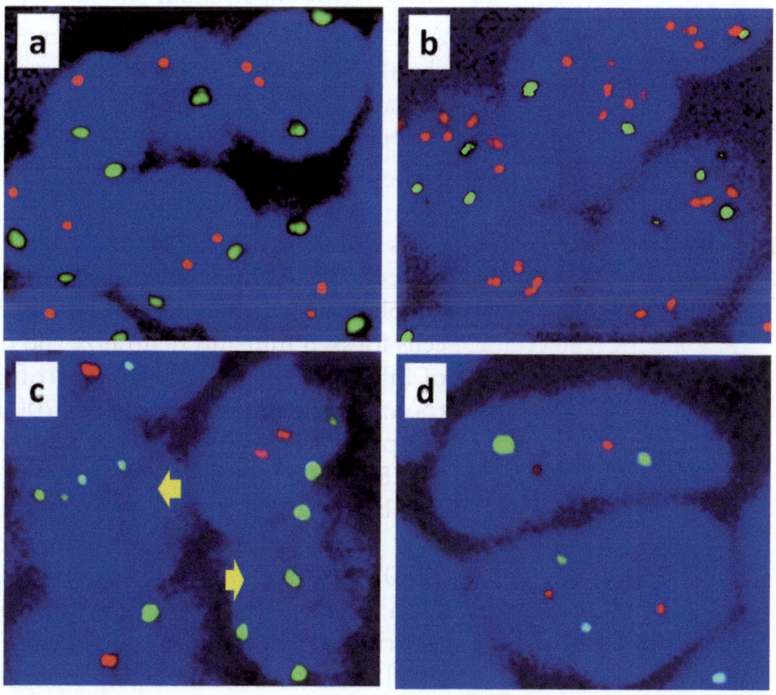


**Fig. 2** Examples of FISH on FFPE prostate tissue sections. (**a**) Control benign prostatic hyperplasia (BPH) tissue with normal, equal copy number of both, *CCND1* (red probe) and control (centromere 1, green probe) signals. (**b**) *CCND1* copy number gain (red) in all visible nuclei of prostate bone metastasis. (**c**) Loss of 17q11.2 region (red) in some nuclei (marked with arrows) in prostate cancer bone metastasis. (**d**) Monosomy of 17q11.2 (red) and control centromere probe (green) in prostate bone metastasis

7. Clean the labeled probe. Prepare Microspin G50 columns: loosen the cup of the column, break off the bottom plug, place the column in collection tube and spin at 3500 rpm (1400 × *g*) for 1 min. Wash the column twice with 200 μl of TE buffer and spin at 3500 rpm (1400 × *g*) for 1 min. Transfer the column to a new eppendorf tube, add the labeled probe solution to the center of the column and spin at 3500 rpm for 2 min.
8. Check the concentration of the cleaned probe with Nanodrop spectrophotometer (expect 90–150 ng/μl which is 10–40× increase in DNA yield).
9. Precipitate the probe: add 30 μl of COT-1 DNA, 1 μl of salmon sperm, 4 μl of 4 M NaCl and 222 μl of 100% ethanol. Mix well and put the samples in −80 °C for 20–30 min.
10. Centrifuge the samples at 13,000 rpm (19000 × *g*) for 20 min at 4 °C.
11. Remove the supernatant and rinse the pellet with 200 μl of fresh 80% ethanol. Spin at 13,000 rpm (19000 × *g*) for 10 min at 4 °C.
12. Remove the supernatant and air-dry the pellet.
13. Add 20 μl of nuclease-free water (final probe concentration should be about 200 ng/μl), mix well and place the samples on a heat-block at 50 °C for 5 min to completely resuspend the probe.
14. Store the probes at −20 °C.

### 3.4 Cell and Tissue Preparation

1. Cells from culture: detach the cells from the flask (only adherent cells. Skip this step for cells grown in suspension such as lymphocytes) by adding trypsin-EDTA solution, incubate the cells for 3–5 min and neutralize the trypsin with equal volume of full cell growth medium. Centrifuge the suspended cells at room temperature for 5 min at 1200 rpm (160 × *g*). Wash the cell pellet with 5 ml PBS and resuspend the cells in 15 ml of 0.075 M prewarmed (37 °C) KCl and add 1 ml of fixative. Incubate the cells for 15 min at room temperature. Pellet the cells by centrifugation at room temperature for 5 min at 1200 rpm (160 × *g*). Resuspend the cell pellet in 15 ml of fresh fixative. Repeat twice the 15 min fixation step followed by centrifugation. After last centrifugation, resuspend the cells in a small volume (200–500 μl) of fixative (*see* **Note 5**). Using plastic pipette, drop two drops of the cell suspension on the slides and allow the cells to air-dry. The slides can be used immediately or stored long-term at −20 °C. Before use, treat the cell slide preparation with 70% acetic acid for 10 min to remove the remains of the cytoplasm from the cells. Wash off the acetic acid by incubating the cell slides in PBS at room

temperature, three times for 5 min. Dehydrate the cells with 70%, 90% and 100% ethanol (3 min each) and air-dry the slides.

2. Fresh-frozen tissue touch imprints: remove fresh-frozen tissue from the −80 °C and place on ice. Gently touch fresh-frozen tissue onto glass slides at room temperature. Air-dry the slides at room temperature overnight (*see* **Note 6**). Any unused touch imprint slides should be stored at − 80 °C. Fix the slides in fixative solution for 15 min at room temperature. Wash in PBS for 5 min. Air-dry.
3. Fresh-frozen tissue sections: pre-fix freshly cut fresh-frozen tissue slides in fixative solution for 15 min at room temperature. Wash in water for 5 min. Treat with 25 U/ml collagenase for 30 min at 37 °C (*see* **Note 7**). Wash in water for 5 min. Allow slides to dry. Digest the tissue with pepsin: place a few drops of pepsin solution on the tissue (digestion enzyme solution from SPOT-Light Tissue Pre-treatment kit) and allow digestion for 5 min at room temperature. Ensure that all tissue is covered in solution. Wash the slides with PBS, 2 × 5 min. Fix digested tissue again in fixative solution for 2 × 10 min at room temperature. Wash the slides with PBS, 2 × 5 min. Dehydrate the slides in 70%, 90% and 100% ethanol, 3 min each. Air-dry.
4. For FFPE tissue slides pretreatment prepare the following: Dewax FFPE tissue slides with two changes of xylene, 5 min each and wash twice in ethanol for 3 min. Bake the slides on 100 °C hot plate to fix the tissue to the slide: place the slides on the hot plate and immediately pipette 200 μl of ethanol on each of them, covering them with cover slips. Wait for ethanol to evaporate from under the cover slips and immediately place the dried slides in the coplin jar with distilled water (*see* **Note 8**). Rehydrate the slides in water for 3 min. Quickly boil the pretreatment solution (SPOT-Light Tissue Pre-treatment kit) in a microwave and transfer it into coplin jars placed in 100 °C water bath. Immediately place the slides in boiling tissue pretreatment solution and boil them for 15 min (*see* **Note 9**). Rinse the slides in water twice for 3 min and spread them on a plastic rack/floating foam tube rack in 28 °C water bath for subsequent enzymatic digestion. Let the slides to warm up for 5 min and apply 4–6 drops of enzyme mix (SPOT-Light Tissue Pre-treatment kit) to each slide, cover the tissue with cover slips and incubate for 5 min. Wash the slides twice in water, 3 min each. Air-dry.

### 3.5 Hybridization with FISH Probes

1. Take out the hybridization buffer and FISH probes from −20 °C and keep them on ice. Place the hybridization buffer on the heat-block for 5 min at 50 °C to ensure that all ingredients are completely dissolved. Prepare the humidity box by placing wet tissue at the bottom of the slide box. Place the box in the 37 °C incubator.

2. Mix 1.5 μl (~300 ng) of the red and 1.5 μl of the green labeled probes (or 1.5 μl of DIG-labeled and 1.5 μl of biotin-labeled probes if using indirectly labeled probes) with 9.6 μl of hybridization buffer per each slide. Pipette the mixture up and down to ensure the probes are well resuspended in the buffer. Pipette the hybridization mix onto the slide and cover the hybridization mix with 22 × 22 mm coverslip. Seal the edges of the coverslip with rubber cement (*see* **Note 10**).
3. Denature the slide with the probes on hot plate at 98 °C for 10 min.
4. Put the denatured slides into humidity box for overnight hybridization. Make sure that the lid is closed properly and there is enough moisture to keep the slides humid during overnight incubation.

### 3.6 Posthybridization Washes and Antibody Incubation

The first three steps apply to both: directly and indirectly labeled probes. For directly labeled probes skip **steps 4–9** and go directly to **step 10**. For indirectly labeled probes follow all the steps in the protocol.

1. In the fume hood prepare 42 °C water bath with two coplin jars of 50% formamide/2× SSC and three coplin jars of 2× SSC. Prepare SSCTM solution in 50 ml falcon tube and filter it with the 0.45 μm syringe filter. Keep it in the 42 °C water bath.
2. Remove the slides from the humidity box. Return the box to 37 °C incubator. Carefully remove the rubber cement and slide off the cover slips from the slides. Perform the subsequent washes in the 42 °C water bath. Rinse off the remaining hybridization mix in 2× SSC for 1 min. Wash the slides in 50% formamide/2× SSC solutions two times 5 min, followed by another two washes in 2× SSC, 5 min each (*see* **Note 11**).
3. Wash the slides with SSCT at room temperature—1 × 3 min. For directly labeled probes skip **steps 4–9** and proceed to **step 10**.
4. Meanwhile prepare antibody solutions for indirectly labeled probes: centrifuge the stock of streptavidin—Cy3 (red fluorescence) and anti-DIG-FITC antibody (green fluorescence) at 13,000 rpm (19000 × *g*) for 5 min. Take 1 μl of the streptavidin—Cy3 stock off the surface and dilute in 200 μl of SSCTM per slide. The same way prepare separate working solution of anti-DIG-FITC antibody. Keep both working solutions in 37 °C incubator.
5. Dry the slides by tapping on paper tissue. Spread the slides on the wet tissue in the humidity box. Apply 200 μl of SSCTM (blocking solution) on each slide and coverslip. Incubate for 15 min at 37 °C.

6. Let the cover slips slide off the slides. Dry the excess of liquid on the paper towel. Spread the slides again on the wet tissue in the humidity box. Add working solution of anti-DIG-FITC. Coverslip. Incubate for 10 min at 37 °C (*see* **Note 12**).
7. Wash the slides in SSCT in coplin jar at room temperature 3 × 2 min.
8. Remove the slides from coplin jar and dry the excess of liquid on the paper towel. Spread the slides on the wet tissue in the humidity box. Add 200 μl of Streptavidin-Cy3 working solution. Coverslip. Incubate for 10 min at 37 °C.
9. Wash the slide in SSCT in coplin jar at room temperature 1 × 2 min.
10. Wash the slide in PBS at room temperature 2 × 5 min.
11. Place the slides in 70% ethanol, 2 × 20 s (*see* **Note 13**).
12. Air dry.
13. Add 20 μl of mounting medium containing DAPI and coverslip. Keep the slides in the fridge at 4 °C. Protect from light. For long-term storage slides can be kept at −20 °C.

### 3.7 Removing the Hybridized Probes for FISH Rehybridization

1. Prepare 68 °C water bath in the fume hood containing a coplin jar with 2× SSC solution and another with 70% formamide/2× SSC solution.
2. Meantime, place the slides with coverslips in SSCT for 5–10 min at room temperature to remove the cover slips. Use gentle agitation to facilitate coverslips removal.
3. Incubate the slides in 2× SSC, 2 × 10 min at room temperature with agitation to remove mounting medium with DAPI.
4. To remove old FISH signal, incubate the slides at 68 °C in 2× SSC for 2 min, followed by 70% formamide/2× SSC (35 ml of formamide + 5 ml of 20× SSC + 10 ml of water) solution for 4 min (*see* **Note 14**).
5. Wash the slides in 2× SSC for 2 min at room temperature, followed by two washes with distilled water for 3 min at room temperature. Air-dry the slides. Slides can be now hybridized with another set of FISH probes as described above.

## 4 Notes

1. The supernatant contains pieces of bacterial cell wall which can inhibit enzyme action during downstream applications. Therefore it is important to remove all remaining supernatant by blotting the tubes upside down on the paper tissue.

2. Alkaline lysis is the most common method for circular plasmid DNA isolation. It depends on a unique property of plasmid DNA of rapidly annealing following denaturation. This makes a basis for plasmid DNA separation from the bacterial chromosome. Bacterial cells after resuspension (solution P1) are lysed with solution P2, containing SDS detergent, which solubilizes the cell membranes, and an alkali (NaOH), which denatures DNA. It is important to be very gentle during this step and not to exceed the incubation time as it may lead to permanent DNA damage and also result in shearing of small fragments of bacterial genomic DNA which then will contaminate your plasmid DNA extraction. The lysed sample is then neutralized with potassium acetate (solution P3), which causes the DNA to renature and cell debris to precipitate. This is the selective step, which allows for easy renature of the small circular plasmid DNA but not large bacterial genomic DNA.
3. It is important to keep the solutions on ice at all times as the master mix contains all the components required for DNA amplification and will generate amplification products if exposed to temperatures over 4 °C for sufficient time.
4. Heating the sample to 65 °C is necessary to inactivate the exonuclease activity of the Phi29 DNA polymerase enzyme which may otherwise begin to degrade the amplification product.
5. It is important to ensure that while having good cell density on the slide, cell nuclei do not overlap so the FISH signal can be clearly assessed per nucleus. Cell density and spread should be checked under the microscope for the first prepared slide and the fixative volume adjusted until desired cell density is achieved. Thorough pipetting of the cells suspension will also help to avoid cell aggregates on the slide.
6. Chromosomes from freshly prepared material tend to become distorted during denaturing ('puffy' chromosomes) which significantly affects quality of the FISH signal. The structure of the chromosomes (so called 'hardness' of the chromosomes) can be stabilized by 'aging' the slides, simply by leaving them on the bench at room temperature for 3 days or longer. Slides can be additionally aged chemically by placing them on 99 °C heat block, adding 200 μl of ethanol and covering them with coverslips. Slides are removed from the hot plate and ready for next FISH step after the ethanol completely evaporates from under the coverslip, which usually takes 30–60 s.
7. Collagenase treatment may vary depending on thickness of the tissue sections and stromal content. Therefore this step should be adjusted depending on the type of tissue used. Our protocol was optimized on prostate tissue sections 10 μm thick. More fibrous tissues such as skin may need longer digestion time or higher collagenase concentration.

8. Lightly pressing the cover slip to the slide will facilitate ethanol evaporation. Do not keep the slides on the hot plate for longer than 1 min. When ethanol evaporates the tissue turns white and the slide should be immediately taken off the hot plate and placed in the coplin jar with water for rehydration.
9. Use thick gloves to avoid burns. Make sure that after placing the slides inside the coplin jars the lids are properly placed to avoid evaporation of tissue pretreatment solution and accidental contamination with boiling water, which would dilute the solution. Start counting 15 min incubation from the point at which the water is boiling again.
10. Apply the rubber cement generously and ensure that all the cover slip corners are completely covered. This will protect the hybridization mix from drying during denaturation and hybridization steps. The rubber cement should still be soft and easy to remove after overnight hybridization in the humidity box. If the tissue section is bigger than 22 × 22 mm, the amount of hybridization mix should be doubled and slide covered with 22 × 50 mm cover slip. It is important to remove all the bubbles from the slide by gently pressing the cover slip. Bubbles will prevent hybridization mix from accessing affected tissue area resulting in lack of FISH signal. The affected tissue areas will appear under the fluorescence microscope as black spots.
11. These washes are designed to remove unbound and weakly bound probes to reduce the background from nonspecifically bound probes. To remove more background, the temperature can be increased to 45–48 °C; however, care must be taken not to remove all of the signal. Balance has to be obtain resulting in the best quality FISH signal combined with minimum background.
12. Slides should be leveled on the wet tissue so the antibody solution stays in place and evenly covers the whole slides. Bubbles should be pushed away from under the cover slips, as they will prevent antibody binding to the FISH probe in the affected area. In case of weak signal, the incubation time with antibodies can be extended/doubled; however, this may increase the background.
13. Ethanol treatment is not necessary; however, it may slightly shrink the nuclei, which is helpful when analyzing tissue sections, where nuclei commonly overlap.
14. The incubation in 70% formamide/2× SSC can be extended if not all FISH signal was successfully washed off. It is always necessary to confirm under fluorescence microscope that all FISH signal was successfully removed before starting with hybridization of a new set of FISH probes.

## References


1. Trask BJ (2002) Human cytogenetics: 46 chromosomes, 46 years and counting. Nat Rev Genet 3(10):769–778. https://doi.org/10.1038/nrg905
2. Cui C, Shu W, Li P (2016) Fluorescence in situ hybridization: cell-based genetic diagnostic and research applications. Front Cell Dev Biol 4:89. https://doi.org/10.3389/fcell.2016.00089
3. Stankiewicz E, Mao X, Mangham DC, Xu L, Yeste-Velasco M, Fisher G et al (2017) Identification of FBXL4 as a metastasis associated gene in prostate cancer. Sci Rep 7(1):5124. https://doi.org/10.1038/s41598-017-05209-z
4. Lu YJ, Dong XY, Guo SP, Ke Y, Cheng SJ (1996) Integration of SV40 at 12q23 in SV40-immortalized human bronchial epithelial cells. Carcinogenesis 17(9):2089–2091
5. Lu YJ, Birdsall S, Summersgill B, Smedley D, Osin P, Fisher C et al (1999) Dual colour fluorescence in situ hybridization to paraffin-embedded samples to deduce the presence of the der(X)t(X;18)(p11.2;q11.2) and involvement of either the SSX1 or SSX2 gene: a diagnostic and prognostic aid for synovial sarcoma. J Pathol 187(4):490–496. https://doi.org/10.1002/(SICI)1096-9896(199903)187:4<490::AID-PATH274>3.0.CO;2-X
6. Tonnies H (2002) Modern molecular cytogenetic techniques in genetic diagnostics. Trends Mol Med 8(6):246–250
7. Schrock E, du Manoir S, Veldman T, Schoell B, Wienberg J, Ferguson-Smith MA et al (1996) Multicolor spectral karyotyping of human chromosomes. Science 273(5274):494–497
8. Speicher MR, Gwyn Ballard S, Ward DC (1996) Karyotyping human chromosomes by combinatorial multi-fluor FISH. Nat Genet 12(4):368–375. https://doi.org/10.1038/ng0496-368
9. Lu YJ, Morris JS, Edwards PA, Shipley J (2000) Evaluation of 24-color multifluor-fluorescence in-situ hybridization (M-FISH) karyotyping by comparison with reverse chromosome painting of the human breast cancer cell line T-47D. Chromosom Res 8(2):127–132
10. Kearney L (2001) Molecular cytogenetics. Best Pract Res Clin Haematol 14(3):645–669. https://doi.org/10.1053/beha.2001.0159
11. Kallioniemi A, Kallioniemi OP, Sudar D, Rutovitz D, Gray JW, Waldman F et al (1992) Comparative genomic hybridization for molecular cytogenetic analysis of solid tumors. Science 258(5083):818–821
12. Wang R, Lu YJ, Fisher C, Bridge JA, Shipley J (2001) Characterization of chromosome aberrations associated with soft-tissue leiomyosarcomas by twenty-four-color karyotyping and comparative genomic hybridization analysis. Genes Chromosomes Cancer 31(1):54–64. https://doi.org/10.1002/gcc.1118
13. Jones C, Foschini MP, Chaggar R, Lu YJ, Wells D, Shipley JM et al (2000) Comparative genomic hybridization analysis of myoepithelial carcinoma of the breast. Lab Investig 80(6):831–836
14. Lu YJ, Birdsall S, Osin P, Gusterson B, Shipley J (1997) Phyllodes tumors of the breast analyzed by comparative genomic hybridization and association of increased 1q copy number with stromal overgrowth and recurrence. Genes Chromosomes Cancer 20(3):275–281
15. Lu YJ, Osin P, Lakhani SR, Di Palma S, Gusterson BA, Shipley JM (1998) Comparative genomic hybridization analysis of lobular carcinoma in situ and atypical lobular hyperplasia and potential roles for gains and losses of genetic material in breast neoplasia. Cancer Res 58(20):4721–4727
16. Lu YJ, Williamson D, Clark J, Wang R, Tiffin N, Skelton L et al (2001) Comparative expressed sequence hybridization to chromosomes for tumor classification and identification of genomic regions of differential gene expression. Proc Natl Acad Sci U S A 98(16):9197–9202. https://doi.org/10.1073/pnas.161272798
17. Bharaj BB, Luo LY, Jung K, Stephan C, Diamandis EP (2002) Identification of single nucleotide polymorphisms in the human kallikrein 10 (KLK10) gene and their association with prostate, breast, testicular, and ovarian cancers. Prostate 51(1):35–41
18. Lu YJ, Williamson D, Wang R, Summersgill B, Rodriguez S, Rogers S et al (2003) Expression profiling targeting chromosomes for tumor classification and prediction of clinical behavior. Genes Chromosomes Cancer 38(3):207–214. https://doi.org/10.1002/gcc.10276
19. Mao X, James SY, Yanez-Munoz RJ, Chaplin T, Molloy G, Oliver RT et al (2007) Rapid high-resolution karyotyping with precise identification of chromosome breakpoints. Genes Chromosomes Cancer 46(7):675–683. https://doi.org/10.1002/gcc.20452

20. Shen Y, Wu BL (2009) Microarray-based genomic DNA profiling technologies in clinical molecular diagnostics. Clin Chem 55 (4):659–669. https://doi.org/10.1373/clinchem.2008.112821
21. Xue L, Mao X, Ren G, Stankiewicz E, Kudahetti SC, Lin D et al (2012) Chinese and Western prostate cancers show alternate pathogenetic pathways in association with ERG status. Am J Cancer Res 2(6):736–744
22. Bastus NC, Boyd LK, Mao X, Stankiewicz E, Kudahetti SC, Oliver RT et al (2010) Androgen-induced TMPRSS2:ERG fusion in nonmalignant prostate epithelial cells. Cancer Res 70(23):9544–9548. https://doi.org/10.1158/0008-5472.CAN-10-1638
23. Mao X, Yu Y, Boyd LK, Ren G, Lin D, Chaplin T et al (2010) Distinct genomic alterations in prostate cancers in Chinese and Western populations suggest alternative pathways of prostate carcinogenesis. Cancer Res 70 (13):5207–5212. https://doi.org/10.1158/0008-5472.CAN-09-4074
24. Giraldo P, Montoliu L (2001) Size matters: use of YACs, BACs and PACs in transgenic animals. Transgenic Res 10(2):83–103

# Chapter 17

# Upconversion Nanoparticle-Based Strategy for Crossing the Blood-Brain Barrier to Treat the Central Nervous System Disease

**Libing Fu, Roger Chung, and Bingyang Shi**

## Abstract

The blood-brain barrier (BBB) is a major challenge for the treatment of central nervous system (CNS) diseases. The BBB strictly regulates the movement of molecules into and out of the brain, and therefore protects the brain from noxious agents. However, for this reason the BBB also acts as a major obstacle that prevents most therapeutic molecules from getting into the target site of the brain. Therefore, it is essential to develop an efficient and general approach to overcome the BBB and transport the drug to the targeted region. Nanoparticle-based drug delivery systems are emerging as a promising drug delivery platform, due to their distinct advantages of tunable biophysical properties such as surface chemistry, size, and shape leading to various biological actions (like clearance, biodistribution, and biocompatibility) in the body. Therefore, it was hypothesized that the surface and shape of nanoparticles will influence their BBB permeation efficiency. Here, we describe a series of upconversion nanoparticles with different surfaces (oleic acid-free, DNA-modified, Silica coating, and PEG-encapsulated), PEGylated UCNPs with various shapes were generated (including sphere and rod). The cellular uptake ability, biodistribution, and BBB penetration of those UCNPs were assessed in cultured cells (NSC-34 neuron- like cells) and in vivo (zebrafish models).

**Key words** Upconversion Nanoparticles, Surface, Shape, Blood-brain Barrier, Central Nervous System Disease

## 1 Introduction

Neurodegenerative diseases including dementia and motor neuron disease (MND) destroy the nerve cells that control cognitive and/or motor function, with no effective long-term therapies available. To develop pharmacological treatments for central nervous system (CNS) diseases, a major challenge is the efficient delivery of therapeutic agents across the blood-brain barrier (BBB). The BBB acts to protect the brain from harmful and poisonous agents; however, it also significantly hinders most drugs from entering the brain. In order to enhance BBB penetration,

Jyotsna Batra and Srilakshmi Srinivasan (eds.), *Theranostics: Methods and Protocols*, Methods in Molecular Biology, vol. 2054, https://doi.org/10.1007/978-1-4939-9769-5_17, 

the drug delivery system should escape clearance through the reticuloendothelial system to ensure sustained presence at effective concentrations [1]. In principle, there are three pathways that support delivery past the BBB. Firstly, surface ligands of carrier-systems which bind to surface receptors abundant on the surface of endothelial cells that form the BBB facilitate binding and uptake into the brain [2–4]. Specific peptides targeting certain receptors such as the transferrin receptor or lipoprotein receptors might help to deliver cargoes across the BBB via receptor-mediated endocytosis. Secondly, certain physicochemical characteristics of delivery systems are suitable for crossing BBB. For example, it was reported that gold nanoparticles (20 nm) were found in the retina after intravenous injection [5]. In addition, some pathological conditions in the brain (such as brain tumors, or injury to the brain such as traumatic brain injury or stroke) lead to an abnormal increase in the permeability of the BBB vasculature. Thirdly, surface modification can improve the BBB permeability of therapeutic compounds. For instance, partial capping of delivery systems with polysorbate 80 or chitosan increased the targeting transport of drugs into the deep brain [6–8]. This combination of characteristics of the delivery system and pathological vasculatures would make BBB more ready for drug-loaded delivery system penetration.

Employing nanoparticles as carriers to transport drugs across the BBB is an emerging approach in the nanotechnology field. The distinct advantage of nano-enabled drug delivery systems is that the biophysical properties of nanoparticles can be readily modified based upon size, shape, and surface, resulting in striking changes in the clearance, biodistribution, and biocompatibility of nanoparticles in vivo [9]. In terms of that, some studies have shown that the effect of size and surface modification of nanoparticles on biological actions in vivo [10, 11] and in vitro [12, 13]. For instance, Chan's group reported the membrane receptor internalization of silver nanoparticles and Au particles with diameters ranging from 2 to 100 nm, and found the preferable efficient uptake was within 25–50 nm size range. Moreover, Zuhorn et al. exploited uncoated, cationic polymer polyethyleneimine (PEI)-attached and prion proteins-functionalized 500 nm beads to investigate their intracellular processing in an in vitro BBB model. They demonstrated that surface modifications of nanoparticles (like charge and protein ligands) affect their mode of internalization by brain endothelial cells and further subcellular fate and transcytosis potential. Meanwhile, the shape is also most likely a key factor of NPs for BBB penetration. However, there are few reports related to the effect of nanoparticles shape on cellular internalization and subsequent intracellular transport across the BBB [14]. This might be because of the limited availability of approaches and the complexity to synthesize nonspherical biocompatible nanoparticles [3].

Upconversion nanoparticles (UCNPs) consist of a crystalline host (like $NaYF_4$ or $NaGdF_4$) together with a low concentration metal-dopant (generally lanthanide ions). UCNPs possess unique optical properties, which can emit shorter visible light under near-infrared long-wavelength excitation through a process known as the anti-Stokes principle. This is particularly valuable for biological applications, as long-wavelength excitation penetrates more deeply into tissue and yields lower phototoxicity than more common forms of excitation. Moreover, UCNPs exhibit other technical advantages, including high photostability, low autofluorescence background, narrow emission bandwidths, high penetration depth, and low toxicity [5, 15–17]. Therefore, recent decades have witnessed substantial growth in the UCNP field, particularly exploring the fabrication and bio-applications of UCNPs as a new class of luminescent optical labels and drug delivery vehicles for therapeutic applications and biological imaging [18–21]. - Importantly, the surface of UCNPs can be modified with various approaches, such as chemical reaction of surface ligands [9, 22, 23], salinization [24–26], ligand exchange [1, 27], and ligand interaction [28–31]. Furthermore, the size, lifetime, and shape of UCNPs are adjustable via controlling the synthesis conditions (heating up speed, reaction temperature, and chemical ratio) and composition parameters (percentage of host and lanthanide ions).

In this study, we developed a library of UCNPs with various shapes to evaluate the effect of surface modification and shape upon BBB penetration. First, we fabricated a batch of high-doped UCNPs (ß-$NaYF_4$:20%Yb,4%Tm) with different shapes, such as sphere and rod. And then we functionalized spherical UCNPs to form four types of UCNPs (such as OA-free, DNA-modified, silica-coated, and PEG-encapsulated) (*see* Fig. 1 and 2) for surface evaluation. Furthermore, we measured the cellular uptake and BBB

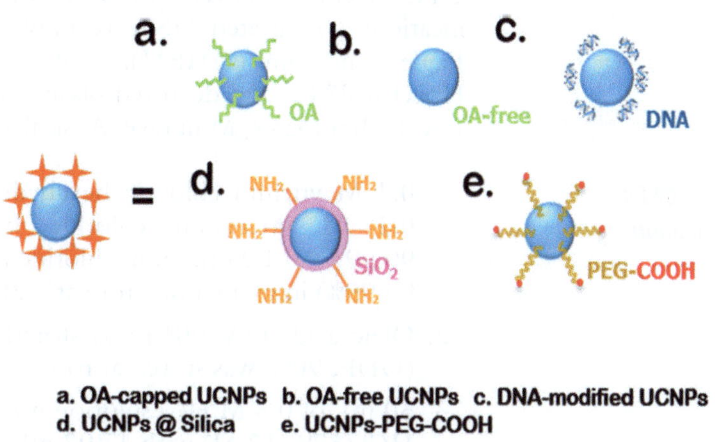


**Fig. 1** Scheme of UCNPs with different surfaces: (**a**) OA-capped, (**b**) OA-free, (**c**) DNA-modified, (**d**) silica-coated, and (**e**) PEG-encapsulated UCNPs

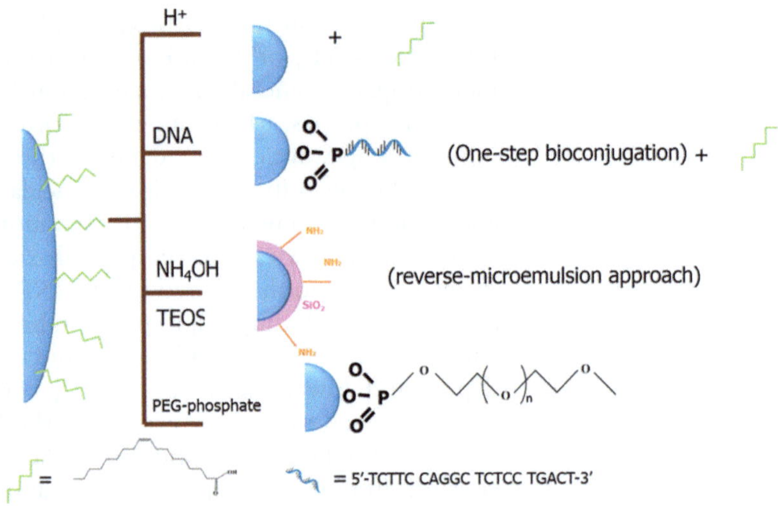


**Fig. 2** Schematic illustration showing the fabrication of the surface-modified UCNPs: OA-free (employing acid to remove OA on the surface), DNA-modified (using one-step bioconjugation approach), silica-coated (utilizing reverse-microemulsion method), and PEG-COOH functionalized (ligand exchange approach) UCNPs

penetration of these transferrin-UCNPs with cultured brain endothelial cells (BBB hCMEC/D3) and injection into the bloodstream of living zebrafish, utilizing confocal microscopy and Imaris software.

## 2 Materials

### *2.1 Experimental Materials*

All chemicals were purchased from Sigma-Aldrich and used as received without further purification. The DNA linkers (5′-TTC CTC CAG ACT GAG GTC TTC-3′) with amine and thiol modifications (Integrated DNA Technologies). Dulbecco's modified Eagle's medium (DMEM), fetal bovine serum (FBS), DAPI, Hoechst 3342, and deep red plasma membrane stain (Invitrogen Life Technologies, Mulgrave, Australia).

### *2.2 UCNPs Fabrication*

1. 0.4 M yttrium chloride hexahydrate ($YCl_3 \cdot 6H_2O$, 99.99%), 0.2 M ytterbium chloride hexahydrate ($YbCl_3 \cdot 6H_2O$, 99.99%), 0.1 M thulium chloride hexahydrate ($TmCl_3 \cdot 6H_2O$, 99.99%) in methanol. Store at −20 °C (*see* **Note 1**).
2. Oleic acid (OA, 90%) was stored at 4 °C and 1-octadecene (ODE, 90%) was stored at room temperature (*see* **Note 2**).
3. 50 mL of 0.5 M HCl solution was made by dilute 2.08 mL of HCl (37%, 12 M) with 47.92 mL MiliQ water.
4. Tetraethyl Orthosilicate (TEOS) and (3-aminopropyl) triethoxysilane (APTES, 99%) should be stored at 4 °C (*see* **Note 3**).

5. 15 mL of 0.01 M $NOBF_4$ solution in dichloromethane (DCM) was made by adding 5 g $NOBF_4$ solid into 15 mL DCM liquid for sonication about 10 min. Then take out the supernatant liquid for further use. Because only a few of $NOBF_4$ can solve in the DCM liquid.
6. The phosphate-PEG-COOH (MW 3500) powder was stored at 4 °C and dissolved into MilliQ water before use (*see* **Note 4**).
7. NSC-34 cell culture media: DMEM cell culture media including 10% FBS and 1% p/s was stored at 4 °C.

### 2.3 Cell Membrane Staining

1. Prepare a fresh deep red plasma membrane staining solution (C10046, 649/666 nm) in warm PBS solution from the provided 1000× concentrated stain solution. For example, to prepare 5 mL of 1× working solution, add 50 μL the deep red plasma membrane stain to 5 mL of warm PBS solution (pH 7.2) (*see* **Note 5**).
2. Seed $1 \times 10^5$ NSC-34 neuron-like cells on coverslips inside a 24-well cell culture dish with 1 mL DMEM cell culture media (including 10% FBS and 1% p/s).
3. After overnight incubation at 37 °C, the cells reach the desired confluency, remove the cell culture media and incubate the cells with 400 μL 1.5× staining solution each well for 10 min at 37 °C (*see* **Note 6**).
4. Remove the staining solution and rinse the coverslip with PBS solution three times.

### 2.4 Cell Nucleus Staining

1. Hoechst 33342 (62249, 5 mL/20 mM) should be stored at 4 °C protected from light. Prepare a fresh staining solution (361/486 nm) in warm PBS solution from the provided 10,000× concentrated stain solution. For example, to prepare 5 mL of 1× working solution, add 5 μL the deep red plasma membrane stain to 5 mL of warm PBS solution (pH 7.2) (*see* **Note 7**).
2. Seed $1 \times 10^5$ NSC-34 neuron-like cells on coverslips inside a 24-well cell culture dish with 1 mL DMEM cell culture media (including 10% FBS and 1% p/s).
3. The cells reach the desired confluency after overnight incubation at 37 °C, remove the cell culture media and incubate the cells with 400 μL 1× staining solution each well for 10 min at 37 °C, protected from light.
4. Remove the excess staining solution and rinse the coverslip with PBS solution three times.
5. Mount the coverslip and image immediately.

#### 2.5 Cell Cytotoxicity

1. $1 \times 10^4$ NSC-34 cells were seeded into a 96-well plate (200 μL medium per well) and incubated at 37 °C with 5% $CO_2$.
2. The MTT assay with minor modifications was conducted by adding 10 μL of the 3-[4,5-dimethylthiazol-2-yl]-2,5-diphenyltetrazoliumbromide (MTT, 5.0 mg/mL in PBS) to each well (*see* **Note 8**).
3. Cells were further incubated at 37 °C for 4 h to allow precipitation of insoluble purple formazan crystals.
4. The supernatant was carefully removed, and 150 μL dimethyl sulfoxide (DMSO) was added to the wells and left for 2 h in the dark at room temperature (*see* **Note 9**).
5. Lastly, the absorbance was measured using the Biotek Microplate Reader (Biotek, USA) at a wavelength of 570 nm. The control well was assumed to have 100% viability. The viable cells number was calculated based on the absorbance of the control cultures.
6. LDH assay followed the LDH Assay Kit (601170): $1.0 \times 10^4$ of cells at a density of $5 \times 10^5$ cells/mL (DMEM containing 10% FBS) were seeded in each well of 96-well plates and grew for 24 h before NPs incubation (*see* **Note 10**).
7. The cells were washed with PBS three times.
8. After 24 h incubation, 20 μL of Triton X-100 (10%) solution was added to three wells and 20 μL of Assay Buffer was added to another three wells before incubation at room temperature for 1 h.
9. Transfer 100 μL of cell supernatant to a new 96-well assay plate. 100 μL of Reaction Solution was added to each well.
10. Afterward, the plate was incubated with gentle shaking on an orbital shaker (150 rpm) for 30 min at 37 °C.
11. Finally, the absorbance of the plate was measured at 490 nm with Biotek Microplate Reader (Biotek, USA). The LDH activity of the samples was obtained by measuring the decreasing rate of NADH absorbance over time.

#### 2.6 Zebrafish Microinjection

1. Five-day-old zebrafish larvae were mounted in 1% low melting agarose before microinjection and imaging as previously described [32, 33] (*see* **Note 11**).
2. 4% paraformaldehyde (PFA) was used to fix zebrafish for 30 min at room temperature before imaging (*see* **Note 12**).
3. PBS solution (pH 7.2) was utilized to emerge zebrafish in the 1% low melting agarose (*see* **Note 13**).

4. Luminescence images were acquired at 0 and 2 h after injection (fed into the fish water in a 28 °C incubator).
5. Transgenic zebrafish model: labeled with blue fluorescent protein; Tg (-3mnx1:TagBFP)) of 5-day-old zebrafish larvae (*see* **Note 14**).

## 3 Methods

All procedures were carried out at room temperature unless otherwise specified.

### *3.1 UCNPs Preparation*

1. The spherical UCNPs ($NaYF_4$:20%Yb,4%Tm) were synthesized following an oxygen-free protocol (shown in Fig. 3). In a typical procedure, 3.5 mL of methanol solution (0.76 mmol $YCl_3$, 0.2 mmol $YbCl_3$ and 0.04 mmol Tm) was magnetically mixed with OA (6.5 mL) and ODE (15 mL) in a 100 mL three-neck round-bottom flask (*see* **Note 15**).
2. The mixture was degassed under Ar flow and heated to 150 °C for 30 min to form a clear solution.
3. Then cooled to room temperature. 5 mL of methanol solution containing $NH_4F$ (4 mmol) and NaOH (2.5 mmol) was added into the solution of $YCl_3$ in ODE and stirred for 30 min at room temperature. The mixture of methanol sodium hydroxide (NaOH, 99%) and ammonium fluoride ($NH_4F$, 99%) should be made freshly before use every time (*see* **Note 16**).
4. The mixture solution was slowly heated up to 110 °C and kept at 110 °C for 30 min to remove methanol and water completely (*see* **Note 17**).

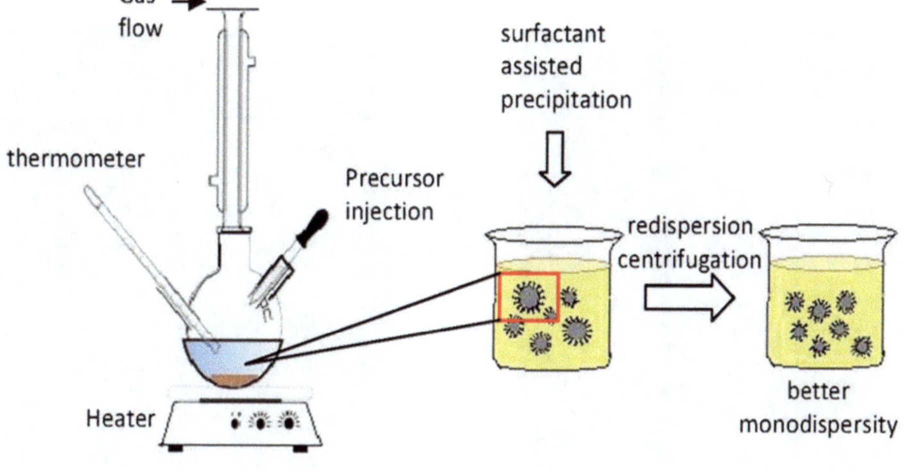


**Fig. 3** Illustration of the synthesis of UCNPs

5. Then the mixture solution was quickly heated to reaction temperature 310 °C and aged for 1.5 h. After the solution cooled down to room temperature, ethanol was added to precipitate the nanocrystals (*see* **Note 18**).
6. After the product was washed with cyclohexane, ethanol, and methanol for three times, the final $NaYF_4$:Yb,Tm nanocrystals were redispersed in 10 mL cyclohexane at −20 °C for further use (*see* **Note 19**).

### 3.2 Preparation of Upconversion Nanorods

1. The longitudinal growth of $NaYF_4$:Yb,Tm onto the core of $NaYF_4$:Yb,Tm upconversion nanoparticles was conducted with a one-pot hot-injection approach. Potassium hydroxide (KOH) was utilized to control the longitudinal growth of UCNPs with a one-pot successive layer-by-layer hot-injection approach (*see* **Note 20**).
2. First, shell precursors were prepared (shown in Fig. 4) 0.76 mmol $YCl_3{\cdot}6H_2O$, 0.2 mmol $YbCl_3$ and 0.04 mmol $TmCl_3{\cdot}6H_2O$ were added to a 50 mL flask containing 6.5 mL oleic acid and 15 mL 1-octadecene.
3. The mixture solution was heated to 160 °C with the protection of argon flow for 30 min to obtain a clear solution and then cooled down to room temperature (RT).
4. Followed by the addition of 4.0 mmol $NH_4F$, 2.5 mmol NaOH and 2.0 mmol KOH in 8 mL methanol.
5. After stirring for 30 min at RT, the solution was heated to 80 °C with argon flow for 20 min to remove methanol.
6. Then the reacting system was continually heated to 150 °C for another 30 min to remove water in the solution (*see* **Note 21**).
7. Finally, the reaction solution was cooled down to room temperature and labeled as 1 mmol $NaYF_4$:Yb,Tm shell precursors for further use.

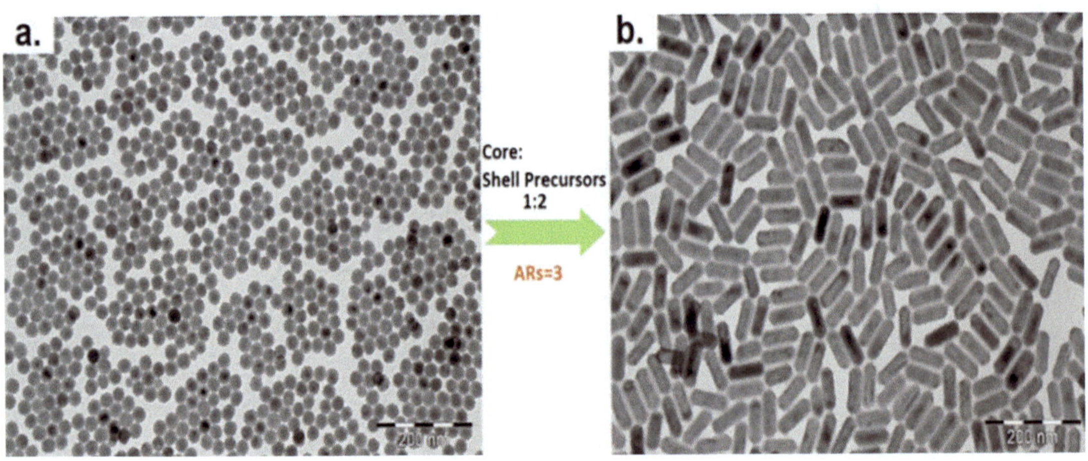


**Fig. 4** TEM images of spherical UCNPs ($NaYF_4$:20%Yb,4%Tm) as the core (**a**) and the as-prepared UCNR (ARs = 3) (**b**) (Scale bar of images is 200 nm)

8. Synthesis spherical UCNPs as core via an oxygen-free method described in our previous work [22, 26]. 0.2 mmol $NaYF_4$:Yb, Tm core particles in cyclohexane were added to a 50 mL three-neck flask containing 1.3 mL OA and 3 mL ODE (*see* **Note 22**).
9. The mixture solution was heated to 160 °C under the protection of argon flow for 30 min to remove cyclohexane, and then the solution was continually heated to 310 °C.
10. In order to control the length of nanorod, the as-prepared $NaYF_4$:Yb,Tm shell precursors should be injected into the reaction mixture slowly at 310 °C with a long needle (*see* **Note 23**).
11. The ratio between the $NaYF_4$:Yb,Tm cores and $NaYF_4$:Yb,Tm shell precursors determines the length of the nanorods (*see* **Note 24**).
12. 0.2 mmol of as-prepared $NaYF_4$:Yb,Tm shell precursors were injected into the reaction mixture at the speed of 0.1 mL per 10 min and followed by another 0.5 h reaction time at 310 °C (*see* **Note 25**).
13. Finally, the formed solution was gradually cooled down to room temperature and the formed upconversion nanorods (ARs = 2) were purified according to the procedures employed for the purification of $NaYF_4$:Yb,Tm spherical particles (*see* Subheading 3.1).

### *3.3 OA-Removal of UCNPs*

1. OA ligand on the UCNPs surface was removed according to literatures [34, 35] with modifications (*see* **Note 26**, shown in Fig. 3a).
2. 1 mL of cyclohexane containing oleic acid-capped UCNPs (20 mg/mL) was added into 4 mL absolute ethanol and performed with sonicating for 10 min.
3. The solution turned to be cloudy, because the UNCPs participated.
4. Afterward, the solution was centrifuged for 10 min at 9501.8 × *g* to separate UCNPs.
5. The as-prepared UCNPs were dispersed in aqueous solution (5 mL) and the pH was adjusted to 2 by adding 0.5 m HCl solution.
6. The reaction was performed with sonication for 10 min. During this reaction, the carboxylate groups of the oleic ligand were protonated (to yield oleic acid).
7. The UCNPs in the solution were recuperated and treated with 5 mL HCl solution (pH 4) by sonicating for 20 min.
8. Last, the particles were recuperated and washed with ethanol for three times. Finally, the particles were dispersed in water for further use.

### 3.4 Fabrication of DNA-UCNPs

1. DNA-modified UCNPs were synthesized via ligand-exchange bio-conjugation [22] (*see* Fig. 2) 10 μL of 10 mg/mL original UCNPs were added into 400 μL chloroform, and then mixed with 300 μL ultrapure $H_2O$ solution containing purified DNA–protein conjugates (*see* **Note 27**).
2. After 2-h incubation on a mixer, the upper layer solution was taken out and then centrifuged twice at 18623.5 × *g* for 5 min.
3. The final settlement was dissolved in 100 μL ultrapure $H_2O$ with sonication for 5 s.
4. The UCNPs-DNA were dispersed in MilliQ-water for further use.

### 3.5 Preparation of Silica-Coated UCNPs

1. We used a modified water-in-cyclohexane (W/C) reverse micro-emulsion coating approach [27]. Firstly, a cyclohexane dispersion of $NaYF_4$:Yb,Tm nanocrystals (0.1 mmol/mL, 1 mL), IGPAL CO-520 (1 mL) and cyclohexane (20 mL) was magnetically mixed in a glass flask for 30 min.
2. Then, 35 mL of ammonia (30 wt%) was injected into the solution and a transparent emulsion formed after sonicating for 40 min.
3. Thereafter, 100 μL of TEOS was added as the silica shell precursor and the solution was kept under gentle stirring (controlled at 600 rpm) at room temperature for 2 days (*see* **Note 28**).
4. After that, 10 mL acetone was added to precipitate the silica-coated UCNPs from the microemulsion.
5. The nanoparticles were centrifugally washed with ethanol two times to remove the excess surfactant, and then washed with water to remove ethanol, and finally redispersed in water for use.

### 3.6 Synthesis of PEG-UCNPs and PEG-UCNRs

1. Firstly, a dichloromethane solution of $NOBF_4$ was used to turn UCNPs hydrophilic according to the literature [28]. In a typical process, 5 mL of cyclohexane (containing UCNPs or UCNRs ~5 mg/mL) was combined with 5 mL of dichloromethane (DCM) solution of $NOBF_4$ (0.01 M) at room temperature.
2. The mixture was shaken overnight to transfer UCNPs (or UCNRs) from the upper cyclohexane layer to the bottom DCM layer.
3. After centrifugation to remove the supernatant, the pellet (UCNPs or UCNRs) was redispersed in DMF at the concentration of 10 mg/mL.
4. Then, 200 μL of solution (~2 mg UCNPs or UCNRs) was taken into a bottle and 2 mL of 100% ethanol was added and stirred, followed by 10 mg PEG dispersed in 1 mL water. The mixture solution became cloudy.
5. Then water (~3 mL) was added until the solution became clear and stirred for 48 h.
6. Finally, the UCNPs-PEG-COOH (or UCNRs-PEG-COOH) were washed with water five times (*see* **Note 29**).

### 3.7 Cell Imaging with Confocal Microscopy

1. Surface-modified UCNPs at the concentration of 0, 20, 50, 100, and 500 μg/mL were incubated with NSC-34 cells at 37 °C for 4 h under 5% $CO_2$ according to our well-established method [36] (*see* **Note 30**).
2. After washing with PBS three times, cells were fixed in 4% paraformaldehyde for 30 min at room temperature. The fixed cells were washed with PBS and incubated with 1× deep red plasma membrane stain and 2 μM Hoechst 33342 for 10 min at 37 °C (*see* **Note 31**).
3. The labeled cells were rinsed with PBS and mounted with Vector Vectashield mounting media (USA) before the coverslips were sealed (*see* **Note 32**).
4. The cells without UCNPs served as the blank control.
5. Furthermore, commercial fluorescent microspheres (FM, 580/605 nm) were employed as the positive control and the OA-capped UCNPs served as the negative control in this process (*see* **Notes 33** and **34**).
6. Confocal microscopy measurement was performed to confirm internalization and quantification of the localized UCNPs using a laser-scanning fluorescence confocal microscope (ZEISS LSM 880 with Airyscan) illuminated with a custom-built 980 nm laser module (*see* **Note 35**).
7. The oil-immersions of 40× and 100× were used in this imaging process. The excitation intensity was set to $9.8 \times 10^7$ W/cm$^2$.
8. For each cell culture, a minimum of three different cell areas was imaged.

### 3.8 Quantification of UCNPs Uptake in NSC-34 Cells

1. Nanoparticle uptake was quantitatively calculated using the Imaris software. For each cell culture slide, a minimum of three different cell areas was imaged [37] (shown in Fig. 5).
2. For the image analysis, stains for the nucleus and membranes were used to outline single cells, cell clusters, or a monolayer region within the excitation spot (*see* **Note 36**).
3. The signal from the pixels of the same area in the PL images was processed by subtracting the background and correcting for the exposure time.
4. The resulting UCNP signals were quantified as a mean UCNP signal per area, which was a measure of the mean cellular uptake of UCNPs.
5. The relative intensity of the UCNP PL signal was recorded and normalized with respect to the largest cellular uptake level observed (that of UCNPs-PEG-COOH; *see* Figs. 6, 7, and 8), which was set to 100%.

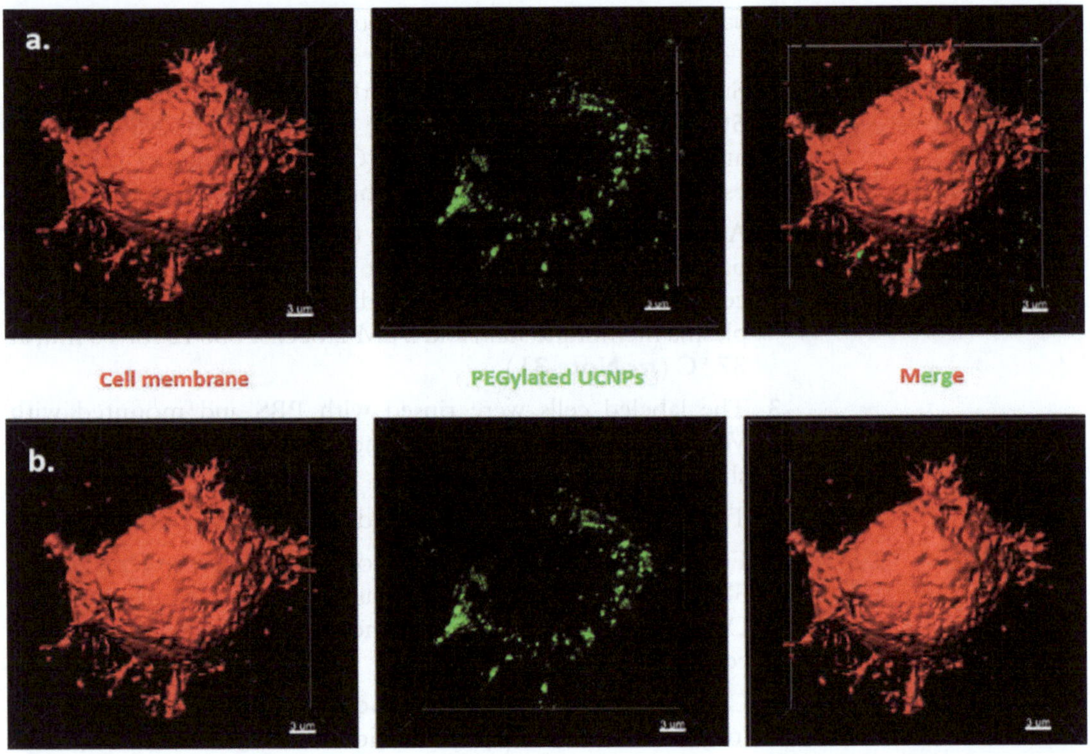


**Fig. 5** 3D rendering of confocal z-stacks to quantitate the cellular uptake of surface-modified UCNPs. These images represent 3D rendering using Imaris software of PEG-UCNP (green) distribution in an individual NSC-34 cell. Both UCNP and membrane signals were individually 3D-rendered (before, row **a**), and then the masks overlaid to isolate and exclude any UCNP signal (after, row **b**) that is outside of the cell membrane mask (and therefore outside of the cell)

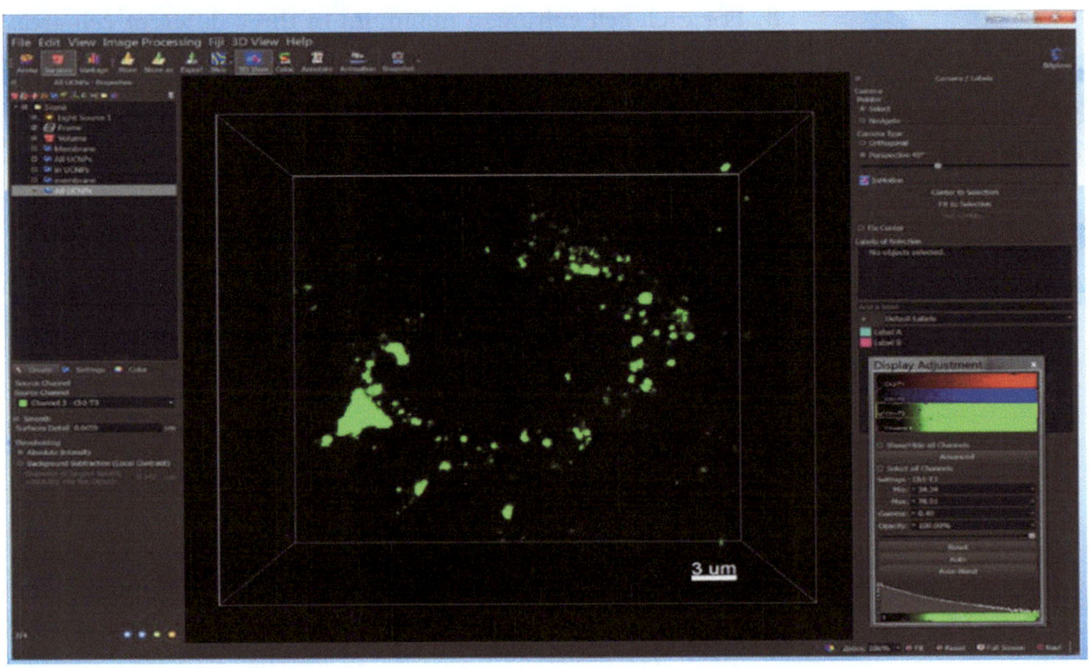


**Fig. 6** 3-D confocal microscope image of all UCNPs in or outside one single NSC-34 cell

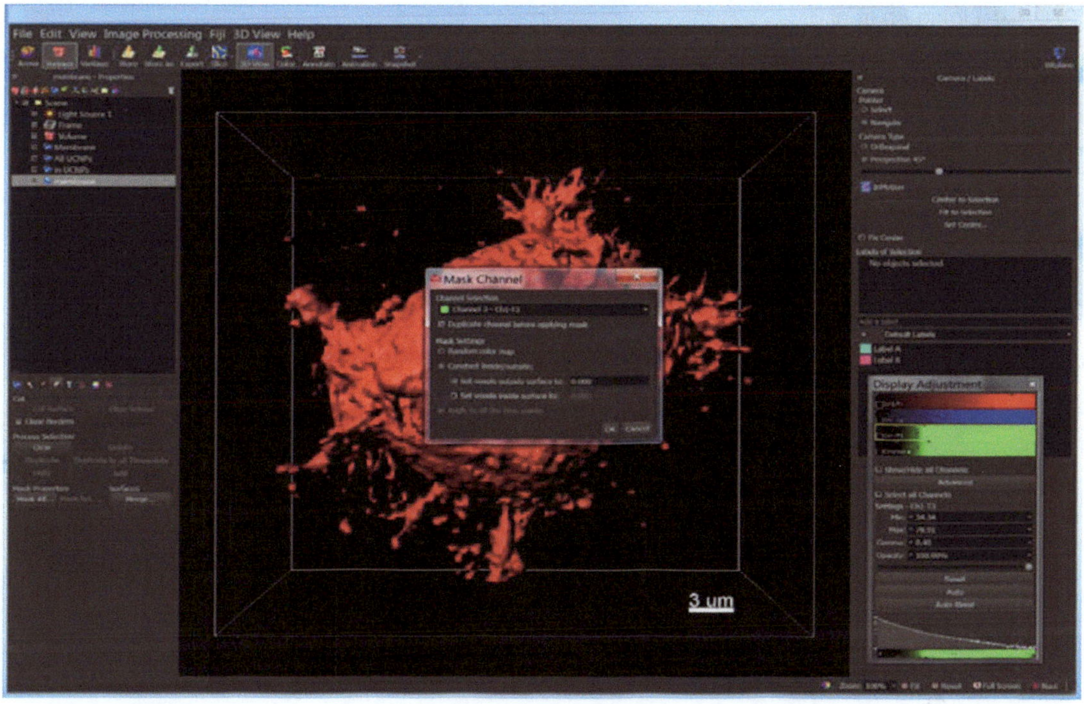


**Fig. 7** 3-D confocal microscope image of cell membrane rendering used to mask selection for UCNPs inside one single NSC-34 cell

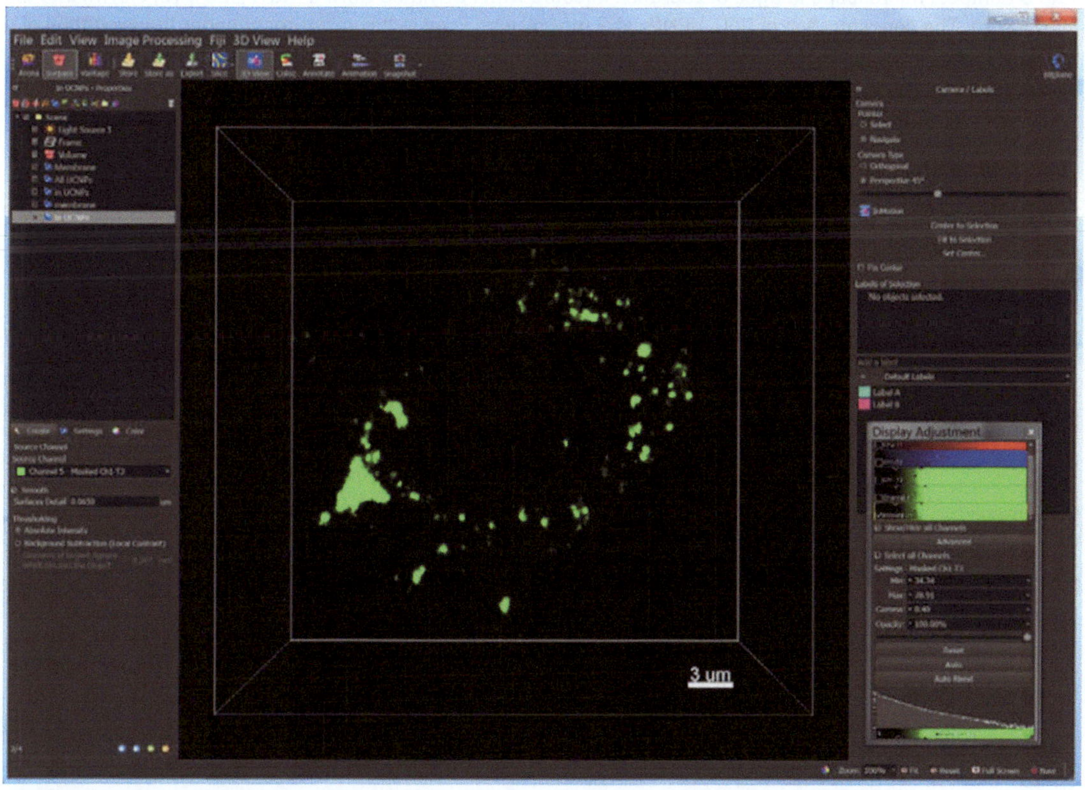


**Fig. 8** 3-D confocal microscope image of all UCNPs inside one single NSC-34 cell after mask selection (i.e.: all UCNPs outside of the cell have been subtracted as per Fig. 10)

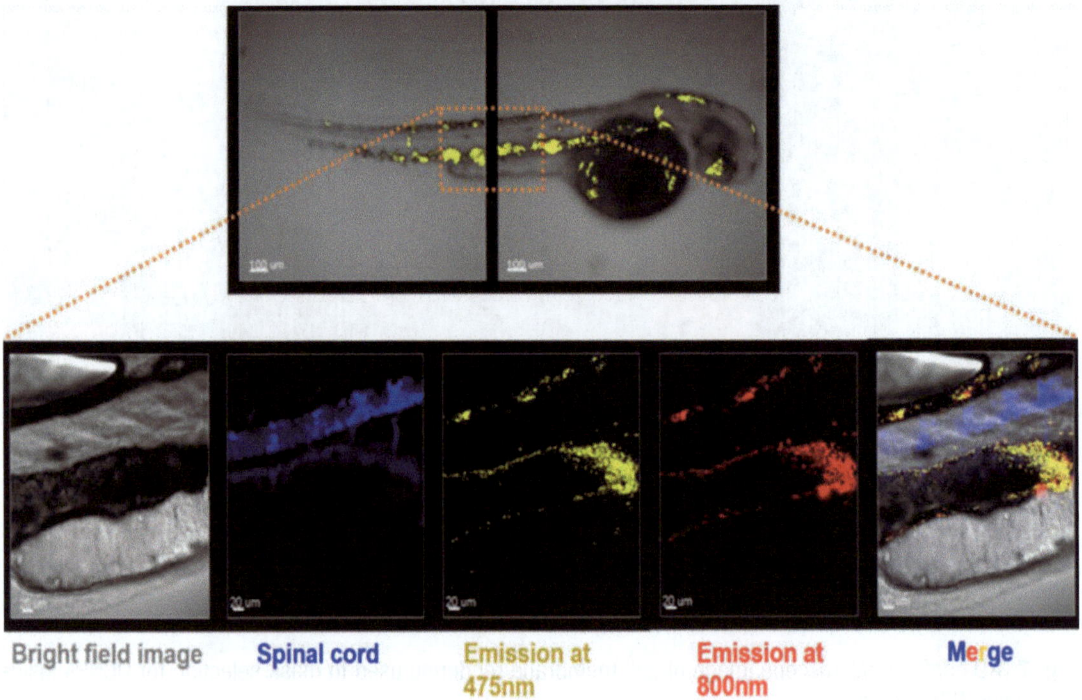


**Fig. 9** Confocal microscope images of zebrafish injected with PEGylated UCNPs ($NaYF_4$:20%Yb,4%Tm) into the spinal cord. The concentration of PEG-UCNPs at 3 mg/mL was used in this experiment under the excitation of a 980 nm laser

#### 3.9 Zebrafish Microinjection

1. The biodistribution of PEG-encapsulated UCNPs were then investigated in vivo using optically transparent zebrafish larvae (*see* Fig. 9). Five-day-old zebrafish larvae were mounted in 1% low melting agarose before microinjection and imaging. And to compare the different blood-brain barrier penetration efficiency of spherical and rod-like UCNPs, they were both injected into blood vessel of zebrafish (*see* Fig. 10).
2. A suspension of UCNPs with various surfaces mixed with microbeads (1 mg/mL) was injected into the spinal cord (labeled with blue fluorescent protein; Tg (-3mnx1:TagBFP)) of 5-day-old zebrafish larvae [8] (*see* **Note 37**).
3. Approximately, 5 nL solution was injected into the spinal cord and luminescence images were acquired at 0 and 2 h after injection (fed into the fish water in a 28 °C incubator).
4. Upon excitation at 980 nm, the distribution of the nanoparticles in the zebrafish can be clearly visualized.

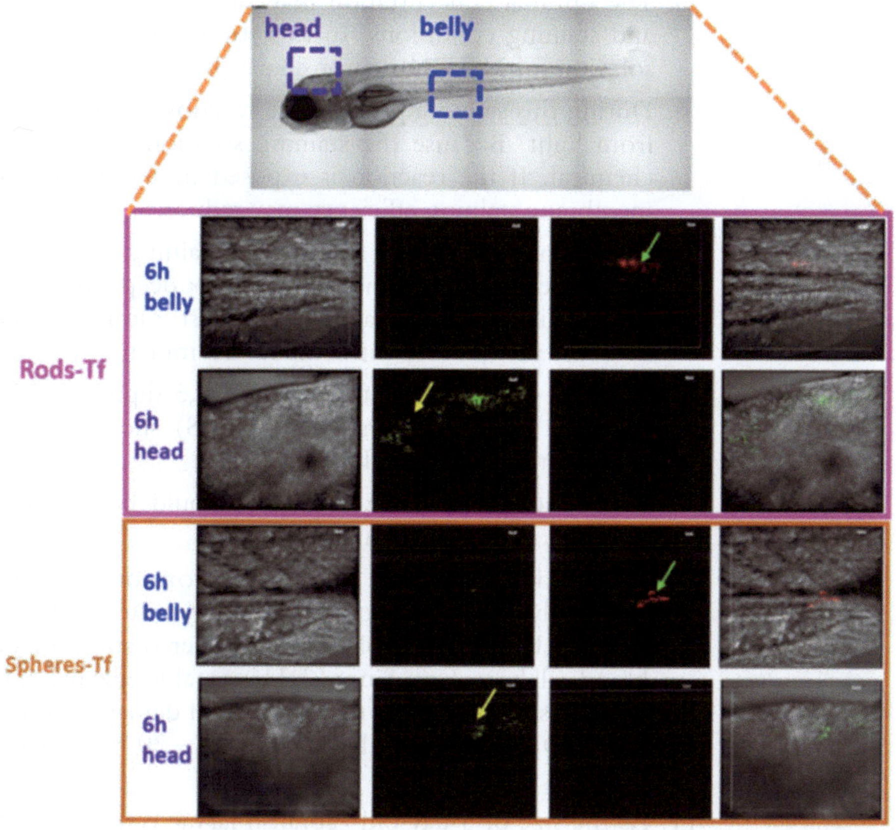


**Fig. 10** Confocal microscope images of zebrafish injected with rod-like and spherical Tf-UCNPs ($NaYF_4$:20% Yb,4%Tm) into the blood vessel and were observe after 6 h. The concentration of Tf-UCNPs at 1 mg/mL was used in this experiment under the excitation of a 980 nm laser

## 4 Notes

1. It is easy to take a specific amount of methanol solution including $YCl_3{\cdot}6H_2O$, $YbCl_3{\cdot}6H_2O$, and $TmCl_3{\cdot}6H_2O$, respectively. Otherwise, we need to measure little amount of chemical powder, which is ready to cause an error.
2. Oleic acid (OA) was solid at 4 °C. Therefore, we need to thaw the solid OA before we use it.
3. The tetraethyl orthosilicate (TEOS) was silica precursor and the APTES here was used to modify the silica layer with amino groups.
4. The phosphate group end of phosphate-PEG-COOH (MW 3500) was utilized to ligand exchange with the OA groups on the surface of UCNPs. After this ligand exchange, the PEG will encapsulate the UCNPs and transfer the hydrophobic UNCPs to hydrophilic ones.

5. The optimal concentration may vary depending on cell type and staining conditions. In this study, we use 1.5× working solution for NSC-34 neuron-like cells.
6. During this staining process, the reaction should be protected from light. Because the staining solution is a light-sensitive chemical. If the reaction is exposed under strong light, the membrane staining efficiency may reduce.
7. The membrane staining and nucleus staining can be conducted simultaneously. The same volume of 2× deep red plasma membrane staining solution and 3× Hoechst solution is used. Then incubate the mixture with cells for 10 min at 37 °C.
8. The 3-[4,5-dimethylthiazol-2-yl]-2,5-diphenyltetrazolium-bromide (MTT, 5.0 mg/mL in PBS) should be stored at 4 °C protected from light.
9. The dimethyl sulfoxide (DMSO) should be protected from light at room temperature.
10. Lactate dehydrogenase (LDH) is an oxidoreductase enzyme that catalyzes the interconversion of pyruvate and lactate. Cells release LDH into the bloodstream after tissue damage or red blood cell hemolysis. Since LDH is a stable enzyme, it has been widely used to evaluate the presence of damage and toxicity of cells. LDH Kit set should be stored at −20 °C protected from light.
11. As the size of 5-day-old zebrafish larvae is only 2–3 mm, it is important to stabilize the position when it is microinjected and observed under confocal microscopy with 1% low melting agarose. Typically, the agarose doesn't affect the imaging process as it is transparent.
12. As the imaging process with confocal microscopy is complicated, in order to confirm the stable of biological samples, it is better to fix the zebrafish sample with 4% paraformaldehyde (PFA) before imaging.
13. It is necessary to add PBS solution (pH 7.2) to emerge zebrafish in the 1% low melting agarose, avoiding the agarose becoming dry. The structure of zebrafish will be damaged when the agarose is dry.
14. This transgenic 5-day-old zebrafish model is labeled with blue fluorescent protein in the spinal cord (Tg (-3mnx1:TagBFP)), so it is an ideal model to study the biodistribution of our UCNPs with various surface modifications in the spinal cord.
15. As mentioned in the Subheading 2.2, here we need to take 1.9 mL 0.4 M $YCl_3$, 1 mL 0.2 M $YbCl_3$ and 0.4 mL 0.1 M $TmCl_3$ in a100 mL three-neck round-bottom flask.

16. The cooled mixture solution should be transparent. If the solution turned cloudy, it indicates the chemicals have not dissolve completely at 150 °C. It will profoundly affect the result of UCNPs fabrication. Furthermore, $NH_4F$ (4 mmol) and NaOH (2.5 mmol) should be completed dissolved into 5 mL of methanol before adding into the mixture solution.
17. It always takes 10–15 min to remove all methanol completely.
18. When the reaction complete, the solution will turn yellow, cloudy and oily. After adding ethanol to precipitate the nanocrystals, a great amount of UCNPs will form. So UCNPs can be separated with the supernatant liquid by centrifugation.
19. After washing steps, the cyclohexane including the clean $NaYF_4$:Yb,Tm nanocrystals is clear. That means the nanocrystals disperse well in cyclohexane. Otherwise, the solution will not be transparent if the nanocrystals are not clean enough.
20. We increased the ratio of $OA^-$ to OAH by adding KOH to lead to the conversion of more OAH into $OA^-$ and to lead to form UCNRs of various lengths.
21. It is important to remove water in the solution before the temperature increasing to 310 °C. Because if the reaction system includes some water in the solution, the reaction system will be to stable at the reaction temperature (310 °C). It will cause the nonuniform UCNPs formation.
22. In order to keep consistent, the solution of OA (1.3 mL) and ODE (3 mL) reduce to 20% as the UCNPs decreased to 0.2 mmol.
23. The reason why we use needle is to avoid oxygen during the reaction at 310 °C.
24. For example, when the ratio between UCNPs core and shell precursors is 1:2, then it will lead to forming UCNRs with ARs (length/width) of 3.
25. In order to synthesize the UCNRs with an aspect ratio (ARs, length/ width) of 2, 0.2 mmol of as-prepared $NaYF_4$:Yb,Tm shell precursors were injected into the reaction mixture slowly.
26. As we know, the as-synthesized UCNPs have OA on the surface, so it is not water-soluble. After OA removal, the nearly neutral UCNPs turn to hydrophilic.
27. DNA-modified UCNPs were synthesized via ligand-exchange bioconjugation. As the phosphate groups of DNA molecules have stronger binding ability than OA with UCNPs. Thus, DNA binds on the surface of UCNPs, transferring UCNPs to be hydrophilic.
28. The thickness of the silica shell on the surface of UCNPs can be adjusted by controlling the amount of TEOS. However, if the

amount of TEOS exceeds, the silica may form silica nanoparticles itself, rather than coating on the surface of UCNPs.

29. We introduced a PEGylation surface modification to the UCNPs to enhance the biostability of the nanoparticles.

30. Four types of UCNPs with different surfaces, including OA-free, DNA-modified, silica-coated, and PEG-encapsulated, were incubated with NSC-34 cells at the concentration of 0, 20, 50, 100, and 500 μg/mL, respectively.

31. The 4% PFA was acquired by diluting the 16% PFA solution with PBS solution (pH 7.2).

32. The VECTASHIELD® Mounting Media are unsurpassed in preventing photobleaching of fluorescent dyes and fluorescent proteins. These ready-to-use antifade mounting media are stored at 4 °C.

33. As the commercial fluorescent microbeads (with carboxyl groups on the surface, 40 nm) can easily uptake by NSC-34 cells as noticed. Thus, in this study, we choose this similar size beads compared to our UCNPs to confirm the uptake of UCNPs.

34. The commercial fluorescent microspheres (Molecular Probes' intensely fluorescent FluoSpheres® beads, 40 nm, 580/605 nm) are manufactured using high-quality, ultraclean polystyrene microspheres. These microspheres are loaded with proprietary dyes, making them the brightest fluorescent microspheres available. Our carboxylate-modified microspheres are coated with a hydrophilic polymer containing multiple carboxylic acids for covalent attachment of ligands.

35. To further confirm whether UCNPs were internalized by NSC-34 cells, we employed high-resolution confocal imaging and three-dimensional (3D) rendering to precisely analyze the distribution of the surface functionalized UNCPs in the motor neuron-like cells.

36. The z-stack confocal images of individual NSC-34 cells were processed through Imaris (Bitplane) to generate individual 3D renderings of the cell membrane and UCNP distribution. Firstly, the volume and intensity of the fluorescence signal of UCNPs inside or outside of a single cell were collected, analyzed, and 3D-rendered using Imaris software. Subsequently, the fluorescence signal of the cell membrane dye was 3D rendered and used as a mask within Imaris software, which allowed the determination of the volume of the UCNP signal that was either inside the cell or outside of the cell membrane. This information was used to determine the volume ratio of UCNPs within NSC-34 cells ($V_{inside}$) versus the total distribution of UCNPs detected in proximity to the cell ($V_{entire}$).

37. A further advantage is that we have generated a transgenic zebrafish that specifically expresses blue fluorescent protein (BFP) in spinal motor neurons, this allows us to visualize these motor neurons and track UCNP uptake in real time in living animals. The surface modified UCNPs were microinjected directly into the spinal cord of the transgenic zebrafish expressing blue fluorescent protein specifically within motor neurons.

## Acknowledgments

This project has been supported by NHMRC grants awarded to R.C. (APP1095215) and B.S. (APP1111611).

### References

1. Petros RA, Desimone JM (2010) Strategies in the design of nanoparticles for therapeutic applications. Nat Rev Drug Discov 9:615–627
2. Bogdan N, Vetrone F, Ozin GA, Capobianco JA (2011) Synthesis of ligand-free colloidally stable water dispersible brightly luminescent lanthanide-doped upconverting nanoparticles. Nano Lett 11:835–840
3. Champion JA, Katare YK, Mitragotri S (2007) Particle shape: a new design parameter for micro- and nanoscale drug delivery carriers. J Control Release 121:3–9
4. Chen C, Duan Z, Yuan Y, Li R, Pang L, Liang J, Xu X, Wang J (2017) Peptide-22 and cyclic RGD functionalized liposomes for glioma targeting drug delivery overcoming BBB and BBTB. ACS Appl Mater Interfaces 9:5864–5873
5. Kim JH, Kim JH, Kim KW, Kim MH, Yu YS (2009) Intravenously administered gold nanoparticles pass through the blood-retinal barrier depending on the particle size, and induce no retinal toxicity. Nanotechnology 20:505101
6. Cheng D, Shami GJ, Morsch M, Chung RS, Braet F (2016) Ultrastructural mapping of the zebrafish gastrointestinal system as a basis for experimental drug studies. Biomed Res Int 2016:1–13
7. Choi HS, Ipe BI, Misra P, Lee JH, Bawendi MG, Frangioni JV (2009) Tissue- and organ-selective biodistribution of NIR fluorescent quantum dots. Nano Lett 9:2354–2359
8. Don EK, Formella I, Badrock AP, Hall TE, Morsch M, Hortle E, Hogan A, Chow S, Gwee SSL, Stoddart JJ, Nicholson G, Chung R, Cole NJ (2017) A Tol2 gateway-compatible toolbox for the study of the nervous system and neurodegenerative disease. Zebrafish 14:69–72
9. Masserini M (2013) Nanoparticles for brain drug delivery. ISRN Biochem 2013:1–18
10. Ge X, Dong L, Sun L, Song Z, Wei R, Shi L, Chen H (2015) New nanoplatforms based on UCNPs linking with polyhedral oligomeric silsesquioxane (POSS) for multimodal bioimaging. Nanoscale 7:7206–7215
11. Gref R, Couvreur P, Barratt G, Mysiakine E (2003) Surface-engineered nanoparticles for multiple ligand coupling. Biomaterials 24:4529–4537
12. Haase M, Schäfer H (2011) Upconverting nanoparticles. Angew Chem Int Ed 50:5808–5829
13. Huang X, Teng X, Chen D, Tang F, He J (2010) The effect of the shape of mesoporous silica nanoparticles on cellular uptake and cell function. Biomaterials 31:438–448
14. Jiang W, Kim BYS, Rutka JT, Chan WCW (2008) Nanoparticle-mediated cellular response is size-dependent. Nat Nanotechnol 9:615–627
15. Kuang Y, An S, Guo Y, Huang S, Shao K, Liu Y, Li J, Ma H, Jiang C (2013) T7 peptide-functionalized nanoparticles utilizing RNA interference for glioma dual targeting. Int J Pharm 454:11–20
16. Kurakhmaeva KB, Djindjikhashvili IA, Petrov VE, Balabanyan VU, Voronina TA, Trofimov SS, Kreuter J, Gelperina S, Begley D, Alyautdin RN (2009) Brain targeting of nerve growth factor using poly(butyl cyanoacrylate) nanoparticles. J Drug Target 17:564–574

17. Lin S, Li M, Dujardin E, Girard C, Mann S (2005) One-dimensional Plasmon coupling by facile self-assembly of gold nanoparticles into branched chain networks. Adv Mater 17:2553–2559
18. Lin YS, Haynes CL (2010) Impacts of mesoporous silica nanoparticle size, pore ordering, and pore integrity on hemolytic activity. J Am Chem Soc 132:4834–4842
19. Liu C, Gao Z, Zeng J, Hou Y, Fang F, Li Y, Qiao R, Shen L, Lei H, Yang W, Gao M (2013) Magnetic/upconversion fluorescent dual-modal molecular probes for imaging tiny tumors in vivo. ACS Nano 7:7227–7240
20. Liu D, Xu X, Du Y, Qin X, Zhang Y, Ma C, Wen S, Ren W, Goldys EM, Piper JA, Dou S, Liu X, Jin D (2016) Three-dimensional controlled growth of monodisperse sub-50 nm heterogeneous nanocrystals. Nat Commun 7:10254
21. Liu L, Venkatraman SS, Yang YY, Guo K, Lu J, He B, Moochhala S, Kan L (2008) Polymeric micelles anchored with TAT for delivery of antibiotics across the blood-brain barrier. Pept Sci 90:617–623
22. Lu J, Chen Y, Liu D, Ren W, Lu Y, Shi Y, Piper J, Paulsen I, Jin D (2015) One-step protein conjugation to upconversion nanoparticles. Anal Chem 87:10406–10413
23. Lu Y, Zhao J, Zhang R, Liu Y, Liu D, Goldys EM, Yang X, Xi P, Sunna A, Lu J, Shi Y, Leif RC, Huo Y, Shen J, Piper JA, Robinson JP, Jin D (2014) Tunable lifetime multiplexing using luminescent nanocrystals. Nat Photonics 8:32
24. Mistry A, Stolnik S, Illum L (2009) Nanoparticles for direct nose-to-brain delivery of drugs. Int J Pharm 379:146–157
25. Morsch M, Radford R, Lee A, Don EK, Badrock AP, Hall TE, Cole NJ, Chung R (2015) In vivo characterization of microglial engulfment of dying neurons in the zebrafish spinal cord. Front Cell Neurosci 9:321
26. Park J, Mattessich T, Jay SM, Agawu A, Saltzman WM, Fahmy TM (2011) Enhancement of surface ligand display on PLGA nanoparticles with amphiphilic ligand conjugates. J Control Release 56:109–115
27. Shi B, Zhang H, Qiao SZ, Bi J, Dai S (2014) Intracellular microenvironment-responsive label-free autofluorescent nanogels for traceable gene delivery. Adv Healthc Mater 3:1839–1848
28. Shi Y, Shi B, Dass AVE, Lu Y, Sayyadi N, Kautto L, Willows RD, Chung R, Piper J, Nevalainen H, Walsh B, Jin D, Packer NH (2016) Stable upconversion nanohybrid particles for specific prostate cancer cell immunodetection. Sci Rep 8:1–11
29. Singh SR, Grossniklaus HE, Kang SJ, Edelhauser HF, Ambati BK, Kompella UB (2009) Intravenous transferrin, RGD peptide and dual-targeted nanoparticles enhance anti-VEGF intraceptor gene delivery to laser-induced CNV. Gene Ther 16:645–659
30. Sun W, Xie C, Wang H, Hu Y (2004) Specific role of polysorbate 80 coating on the targeting of nanoparticles to the brain. Biomaterials 25:3065–3071
31. Tang J, Kemp KW, Hoogland S, Jeong KS, Liu H, Levina L, Furukawa M, Wang X, Debnath R, Cha D, Chou KW, Fischer A, Amassian A, Asbury JB, Sargent EH (2011) Colloidal-quantum-dot photovoltaics using atomic-ligand passivation. Nat Mater 10:765–771
32. Wang F, Deng R, Liu X (2014) Preparation of core-shell NaGdF4 nanoparticles doped with luminescent lanthanide ions to be used as upconversion-based probes. Nat Protoc 9:1634–1644
33. Wang F, Liu X (2009) Recent advances in the chemistry of lanthanide-doped upconversion nanocrystals. Chem Soc Rev 38:976–989
34. Wang H, Zhang S, Liao Z, Wang C, Liu Y, Feng S, Jiang X, Chang J (2010) PEGlated magnetic polymeric liposome anchored with TAT for delivery of drugs across the blood-spinal cord barrier. Biomaterials 31:6589–6596
35. Wong AC, Wright DW (2016) Size-dependent cellular uptake of DNA functionalized gold nanoparticles. Small 12:5592–5600
36. Yi DK, Selvan ST, Lee SS, Papaefthymiou GC, Kundaliya D, Ying JY (2005) Silica-coated nanocomposites of magnetic nanoparticles and quantum dots. J Am Chem Soc 127:4990–4991
37. Fu LB, Shi BY, Jin DY, Chung R (2017) A versatile upconversion surface evaluation platform for bio–nano surface selection for the nervous system. Nanoscale 9:13683–13692

# Chapter 18

# Genetically Engineered Plasma Membrane Nanovesicles for Cancer-Targeted Nanotheranostics

**Pengfei Zhang, Hu Chen, Jingyi Liu, and Gang Liu**

## Abstract

A series of ligand-targeted nanosystems have been rapidly exploited to selectively deliver drug molecules to desired cell populations. The conjugation of protein ligands to the nanoparticle (NP) surface endows nanovehicles with active targeting properties. However, the nonspecific covalent coupling of protein ligands to nanocarriers may compromise the protein targeting due to the uncontrolled ligand orientation as well as the decline in ligand activity during linkage process. With this regard, biomimetic synthetic strategies are employed for the preparation of genetically engineered nanovesicles (GNV) from cellular plasma membrane with targeting moieties on the surface in a ligand-oriented manner. Herein, we introduce the biomimetic synthetic strategy and procedures for GNV preparation. This chapter may guide readers to design analogous NPs for cell-specific targeting by displaying particular protein probes (e.g., antibody, nanobody, and single-chain antibody) on the surface of GNVs.

**Key words** Nanovesicles, Cell membrane vesicles, Ligand-targeted delivery, Drug delivery, Genetically engineered vesicles

## 1 Introduction

Nanotheranostics, which combines diagnosis with therapy of malignant tumors by drug-loaded nanovehicles, continues to thrive to achieve better precision medicine and visualize the therapy process. It can be applied for monitoring the therapeutic effect, detecting the tumor metastasis after treatment, and diminishing the side effects [1, 2]. At present, various types of theranostic nanomedicines have been developed [3–6]. Notably, a wide range of ligand-targeted nanocarriers have been fabricated for cell-specific drug delivery [7–9]. The coupling of targeting peptide/protein moieties to nanovehicles is normally the prerequisite for active targeting. However, the nonspecific covalent conjugation to nanovehicles may lead to the target impairment of vulnerable protein ligands due to uncontrolled ligand orientation, structure damages as well as declined protein activity during the chemical linkage process

Jyotsna Batra and Srilakshmi Srinivasan (eds.), *Theranostics: Methods and Protocols*, Methods in Molecular Biology, vol. 2054, https://doi.org/10.1007/978-1-4939-9769-5_18, 

[10, 11]. Two major challenges impeded the application of these targeted nanomedicines; one is how to preserve the accurate orientation and bioactivities of conjugated ligand and the other is how to engineer the targeting protein/antibody moieties [12].

Biomimetic nanovehicles [13], whose size, morphology as well as surface properties mimic the natural organisms like exosomes [14], viruses [15, 16], and red blood cells [17, 18], have the capacities to perform unique functions. Exosomes, extracellular vesicles secreted by most eukaryotic cells, are 70–150 nm in diameter and have inherent capacities to selectively deliver its cargoes (i.e., mRNA, miRNA, and proteins) to other cells [19]. This selective delivery is indeed mediated through adhesion protein ligands on vesicular surfaces [20]. Inspired from cellular communication functions of natural exosomes [21], therefore, we fabricated genetically engineered nanovesicles (GNV) by utilizing biomimetic synthesis strategies that involve imitating or exploiting biologically synthetic process, pathway, and structures [22–24]. In order to target diverse desired cell populations, this GNV nanoplatform was able to display particular targeting protein ligands on its surface using biosynthetic pathways in genetically engineered cells.

The preparation procedures of GNVs are summarized as follows. Briefly, the synthetic genes of targeted protein moieties are genetically modified and transfected into 293T cells to express targeting ligands on cellular surfaces. Subsequently, during the incubation of transfected 293T cells with chemical surfactants such as sodium deoxycholate, micron-sized vesicles (giant plasma membrane vesicles, GPMV) bearing targeted protein probes rapidly bud from cellular plasma membranes in the way similar to exosome secretion. In order to obtain uniform nanovesicles, sodium deoxycholate and Triton-X100 are mixed with the modified GPMV, meanwhile, the ultrasonic treatment helps the encapsulation of drug molecules into nanoscaled GNVs [25] (*see* Fig. 1). Herein, we describe the biomimetic synthetic strategy and an approach to prepare GNV as analogous nanovesicles displaying different protein probes for cancer cell-specific targeting (*see* **Notes 1, 2**).

## 2 Materials

### 2.1 Reagents

Protease inhibitor (Roche); Phenylmethanesulfonyl fluoride (PMSF) (Sigma); DMEM medium (Hyclone); Penicillin–streptomycin, trypsin–EDTA (Thermo Scientific); Fetal bovine serum (FBS) (Hyclone); Turbofect transfection reagent (Thermo Scientific); Plasmid pLVX-IRES-Puro (Clontech Laboratories, Inc., Takara Bio Company); HEK293T cells (American Type Culture Collection); Triton-X100 (BBI); Sodium deoxycholate (BBI); Doxorubicin (Dox, TCI); Indocyanine green (ICG, Sigma); Matrigel (BD); and Phosphate Buffered Saline (PBS, Sigma).

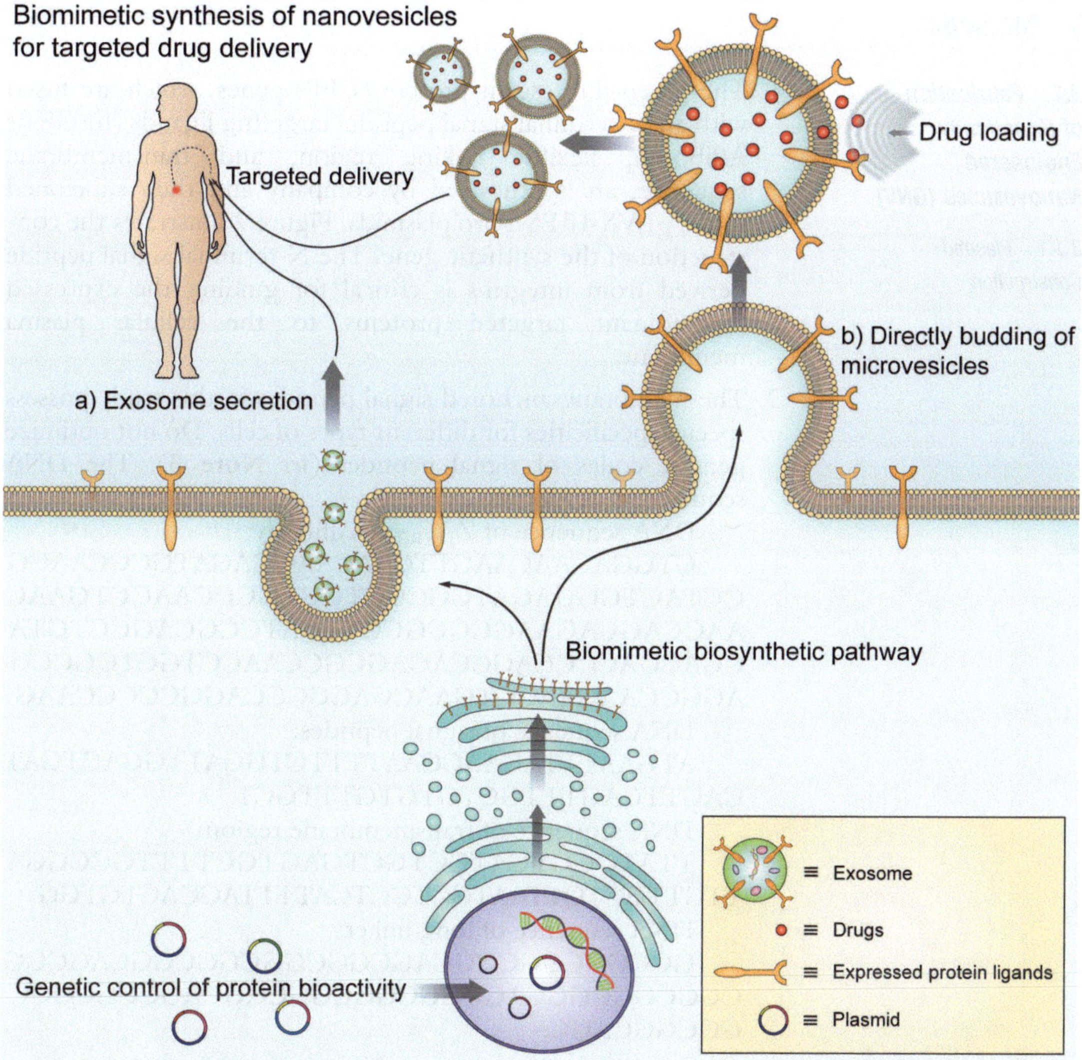


**Fig. 1** Genetically engineered plasma membrane nanovesicles (GNV), generated by biomimetic synthesis that exploit natural intracellular synthetic apparatus, can selectively deliver drugs to diverse cell population of interest through displaying appropriate protein probes (e.g., antibody, nanobody, and single-chain antibody) on vesicular surface. (Reproduced from ref. 25 with permission)

### 2.2 Equipment

Animal IVIS optical imaging system (Caliper Life Sciences Inc.); Photoacoustic (PA) imaging apparatus (Endra Nexus 128, Ann Arbor, MI, USA); Centrifuge (5424R, Eppendorf); Ultrasonic Cell Disruptor (VCX150PB, Sonics Inc.); Ultracentrifuge (BECMAN); Incubator (Thermo Scientific); Near-infrared (NIR) laser generator (808 nm, Stone laser, Shenzhen); and FLIR A×5 camera (FLIR Systems Inc., Wilsonville, USA).

## 3 Methods

### 3.1 Fabrication of Genetically Engineered Nanovesicles (GNV)

#### 3.1.1 Plasmid Construction

1. The green fluorescent protein (GFP) genes, which are fused with an N-terminal signal peptide, targeting ligands (hEGF or Affibody), flexible linking region, and transmembrane sequence, are synthesized by company and then subcloned into a pLVX-IRES-Puro plasmids. Figure 2 illustrates the construction of the synthetic gene. The N-terminal signal peptide derived from integrins is critical for guiding the expressed recombinant targeted proteins to the cellular plasma membrane.
2. The membrane-anchored signal peptides in this study possess species specificities for different types of cells. Do not optimize genetic codes of signal peptides (*see* **Note 3**). The DNA sequence is as follows.

   DNA sequence of $Z_{HER2:342}$Affibody:

   GTGGACAACAAGTTCAACAAGGAGATGCGCAACG CCTACTGGGAGATCGCCCTGCTGCCCAACCTGAAC AACCAGCAGAAGCGCGCCTTCATCCGCAGCCTGTA CGACGACCCCAGCCAGAGCGCCAACCTGCTGGCCG AGGCCAAGAAGCTGAACGACGCCCAGGCCCCCAAG.

   DNA sequence of signal peptides:

   ATGAATTTACAACCAATTTTCTGGATTGGACTGAT CAGTTCAGTTTGCTGTGTGTTTGCT.

   DNA sequence of transmembrane region:

   TTATGGGTCATCCTGCTGAGTGCTTTTGCCGGA TTGTTGCTGTTAATGCTGCTCATTTTAGCACTGTGG.

   DNA sequence of long linker:

   GGCGGCGGCGGCAGCGGCGGCGGCGGCAGCGG CGGCGGCGGCAGCGGCGGCGGCGGCAGCGGCGGC GGCGGCAGC.

#### 3.1.2 Plasmid Transfection

1. Dilute 12 μg of constructed plasmids are in 1.2 mL of FBS-free DMEM culture medium.
2. Then, add 30 μL of turbofect transfection reagents to the treated medium and mix them rapidly by vortexing or pipetting.
3. Place the mixture for 20 min.
4. Add all the mixture dropwise to $3 \times 10^6$ of HEK293T cells in complete growth medium.

**Fig. 2** The construction of synthetic genes encoding plasma membrane-bound targeting protein

5. Gently shake the cell culture plate to achieve even distribution of transfection reagent/plasmids complexes.
6. Culture the transfected cells in normal cultural condition (5% $CO_2$, 37 °C) for 48 h.

#### 3.1.3 Preparation of GPMVs

Carry out the following procedures at 4 °C to prevent vesicular protein from degradation by intracellular proteases.

1. At 48 h following transfection, incubate 0.35–0.6 mM sodium deoxycholate to transfected 8 × $10^6$ HEK293T cells in dishes for 30 min to trigger the production of micron-sized GPMV, after addition of 0.1 mM of protease inhibitor and 1 mM of PMSF.
2. Collect the supernatant in treated dishes and centrifuge it at 1500 × *g* and 3500 × *g* (4 °C, 6 min), respectively. Collect the supernatant once more and discard the precipitate to exclude intracellular organelles.
3. Pellet the supernatant in 2 mL tubes by centrifugation at 15,000 rpm (21,000 × *g*) for 20 min (Eppendorf) to obtain GPMV in precipitate.
4. Disperse the GPMV precipitate with PBS solution. Separate the resuspension through 5–55% sucrose density gradient centrifugation at 200,000 × *g* for 16 h. Harvest the white floccule fraction carefully in the middle location of centrifuge tubes. This step can be omitted according to the practical requirements in the study (*see* **Notes 4, 5**).

#### 3.1.4 Preparation of Drug-Loaded GNVs

1. Resuspend the raw GPMV in PBS. Add 0.05 mM of PMSF to raw GPMV solution.
2. Subsequently, incubate 3 mg/mL total protein of GPMV with 1–1.5 mM sodium deoxycholate, 0.75 mM Triton-X100 and 0.5 mg/mL of Dox or ICG for 10 min under mild sonication at 22–25 W in order to load drug molecules into GNVs, meanwhile, let micron-sized GPMV become nanoscaled drug-loaded GNVs that can display targeting proteins.
3. Transfer aqueous suspension to dialysis bag (MWCO: 100 kDa) and dialyze it for 24 h more than twice at 4 °C to separate the GNVs from free chemical agents.
4. Store the resulting GNVs at 4 °C in the presence of 0.05 mM of PMSF for about 1 week for further usage.
5. Calculate indocyanine green (ICG) concentration by comparing its absorbance values at 784 nm with corresponding standard curve utilizing UV spectrophotometer analysis.
6. The characteristic of resulting GNV can be seen in Fig. 3 (*see* **Notes 6-9**).

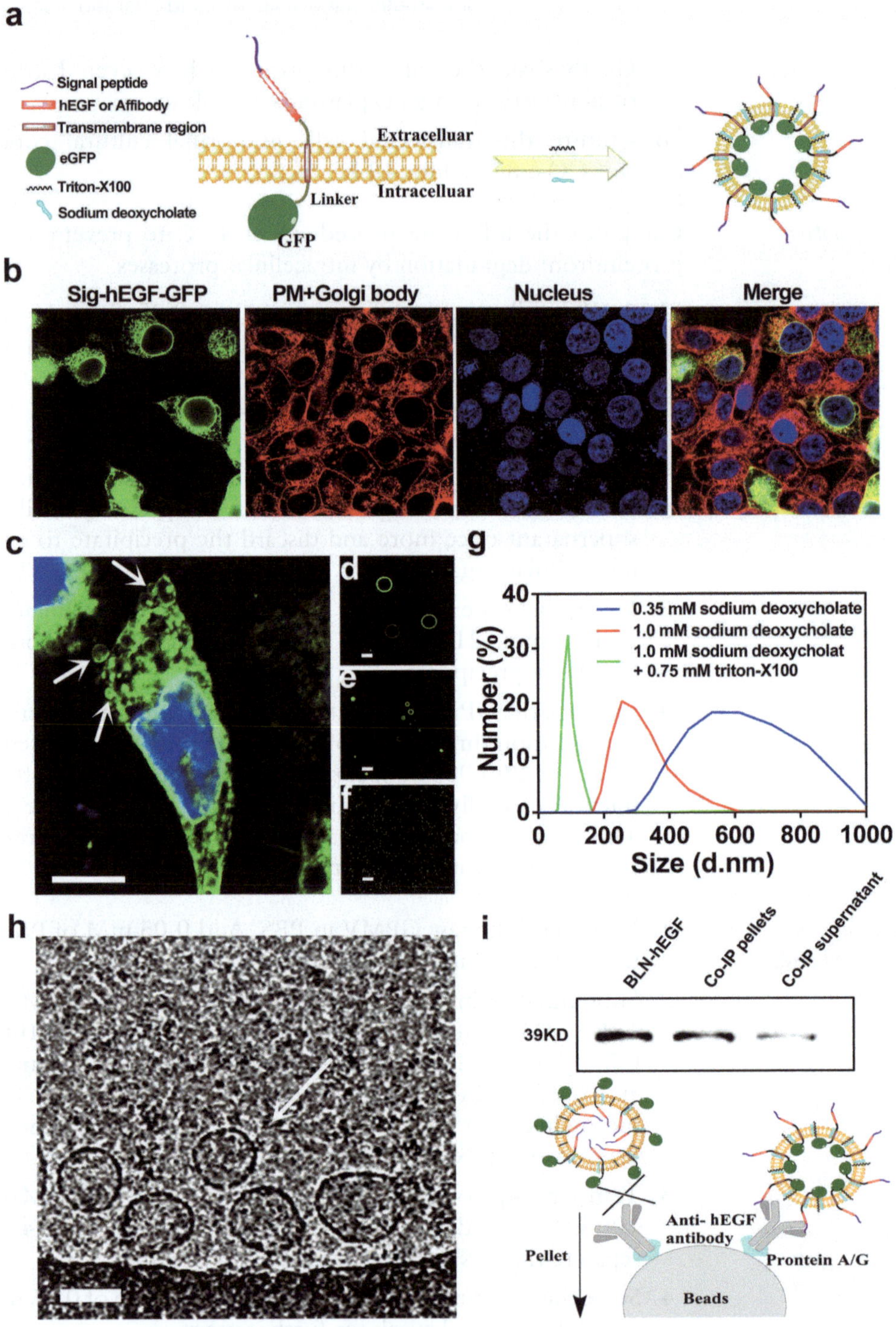


**Fig. 3** Design, fabrication, and characterization of GNV. (**a**) Schematic showing the production process of GNV from cell plasma membrane. (**b**) Location of targeting proteins (Sig–HEGF–GFP) in transfected cells. (**c**) Ligand-displaying GPMVs budded from cellular plasma membrane with the induction of sodium deoxycholate (scale bar: 15 μm). (**d–g**) Controlling of vesicular size under different contents of chemical surfactants. The micron-sized giant plasma membrane vesicles resulted from plasma membranes were incubated with (**d**) $0.35 \times 10^{-3}$ M sodium deoxycholate, (**e**) $1 \times 10^{-3}$ M sodium deoxycholate, or (**f**) $1 \times 10^{-3}$ M sodium deoxycholate and $0.75 \times 10^{-3}$ M Triton-X100, then monitored with confocal laser scanning microscope (**d–f**, Scale bar:1 μm) as well as (**g**) DLS. (**h**) Cryo-elelctron microscopy images showing the morphology and size of GNV. Scale bar: 100 nm. (**i**) Co-IP analysis of GNV through western blotting. (Reproduced from ref. 10 with permission)

#### 3.1.5 Fabrication of PEGylated Liposomes Coupled with Proteins (Liposome-Ligand)

1. Dissolve multiple lipids DSPE-PEG2000:CHOL:HSPC in a molar ratio of 0.15:1:1.83 in 10 mL of chloroform: methanol solution (in a volume ratio of 1:9).
2. Evaporate the lipid mixtures at 62 °C to produce lipid films under vacuum condition.
3. Disperse the performed film in 30 mL of PBS solution to produce neutral PEGylated liposomes under vigorous water-bath sonication for 30 min.
4. Filter the dispersed solution through 0.22 μm filter PES membrane more than 12 times to acquire the nanosized liposomes.
5. Gently mix NHS-DSPE-PEG2000 with protein ligands in MOPS buffer at PH 8.5 in a molar ratio of 1: 1 for 6 h at room temperature.
6. Incubate protein-DSPE-PEG2000 conjugates with prepared liposomes in PBS for 8 h to decorate PEGylated liposomes with targeting protein-PEG ligands.
7. At 4 °C, transfer the mixture to dialysis tubes (MWCO: 200 kDa) and dialyze it against PBS for 18 h more than thrice to isolate PEGylated liposome-protein from free protein-DSPE-PEG2000 conjugates.

### 3.2 Establishment of a Tumor-Bearing Mouse Model

1. After washing with PBS, digest $10^7$ of MDA-MB-468 cells in 1 mL of 0.25% trypsin-EDTA solution at 37 °C for 5 min.
2. Add 4 mL of serum-free DMEM medium to the digested cells and centrifuge it at 1000 rpm/min for 5 min. Collect the precipitate and repeat the **step 2** once more to remove the free serum and trypsin.
3. Gently mix MDA-MB-468 cells (in serum-free DMEM medium) with Matrigel at a volume ratio of 1:1 on the ice (*see* **Notes 10, 11**).
4. Subcutaneously inject the mixture containing $5 \times 10^6$ of tumor cells (100–150 μL of total volume per mouse) into the flank region of nude mice (6 weeks old) prior to the coagulation of Marrigel at room temperature.

### 3.3 Multimodal Imaging In Vivo

#### 3.3.1 Mouse Fluorescent (FL) Imaging and the Biodistribution of GNVs

1. Intravenously administrate tumor-bearing mice with a single dose of free ICG, GNV-ICG, or ICG-loaded nanovesicles (4 mg/kg of equivalent ICG in 150 μg vesicular protein formulation for each mouse).
2. Every time point post-injection, conduct fluorescent (FL) imaging by the use of animal IVIS optical imaging system at 745 nm excitation wavelength and 780 nm emission wavelength.

3. 3 h following injection, excise the tumors and mouse major organs, such as spleen, liver, heart, lung, and kidney, for in vitro NIR FL imaging with a 745 nm excitation wavelength.
4. Calculate the average FL intensities from a region of interest to quantify the biodistribution of different formulations ($n = 5$/group).
5. Measure mean radiation fluorescence efficiency using Living Image 3.1 software (*see* Fig. 4a–c).

#### 3.3.2 Photoacoustic (PA) Imaging

1. When tumors grow up to an average volume of 100 $mm^3$, randomly divide the tumor-bearing mice into three groups: (a) ICG-loaded GNV with ligands, (b) ICG-loaded nanovesicles without targeting ligands, (c) Free ICG group. The selected mice have similar weights.
2. Intravenously inject equivalent free ICG, GNV-ICG, and ICG-loaded nanovesicles without ligands (each containing 80 μg ICG in 100–200 μg of vesicular proteins per mouse) into tumor-bearing mice to acquire PA images. Measure the vesicular protein contents by normal BCA assays.
3. Anesthetize the treated mice using 1–2% isoflurane mixed with pure oxygen in sealed cases for 10 min and then place the anesthetized mouse in a PA scanning instrument.
4. Scan the tumors at a laser wavelength of 800 nm with 25 pulses averaging and keep water temperature in PA scanning instrument at 37 °C. Acquire PA images at predetermined tines post administration without the need to sacrifice mice.
5. Generate three-dimensional PA images using a hemispherical ultrasonic detector in Endra Nexus 128 (*see* Fig. 4d, e).

### 3.4 Photothermal Therapy

1. When tumors grow up to an average volume of 150 $mm^3$, randomly divide the tumor-bearing mice into five groups: (a) PBS, (b) blank GNV with ligands, (c) ICG-loaded GNV with ligands, (d) ICG-loaded nanovesicles without targeting ligands, and (e) Free ICG group. The selected mice have similar weights ($n = 5$/group).
2. Inject above-mentioned different formulations (equal 80 μg of ICG in 150 μg of total vesicular protein for each mouse) into tumor-bearing mice via tail veins, respectively.
3. Adjust laser spot area to cover the whole tumor at an area of 2 $cm^2$ prior to NIR laser irradiation.
4. 2 or 4 h following administration, irradiate tumor sites using 808 nm near-infrared (NIR) laser at a power density of 1 W/$cm^2$ for 10 min at room temperature.

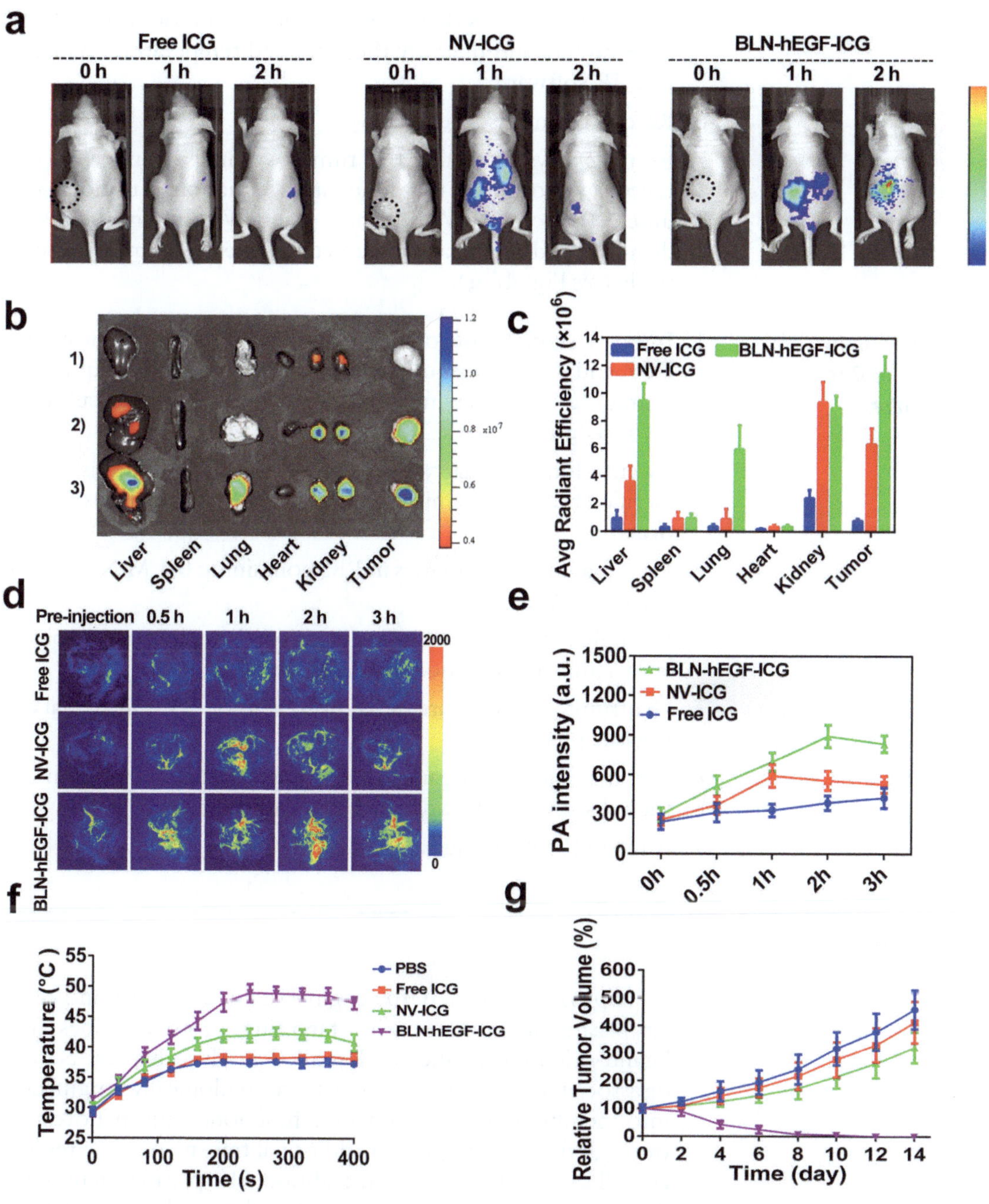


**Fig. 4** Imaging-guided GNV phototherapy. (**a**) FL imaging of nude mice bearing tumors following intravenous administration of GNV at predetermined time. (**b**) FL images and (**c**) average FL radiant efficiency analysis of major organs and tumors of treated mice at 2 h post-injection. (1) free ICG-treated group; (2) NV-ICG-treated group; (3) GNV–treated group. (**d**) PA imaging and (**e**) mean PA intensity analysis of tumor regions following different treatments. (**f**) Temperature profiles of tumors in mice at 2 h following intravenous injection under laser irradiation for 10 min (1 W/cm$^2$, 808 nm). (**g**) Relative volume changes of tumors post-phototherapy. (Reproduced from ref. 10 with permission)

5. Obtain real-time NIR thermal images of tumors with a FLIR A×5 camera and quantify the increased temperature by use of BM_IR software.
6. Record the mouse weights every 3 days.
7. Every 2 days, measure the tumor volumes using a vernier caliper according to following equation: tumor volume $= 0.5 \times L \times (W)^2$ in which $L$ and $W$ represent the largest diameter and the smallest diameter of treated tumors, respectively (*see* Fig. 4f, g).

### 3.5 Quantitative Analysis of Dox Distribution

1. Inject a single dose of Doxil, Dox-loaded nanovesicle, free Dox and GNV-Dox displaying Affibody ligands (equivalent 5 mg/kg of Dox) into tumor-bearing mice via tail veins ($N = 4$/group).
2. Twenty-four and forty-eight hours following administration, sacrifice the treated mice, collect and weigh the mouse major organs.
3. Homogenize these tissues in PBS containing 0.1 M NaOH and SDS at 4 °C.
4. Strongly mix 6 mL of methanol-chloroform (1:4, v/v) solvent with 2 mL of tissue homogenates for 20 min.
5. Centrifuge the mixture at 18,000 × *g* for 15 min to harvest organic phases.
6. Dry the collected organic phases under vacuum condition and then dissolve the powder in 150 μL of methanol.
7. Analyze the resulting samples by HPLC.

## 4 Notes

1. An approach for genetically engineered nanovesicles and cellular plasma membranes with targeting protein probes in a ligand-oriented manner is described in this chapter. Natural disassociated targeting proteins are biologically engineered into membrane-bound forms for functionalizing nanovesicles through the introduction of additional transmembrane regions as well as signal peptides. In addition, the plasma membrane-directed signal peptides are introduced to make sure that targeting protein moieties have been correctly displayed on the exterior of GNVs in protein-oriented ways.
2. The described approach in this chapter can also be employed to fabricate analogous nanocarriers, which enable cell-specific drug delivery through displaying appropriate protein ligands (e.g., affibody, antibody, nanobody, or single-chain antibody) on GNV surfaces.

3. It is very critical to select suitable signal peptide to guide the recombinant targeting proteins to plasma membranes of the transfected cell. The chosen signal peptides have species specificities in different sorts of mammalian cells. Alteration of its native DNA sequences is not allowed for genetic code optimization, because this may lead to the loss of membrane-bound functionalities.
4. In the preparation process of drug-loaded GNVs, make sure that protease inhibitors are added sufficiently and the solution is kept below 4 °C so as to prevent vesicles from degradation by intracellular proteases.
5. The step, which exploits sucrose density gradient centrifugation to further purify the vesicles, can be omitted according to the practical requirements for vesicular purity.
6. The resulting GNVs can be stored in PBS containing 0.1 mM PMSF at 4 °C for about 1 week. Avoid repeated freezing and thawing to prevent leakage of drug molecules from GNVs.
7. Be cautious that PMSF is highly toxic. Wear a face mask and latex gloves to handle it carefully.
8. It is recommended to prepare the GNVs 48–52 h after cellular transfection. The expressed levels of targeting proteins need to be validated by fluorescence microscopes and western blot assays prior to GNV preparation.
9. Prepare all solutions using ultrapure water, if there is no special statement.
10. Digested tumor cells should be centrifuged to remove free serum and trypsin. Importantly, use the serum-free DMEM medium or PBS to suspend the cell pellets and then administrate mice with it. A small amount of serum in injected tumor cells has negative effect on the establishment of tumor-bearing mouse model.
11. To acquire the GNVs that are derived from mesenchymal cell, stem cells, or immune cells, lentivirus infection would be an effective approach to transfer genes encoding recombinant targeting proteins to desired cells.

## Acknowledgements

This work was supported by the National Natural Science Foundation of China (Grant Nos. 81371596, 81422023, 51273165, U1705281, and U1505221), the Major State Basic Research Development Program of China (Grant Nos. 2017YFA0205201 and 2018YFA0107301), the Program for New Century Excellent Talents in University (NCET-13-0502), and the Fundamental Research Funds for the Central Universities, China (20720150141 and 20720150206).

## References


1. Chu C, Lin H, Liu H, Wang X, Wang J, Zhang P et al (2017) Tumor microenvironment-triggered supramolecular system as an in situ nanotheranostic generator for cancer phototherapy. Adv Mater 29:1605928
2. Zhang L, Gao S, Zhang F, Yang K, Ma Q, Zhu L (2014) Activatable hyaluronic acid nanoparticle as a theranostic agent for optical/photoacoustic image-guided photothermal therapy. ACS Nano 8:12250–12258
3. Cai W, Gao H, Chu C, Wang X, Wang J, Zhang P et al (2017) Engineering photo-theranostic nanoscale metal-organic frameworks for multimodal imaging-guided cancer therapy. ACS Appl Mater Interfaces 9:2040–2051
4. Topete A, Alatorre-Meda M, Iglesias P, Villar-Alvarez EM, Barbosa S, Costoya JA et al (2014) Fluorescent drug-loaded, polymeric-based, branched gold nanoshells for localized multimodal therapy and imaging of tumoral cells. ACS Nano 8:2725–2738
5. Lin G, Zhang Y, Zhu C, Chu C, Shi Y, Pang X et al (2018) Photo-excitable hybrid nanocomposites for image-guided photo/TRAIL synergistic cancer therapy. Biomaterials 176:60–70
6. Wang J, Tao W, Chen X, Farokhzad OC, Liu G (2017) Emerging advances in nanotheranostics with intelligent bioresponsive systems. Theranostics 7:3915–3919
7. Ying M, Zhan C, Wang S, Yao B, Hu X, Song X et al (2016) Liposome-based systemic glioma-targeted drug delivery enabled by all-d peptides. ACS Appl Mater Interfaces 8:29977–29985
8. Mikhaylov G, Mikac U, Magaeva AA, Itin VI, Naiden EP, Psakhye I et al (2011) Ferri-liposomes as an MRI-visible drug-delivery system for targeting tumours and their microenvironment. Nat Nanotechnol 6:594–602
9. Lin G, Mi P, Chu C, Zhang J, Liu G (2016) Inorganic nanocarriers overcoming multidrug resistance for cancer theranostics. Adv Sci 3:1600134
10. Zhang P, Zhang L, Qin Z, Hua S, Guo Z, Chu C et al (2018) Genetically engineered liposome-like nanovesicles as active targeted transport platform. Adv Mater 30:1705350
11. Noble GT, Stefanick JF, Ashley JD, Kiziltepe T, Bilgicer B (2014) Ligand-targeted liposome design: challenges and fundamental considerations. Trends Biotechnol 32:32–45
12. Mazzucchelli S, Colombo M, Verderio P, Rozek E, Andreata F, Galbiati E et al (2013) Orientation-controlled conjugation of haloalkane dehalogenase fused homing peptides to multifunctional nanoparticles for the specific recognition of cancer cells. Angew Chem Int Ed Engl 52:3121–3125
13. Zhang P, Liu G, Chen X (2017) Nanobiotechnology: cell membrane-based delivery systems. Nano Today 13:7–9
14. Jang SC, Kim OY, Yoon CM, Choi DS, Roh TY, Park J et al (2013) Bioinspired exosome-mimetic nanovesicles for targeted delivery of chemotherapeutics to malignant tumors. ACS Nano 7:7698–7710
15. Zhang P, Chen Y, Zeng Y, Shen C, Li R, Guo Z et al (2015) Virus-mimetic nanovesicles as a versatile antigen-delivery system. Proc Natl Acad Sci U S A 112:E6129–E6138
16. Somiya M, Kuroda S (2015) Development of a virus-mimicking nanocarrier for drug delivery systems: the bio-nanocapsule. Adv Drug Deliv Rev 95:77–89
17. Wang D, Dong H, Li M, Cao Y, Yang F, Zhang K et al (2018) Erythrocyte-cancer hybrid membrane camouflaged hollow copper sulfide nanoparticles for prolonged circulation life and homotypic-targeting photothermal/chemotherapy of melanoma. ACS Nano 12:5241–5252
18. Hu CM, Zhang L, Aryal S, Cheung C, Fang RH, Zhang L (2011) Erythrocyte membrane-camouflaged polymeric nanoparticles as a biomimetic delivery platform. Proc Natl Acad Sci U S A 108:10980–10985
19. Trajkovic K, Hsu C, Chiantia S, Rajendran L, Wenzel D, Wieland F et al (2008) Ceramide triggers budding of exosome vesicles into multivesicular endosomes. Science 319:1244–1247
20. Wolfers J, Lozier A, Raposo G, Regnault A, Thery C, Masurier C et al (2001) Tumor-derived exosomes are a source of shared tumor rejection antigens for CTL cross-priming. Nat Med 7:297–303
21. Thery C, Zitvogel L, Amigorena S (2002) Exosomes: composition, biogenesis and function. Nat Rev Immunol 2:569–579
22. Yoo JW, Irvine DJ, Discher DE, Mitragotri S (2011) Bio-inspired, bioengineered and biomimetic drug delivery carriers. Nat Rev Drug Discov 10:521–535
23. Hutchens SA, Benson RS, Evans BR, O'Neill HM, Rawn CJ (2006) Biomimetic synthesis of calcium-deficient hydroxyapatite in a natural hydrogel. Biomaterials 27:4661–4670
24. Mantion A, Graf P, Florea I, Haase A, Thunemann AF, Masic A et al (2011) Biomimetic synthesis of chiral erbium-doped silver/peptide/silica core-shell nanoparticles (ESPN). Nanoscale 3:5168–5179
25. Zhao L, Wang J, Lv P, Liu G (2018) Biomimetic synthesis of nanovesicles for targeted drug delivery. Sci Bull 63:663–665

# INDEX



Jyotsna Batra and Srilakshmi Srinivasan (eds.), *Theranostics: Methods and Protocols*, Methods in Molecular Biology, vol. 2054, https://doi.org/10.1007/978-1-4939-9769-5, © Springer Science+Business Media, LLC, part of Springer Nature 2019